Second Edition

Pharmaceutical Formulation Development of Peptides and Proteins

W0234985

Second Edition

Pharmaceutical Formulation Development of Peptides and Proteins

Edited by
Lars Hovgaard
Sven Frokjaer
Marco van de Weert

CRC Press
Taylor & Francis Group
Boca Raton London New York

CRC Press is an imprint of the
Taylor & Francis Group, an **informa** business

First published in paperback 2024

First published 2013 by CRC Press
2385 NW Executive Center Drive, Suite 320, Boca Raton FL 33431

and by CRC Press
4 Park Square, Milton Park, Abingdon, Oxon, OX14 4RN

CRC Press is an imprint of Taylor & Francis Group, LLC

© 2013, 2024 Taylor & Francis Group, LLC

Library of Congress Cataloging-in-Publication Data

Pharmaceutical formulation development of peptides and proteins / editors, Lars
   Hovgaard, Sven Frokjaer, Marco van de Weert. -- 2nd ed.
     p. ; cm.
   Includes bibliographical references and index.
   ISBN 978-1-4398-5388-7 (hardcover : alk. paper)
   I. Hovgaard, Lars, 1962- II. Frokjaer, Sven, 1947- III. Weert, Marco van de, 1972-
   [DNLM: 1. Peptide Biosynthesis. 2. Chemistry, Pharmaceutical--methods. 3. Protein
Biosynthesis. QU 68]

615.1'9--dc23   2012030490

ISBN: 978-1-4398-5388-7 (hbk)
ISBN: 978-1-03-292065-8 (pbk)
ISBN: 978-0-429-10671-2 (ebk)

DOI: 10.1201/b12951

Visit the Taylor & Francis Web site at
http://www.taylorandfrancis.com

and the CRC Press Web site at
http://www.crcpress.com

Contents

Preface

Ever since the first edition of this book was published in 2000, peptides and proteins have become an increasingly more important addition to the treatment of a wide variety of serious, chronic, and/or debilitating diseases. There are at present over 100 marketed pharmaceutical products containing a peptide or protein as the active compound, and annual sales of these products exceeded $100 billion in 2010. Thus, it is important that pharmaceutical scientists are trained in the science of formulating these compounds into stable, active, and safe therapeutic products.

Although conventional (low-molecular-weight) drug compounds should also be stable, active, and safe, peptides and proteins provide a particular challenge. The latter compounds combine relatively low chemical and physical stability with highly suboptimal physicochemical characteristics to allow transport over biomembranes. In addition, peptides and proteins can cause an immune response that may have severe adverse effects on the patient. The formulation of a successful peptide or protein drug is thus a considerable challenge that requires a highly interdisciplinary approach. As significant progress has been made within the area of peptide and protein formulation development, we believed it was time to update this book.

In this second edition, we have again compiled chapters, written by experts in the field, ranging from the production of the active compound via basic preformulation and formulation to registration of the final product. The reader will thus get a broad overview of the whole development process, which will hopefully aid in communication across the various disciplines involved in this process. Most of the chapters also appeared in the first edition but are either updated or fully revised by new authors to allow incorporation of the latest developments; two new chapters complete this edition.

This book can be used as a textbook at an undergraduate and graduate level in courses focusing on various stages of peptide and protein drug development. Moreover, it can also serve as a useful reference book to those already active in this area, in both industry and academia. Chapter 1 discusses the synthesis of peptides. The complex chemistry discussed should garner appreciation of the challenges involved among those who use peptides in their research. Chapter 2 discusses the biotechnological production of proteins through recombinant DNA technology. Chapter 3 on protein purification is the only chapter that has not been updated or revised. However, two references to more recent literature have been added. Chapters 4–6 discuss the physicochemical characteristics and stability of peptides and proteins and the analytical toolbox available to the pharmaceutical scientist. The high-molecular-weight and complex three-dimensional structures of proteins mean that many different analytical techniques need to be used to characterize a protein. Further analytical challenges are added by the many different chemical and physical degradation pathways of proteins, as discussed in Chapters 5 and 6.

In Chapters 7–11, various formulation principles for peptides and proteins are discussed. Chapter 7 describes various methods to modify peptides and proteins to alter their physicochemical characteristics. In general, the aim of such derivations is to improve the pharmacological properties of the protein, for example, the circulation time and compartmentation/tissue distribution. Chapters 8–10 take on the formulation of proteins as suspensions, solutions, and (mostly freeze-dried) solids. These chapters should be considered as the highlights of this book, and hence also take up the most space. In Chapter 11, the authors take a look at the opportunities and challenges of nonparenteral delivery of peptides and proteins. As is made clear, solving the challenges requires a highly integrated approach of reduced costs of the raw product and advanced formulation and administration strategies.

The next two chapters, Chapters 12 and 13, are new to this edition. Chapter 12 discusses the potential of protein drugs to elicit an immune response and its potential consequences. Although we still have a limited grasp on predicting the immune response, there are already some risk factors that will need to be considered when formulating a protein. In Chapter 13, a simulation approach is discussed to describe the fate of peptides and proteins upon administration to a biological system. This computer-based simulation can significantly speed up the development process by predicting areas of concern and opportunity well before the peptide or protein is even produced in significant amounts to allow biological testing. Chapter 14 rounds off the development process by discussing the documentation required to register a protein-based drug.

This book would not have been possible without the authors of the various chapters. We would like to thank them for their very valuable contributions. A note of appreciation should also go to the reviewers of the various chapters for providing important feedback. It is our hope that this second edition will be at least as successful as the first edition and contribute to the further peptide and protein drug development.

Editors

Marco van de Weert is an associate professor in the Department of Pharmacy, Faculty of Health and Medical Sciences, University of Copenhagen, Denmark. His research focuses on the understanding of protein physical instability and protein–excipient interactions and on the development of the analytical toolbox to study protein structure and stability within pharmaceutical formulations.

Sven Frokjaer is a professor in the Department of Pharmacy, Faculty of Health and Medical Sciences, University of Copenhagen, Denmark. His research focuses on peptide and protein formulation with a special emphasis on particulate drug delivery systems—for example, microspheres, liposomes and lipid emulsions, and peptide transport across biological membranes including carrier-mediated mechanisms.

Lars Hovgaard is principal scientist in the Department of Oral Formulation Development at Novo Nordisk A/S and a professor at the University of Copenhagen. His current research covers peptide and protein formulation development with a focus on oral delivery. During the course of his research career, the area of oral peptide and protein drug delivery has been a major and ongoing element. He has addressed the topic from drug delivery, biopharmaceutical, and pharmaceutical technological points of view.

Contributors

Michael J. Akers
Baxter BioPharma Solutions (retired)
Bloomington, Indiana

Tudor Arvinte
University of Geneva and Therapeomic, Inc.
Geneva, Switzerland

Simon Bjerregaard
Novo Nordisk
Copenhagen, Denmark

Vera Brinks
Utrecht University
Utrecht, the Netherlands

Morten Colding-Jørgensen
Novo Nordisk
Copenhagen, Denmark

Michael R. DeFelippis
Bioproduct Research and Development
Eli Lilly and Company
Indianapolis, Indiana

Nanni Din
Novo Nordisk
Copenhagen, Denmark

Sven Frokjaer
University of Copenhagen
Copenhagen, Denmark

Lars Hovgaard
Novo Nordisk
Århus, Denmark

Thomas Høeg-Jensen
Novo Nordisk
Copenhagen, Denmark

František Hubálek
Novo Nordisk
Copenhagen, Denmark

Knud J. Jensen
University of Copenhagen
Copenhagen, Denmark

Wim Jiskoot
Leiden University
Leiden, the Netherlands

Grzegorz Kijanka
Utrecht University
Utrecht, the Netherlands

Niamh Kinsella
NDA Regulatory Science Ltd.
Leatherhead, United Kingdom

Erik Mosekilde
Technical University of Denmark
Kongens Lyngby, Denmark

Michael J. Pikal
University of Connecticut
Storrs, Connecticut

Ulrik Lytt Rahbek
Novo Nordisk
Copenhagen, Denmark

Theodore W. Randolph
University of Colorado at Boulder
Boulder, Colorado

Christian Hove Rasmussen
Novo Nordisk
Copenhagen, Denmark

Melody Sauerborn
Utrecht University
Utrecht, the Netherlands

Huub Schellekens
Utrecht University
Utrecht, the Netherlands

Christian Schöneich
The University of Kansas at Lawrence
Lawrence, Kansas

Teruna J. Siahaan
The University of Kansas at Lawrence
Lawrence, Kansas

Lars Skriver
BioProcess Technology Consultants
Rungsted Kyst, Denmark

Tue Søeborg
Copenhagen University Hospital
Copenhagen, Denmark

Kristian Strømgaard
University of Copenhagen
Copenhagen, Denmark

Bingquan (Stuart) Wang
Genzyme Corporation
Framingham, Massachusetts

Marco van de Weert
University of Copenhagen
Copenhagen, Denmark

1 Peptide Synthesis

Knud J. Jensen

CONTENTS

1.1 INTRODUCTION

Synthetic peptides are ubiquitous in biology, biomedicine, drug discovery, and many other fields. Chemically synthesized peptides serve very diverse purposes, including as biopharmaceutical drugs and for epitope mapping, peptide microarrays, and vaccine development. While proteins are generally prepared by recombinant methods, chemical synthesis is the prevailing method for the preparation of peptides. This is due to the ease, predictability, and flexibility of chemical synthesis, which also allows the convenient incorporation of many nonproteinogenic modifications. Peptide synthesis has allowed the preparation of numerous peptides, both on a laboratory scale and on a ton scale. However, there are also limitations—or current limitations, if you will—and having an understanding of the possibilities as well as the limitations will allow the biomedical users to better incorporate synthetic peptides in their research and applications. This chapter will provide an introduction to some of the most common methods in solid-phase peptide synthesis (SPPS), but will also briefly introduce solution synthesis of peptides and some other methods.

Even a brief history of peptide synthesis will have to include that Emil Fischer reported the first synthesis of a peptide, Leonidas Zervas developed the Z (Cbz) protecting group, and Vincent du Vigneaud was awarded the Nobel Prize in 1955 for the chemical synthesis of the cyclic peptide, oxytocin. In 1963, Bruce Merrifield introduced the concept and the first implementation of solid-phase synthesis for which he was awarded the Nobel Prize in chemistry in 1984 (Merrifield, 1963, 1985). SPPS has for decades been the primary source of synthetic peptides on a laboratory scale.

However, the solution synthesis of peptides has remained a viable option, especially for the large-scale production of peptides, and a very brief introduction to this strategy for the preparation of peptides can also form a useful backdrop for the description of solid-phase synthesis (Figure 1.1). Amino acids carry at least two functional groups, as the name indicates. When two amino acids have to be coupled together by an amide bond, one of the N^{α}-amino and one of the carboxyl groups have to be protected. This is needed to ensure that the correct amide bond, which is often called a "peptide bond," is formed. For the bifunctional amino acids such as Gly, Ala, and Leu, no further protecting groups are needed. However, for trifunctional amino acids such as Lys, Asp, and Cys, side-chain protecting groups are required to prevent undesired reactions at the side-chain functional groups. Side-chain protecting groups are required to protect functional groups such as carboxyls (Asp and Glu), amines (Lys), thiols (Cys), hydroxyls (Ser, Thr, and Tyr), imidazoles (His), and sometimes indoles (Trp). These protecting groups need to be removed at the end of the synthesis.

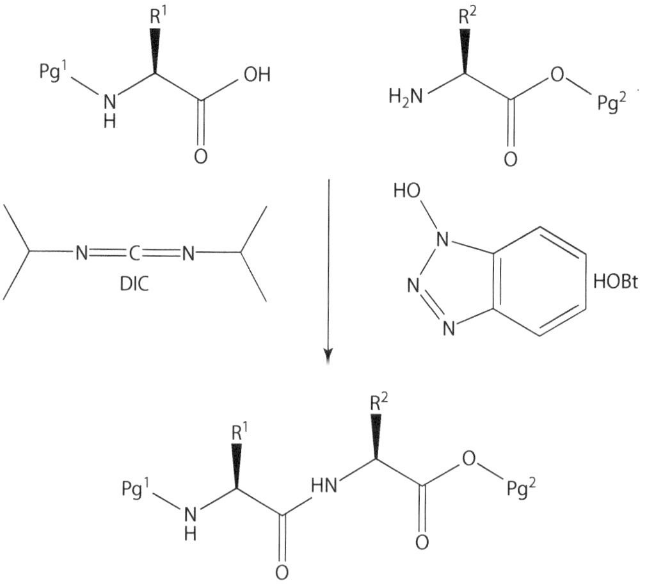

FIGURE 1.1   Solution synthesis of a dipeptide (Pg, protecting group; Pg^1, N^{α}-amino Pg; Pg^2, carboxyl protecting group). Side-chain Pg may also be required. Pg^1 could, for example, be the carboxybenzyl (Cbz) group.

1.2 PROTECTING GROUPS

Some of the protecting groups used in solution synthesis are based on benzyl derivatives, which make them conveniently removable by hydrogenolysis, that is, treatment with hydrogen gas in the presence of a suitable catalyst. One example is the classical carboxybenzyl (Cbz) protecting group for amines, which was reported for peptide synthesis by Bergmann and Zervas (1932). The Cbz group was the first in a long series of protecting groups where the amine forms part of a carbamate. Thus, the N^α-amine of one amino acid could be protected by Cbz, while the carboxylic acid of the other amino acid could be protected by benzyl ester. After coupling of the two amino acids by the formation of the intended amide bond, the two protecting groups can be removed in one step. In syntheses of longer peptides, the N^α-amino protecting group is removed after each coupling, while the C-terminal carboxyl needs to be semipermanent, that is, stable during the removal of the N^α-group.

Amide bond formation requires that the hydroxyl of the carboxylic acid is converted to a better leaving group, such as in a reactive ester or an acyl halide. Typically, the carboxylic acid is reacted with an electrophile, for example, DIC (N,N'-di-isopropylcarbodiimide), which forms an activated species that is then directly reacted with an auxiliary nucleophile, such as HOBt (1-hydroxybenzotriazole), to form an activated ester—in this case an OBt ester. This ester can then react with the amino group of another amino acid or peptide segment.

Two peptides can also be coupled together in solution. This is often referred to as segment condensation. A successful segment condensation requires that all the side chains be protected on each peptide, such that one peptide has a free C-terminal carboxylic acid and the other a free N-terminal amino group. They are then coupled together by the formation of an amide bond, as described above. This means that the two protected segments can be prepared and purified separately before being coupled together. While this can be an attractive approach in some cases, it may need extensive optimization, as protected peptides can suffer from low solubility. Furthermore, carbamate N^α-protecting groups such as Cbz actually reduce the risk of epimerization in activated amino acid derivatives. The nitrogen in the C-terminal amino acid in a peptide is part of an amide and not a carbamate, which increases the risk of epimerization at the C-terminal amino acid that is being activated for segment couplings.

1.3 SOLID-PHASE PEPTIDE SYNTHESIS

The introduction of a functionalized solid support that allows for anchoring of an amino acid revolutionized the field of peptide science and inaugurated the SPPS methodology. Since the initial report by Merrifield in the 1960s, all aspects of SPPS have been developed further and refined, thus extending the reach of synthetic peptide chemistry tremendously. Solid-phase synthesis has evolved into a highly efficient set of techniques for the preparation of numerous peptides and even small proteins. It has been crucial in the development of combinatorial and high-throughput chemistry as well as many other research areas. While the term "solid-phase synthesis" is commonly used, maybe a more precise terminology would be "matrix-assisted synthesis," as the resins most commonly used are anything but solid (Hudson, 1999a,b). The main characteristics of solid-phase synthesis are that (1) the first building block is attached (anchored) to a matrix, which can be filtered; (2) repeated cycles of chemical transformations (especially deprotection and coupling) are performed, also by automation; and (3) most often the final product is released from the matrix and deprotected in the same step. However, by proper choice of protecting groups and linker, the peptide can either be deprotected while it remains attached to the support or it can be released from the support in the fully protected form (Figure 1.2).

SPPS is defined by the set of N^α-protecting groups, side-chain protecting groups, coupling reagents, linkers (handles), as well as resins and other solid supports that can be used. Suitably N^α- and side-chain protected amino acids are coupled sequentially in the $C \rightarrow N$ direction to a growing peptide chain anchored to the resin. Typically, the C-terminal amino acid is first anchored at the carboxy terminus to the solid support through a cleavable handle. Then the N^α-protecting group can be removed without affecting the side-chain protecting groups; thus, the polypeptide chain is prepared for the next coupling cycle. SPPS reactions are driven to completion by the use of soluble reagents in excess, which can be removed by filtration and washing. Following the completion of the desired sequence of amino acids, the

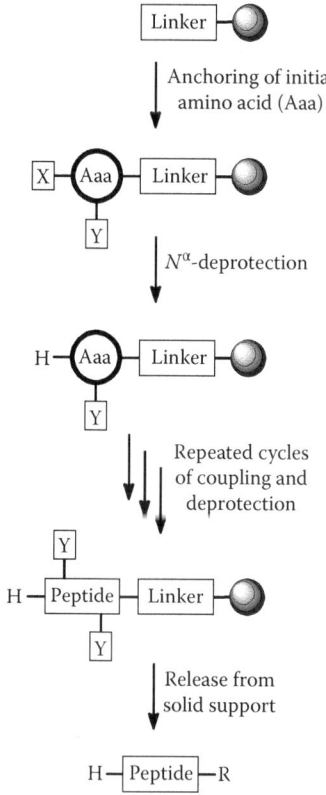

FIGURE 1.2 Principles of solid-phase peptide synthesis (SPPS).

peptide is released from the solid support, and simultaneously the semipermanent side-chain protecting groups are typically removed concomitantly (Figure 1.2, final step). The amino acid protecting groups, coupling reagents, and resins have been refined over the last three decades, and they are now very efficient, which enables them to be used in routine syntheses. However, the peptides that are synthetically accessible, for example, their maximal length, specific post-translational modifications, and unnatural modifications, are defined by the effectiveness of the underlying organic chemistry, which continues to evolve.

At this point, it is important to remember that amino acids are defined as they appear in proteins. Proteins are depicted with the N-terminal to the left and the C-terminal to the right. We thus recommend always drawing amino acids with the nitrogen left and the carboxylate right. When depicting SPPS reactions where the peptide is anchored by its C-terminal anchoring, it is mandatory to place the resin to the *right*.

1.4 N^α-AMINO PROTECTING GROUPS: BOC VERSUS FMOC

The two most general protecting groups in SPPS are fluoren-9-ylmethyloxycarbonyl (Fmoc) (Carpino and Han, 1970, 1972) (Figure 1.3) and *tert*-butoxycarbonyl (Boc), with each N^α-protecting group defining an overall strategy for SPPS. The chemical conditions for the removal of these transient protecting groups, that is, base versus acid, each define a "chemical window" of opportunities for the other chemical steps in the overall SPPS strategy. The solid-phase strategy is defined by the choice of the N^α-protecting group for the amino acid building blocks. This N^α-protecting group will be removed after each coupling and the chemical conditions required for removal thus define what the linker (often also referred to as a "handle") and the side-chain protecting groups have to be compatible with and hence which conditions they are stable to. The choice of semipermanent side-chain protecting group and linker depends on the chosen N^α-protecting group. Conversely, the linkage to the solid support defines the chemistry possible in the following step and the conditions for repetitive removal codetermine the choice of semipermanent side-chain protecting groups and handle.

The Boc strategy, initially introduced by Merrifield, requires trifluoroacetic acid (TFA) for repetitive removal of the Boc groups, while often relying on hydrofluoric acid (HF) for release of the final peptide from the support. Thus the Boc strategy relies on differences in acid-lability of the N^α- and side-chain protecting groups. Due to the use of corrosive and toxic HF and the requirement for a specialized apparatus to handle HF in Boc-SPPS, the Fmoc strategy is often preferred over the Boc strategy for routine synthesis. The Fmoc group can be removed under mild conditions with secondary amines, typically a 1:4 solution of piperidine in DMF (Figure 1.3) (Atherton et al., 1978a).

Boc strategies typically rely on graduated acid-lability between the Boc (removed with TFA) and the linkage to the support (typically cleaved with HF or, alternatively, trifluoromethanesulfonic acid, TFMSA). In contrast, Fmoc strategies often display an orthogonality (Barany and Albericio, 1985) between the conditions for removal of the Fmoc protecting group (e.g., piperidine) and the conditions required for release from the support (often TFA).

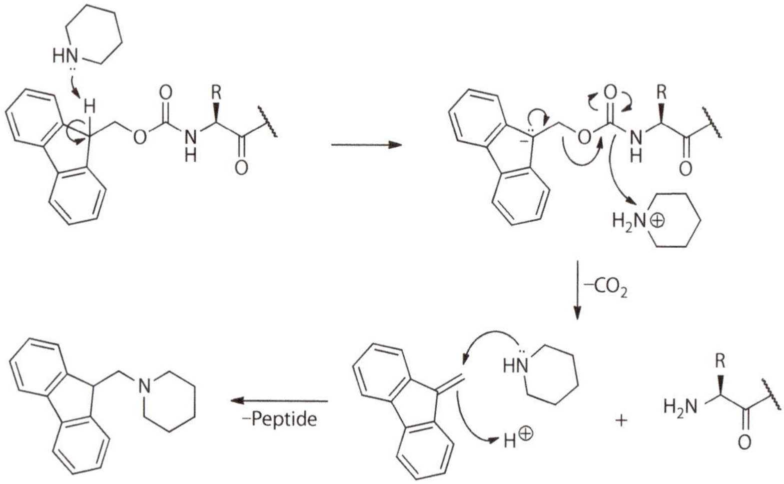

FIGURE 1.3 N^α-Fmoc deprotection by piperidine as base and nucleophilic scavenger.

In the following, we will focus on Fmoc-based solid-phase peptide synthesis (Fmoc-SPPS) and provide a description of side-chain protecting groups as well as other chemistries used in this strategy.

1.5 SIDE-CHAIN PROTECTING GROUPS

The semipermanent side-chain protecting groups for the Fmoc strategy have been developed extensively during the past decades (Isidro-Llobet et al., 2009). For some trifunctional amino acids such as Cys, Asp, Glu, and Lys, side-chain protection is essential for successful peptide synthesis; however, generally all other trifunctional amino acids are also semipermanently side-chain protected. The currently used protecting groups are *tert*-butyl (*t*-Bu) ester for Glu and Asp; *t*-Bu ether for Ser, Thr, and Tyr; 2,2,4,6,7-pentamethyl-dihydrobenzofuran-5-sulfonyl (Pbf) (Carpino et al., 1993) for Arg; and trityl (Trt) for Cys, Asn, Gln, and His (Figure 1.4). The phenyl rings in Trt derivatives can easily be modified with an electron-withdrawing or -donating group, which can be used to fine-tune its properties as a protecting group. For example, the monomethyl and monomethoxy-trityl moieties are acid-labile protecting groups for the N^ε-amine in Lys (Aletras et al., 1995).

There are specialized protecting groups that are typically not removed under the conditions required for the final release of the peptide from the support. They include the acetamidomethyl (Acm)-protecting group for Cys, which can be removed selectively with heavy metal salts such as thallium trifluoroacetate or, environmentally more benign, by iodine. Cys(Acm) can be used both in Boc- and in Fmoc-SPPS. The conditions for Acm removal, which are orthogonal to the conditions for removal of most other protecting groups, enable the use of Acm for the directed and sequential installment of a disulfide bridge in peptides with multiple disulfide bridges. There are also side-chain protecting groups for Lys, which allow the chemoselective deprotection of the N^ε-amine while leaving other side-chain protecting groups on the peptides.

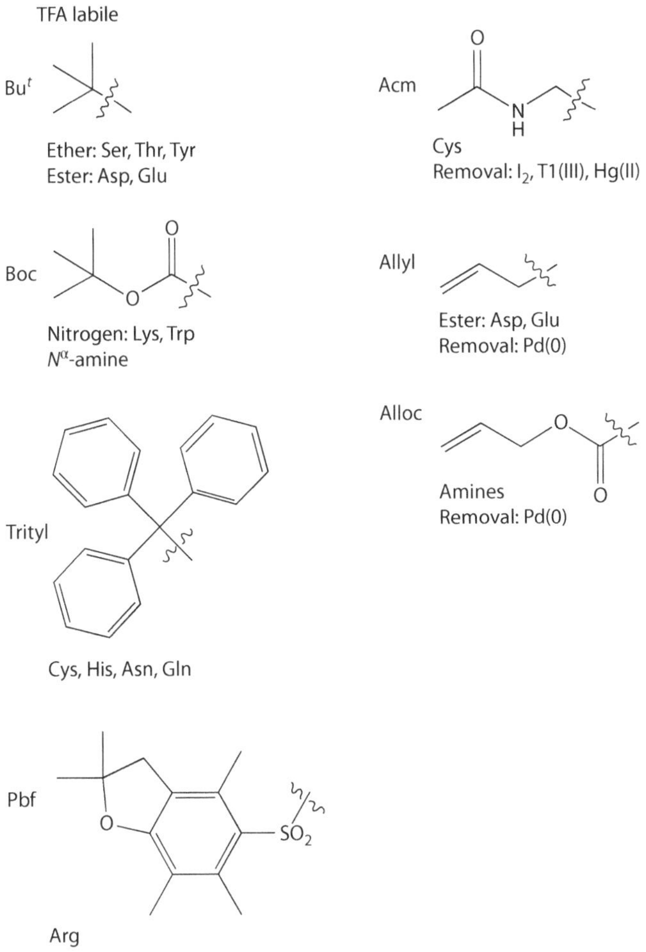

FIGURE 1.4 Some common side-chain protecting groups used in Fmoc-SPPS. The relevant amino acids are mentioned below each group. The protecting groups in the left column can be removed with concentrated trifluoroacetic acid (TFA), while the protecting groups in the right column are stable to TFA and are removed by other conditions.

These Lys-protecting groups include alloc, MMT (Matysiak et al., 1998), as well as Dde and ivDde. Similarly, selectively removable protecting groups for Glu include allyl and 2-phenyl-*iso*-propyl (Ph*i*Pr) (Yue et al., 1993).

1.6   COUPLING REAGENTS

Activation of the carboxylic acid moiety of the amino acid is required to be able to react with the N^α-amino group of the growing peptide chain. Carbodiimide-based coupling reagents, such as DCC or DIC (*N,N'*-diisopropylcarbodiimide), have been used for decades. A potential side reaction with carbodiimide-based reagents is the *O*-to-*N* rearrangement of the *O*-acylisourea intermediate and "overactivation" by the formation of symmetrical anhydride, which can lead to epimerization. Thus, carbodi-imides are used in combination with auxiliary nucleophiles such as 1-hydroxybenzo-triazole (HOBt) (König and Geiger, 1970) or 1-hydroxy-7-azabenzotriazole (HOAt), which form the corresponding activated esters. A newcomer is ethyl 2-cyano-2-(hydroxyimino)acetate (Oxyma) (Subirós-Funosas et al., 2009). Auxiliary nucleo-philes, such as HOBt, ensure that the optical integrity of the stereogenic center at the C-terminal of the activated amino acid residue is maintained throughout the cou-pling step. In automated syntheses, the coupling reagent DIC is preferred over DCC.

Numerous so-called in situ coupling reagents have been developed to reduce coupling time and minimize epimerization. The most important are HBTU (N-[(1H-benzotriazol-1-yl)(dimethylamino)methylene]-N-methylmethanaminium hexa-fluorophosphate N-oxide) (Dourtoglou et al., 1984), HATU (N-[(dimethylamino)-1H-1,2,3-triazole[4,5-b]pyridine-1-ylmethylene]-N-methylmethanaminium hexafluorophosphate N-oxide) (Carpino et al., 2002), PyBOP (1-benzotriazolyloxy-tris-pyrrolidinophosphonium hexafluorophosphate) (Coste et al., 1990), and the novel COMU (1-[(1-(Cyano-2-ethoxy-2-oxoethylideneaminooxy)-dimethylamino-morpholino-methylene)] methanaminium hexafluorophosphate) (El-Faham et al., 2009) reagents.

Although peptides normally are assembled by in situ activation, that is, by activation of the carboxylic acid immediately prior to coupling, preactivated amino acid build-ing blocks are also a viable possibility. Thus, pentafluorophenyl (Pfp) (Kisfaludy and Schön, 1983) esters of some Fmoc-amino acids are commercially available or can be prepared easily. For the couplings, just an auxiliary nucleophile such as HOBt is added, which then generates the transient OBt ester. The N-hydroxysuccinimide (NHS) esters are mainly used in amide bond formations in aqueous solution and not on solid phase.

1.7   AMINO ACIDS AND PEPTIDE SEQUENCES THAT SPELL TROUBLE

Although peptide synthesis most often is successful, some amino acids are prone to side reactions. This can also be viewed as a limitation in the current chemistry for protection and coupling of amino acids, including protecting groups. For example, when the Pbf group is cleaved of the Arg side chain, the resulting Pbf electrophile can alkylate nearby Trp indole rings in an irreversible reaction. This can often be mini-mized by using a Trp building block that is Boc protected at the indole nitrogen. The Asn residue is prone to dehydration and aspartimide formation, although side-chain Trt protection solves some of the problem. The thiol moiety in Cys can be protected by a number of protecting groups during Fmoc-SPPS. While the Trt protecting groups can be removed by acidolysis, the Acm group is stable to acid and base, but can be removed by salts of heavy metals, such as Hg and Tl, or by I_2. This provides a level of orthogonality, which can be utilized in the synthesis of peptides with multiple disulfide bridges, as mentioned. However, the incorporation of Cys in peptides is also likely to cause some level of racemization (Han et al., 1997; Kaiser et al., 1996). Cys can also be epimerized during Fmoc-SPPS; however, the Trt side-chain protecting group is often well suited to prevent this. β-branched amino acids such as Val and Ile can sometimes be difficult to incorporate quantitatively, giving rise to so-called difficult couplings.

There is a growing expectation that SPPS in the near future will be able to reliably provide proteins by direct synthesis. However, low purities and sometimes even failure in achieving the desired peptide sequence is still a frequently occurring problem, especially as the peptide becomes longer. The main reasons for this are believed to be steric hindrance and intra- and intermolecular aggregation. Amino acids that are prone to form β-sheets often lead to aggregation during peptide strand elongation, most likely due to their hydrogen bonding and hydrophobic properties (especially peptides containing a high proportion of Ala, Val, Ile, Asn, or Gln). These problems often lead to premature terminations or deletions of the elongating peptide sequence, which can be tedious to purify afterward. Intermolecular aggregation often leads to poor solvation of the peptidyl-polymer, but it is less pronounced when resins with a low loading capacity are being used.

1.8 RESINS

The most common resins for SPPS are based on polystyrene (PS), typically with 1% cross-linking. A starting point for the production of many other resins is provided by chloromethyl-polystyrene (Merrifield resin), made either by "chloromethylation" of polystyrene or, better, by copolymerization to directly incorporate the chloromethyl moiety. Another important base resin is aminomethyl-polystyrene. Although chloromethyl-polystyrene can be used for Boc-SPPS, in by far most cases a dedicated linker (handle) is inserted between the base resin and the first amino acid.

While polystyrene is an inexpensive resin that has been used widely, especially in Boc-SPPS and in Fmoc-SPPS of shorter sequences, other resins provide certain advantages. An important class of resins is constructed from a polystyrene core onto which PEG chains have been attached. Tentagel (Rapp Polymere, Germany) carries amino groups at the end of the PEG chains. It is important to realize that the polystyrene microparticles have been functionalized with PEG inside out, thus not only on the surface of the resin bead. These supports often provide higher purity of synthesized peptide, especially with longer sequences and with peptides that are prone to aggregate.

Another class of resins consists mainly of PEG and contains no polystyrene. They rely on cross-linking of PEG chains. This class includes the poly(ethylene glycol)-poly-(N,N-dimethylacrylamide) copolymer) PEGA supports, developed by Meldal, and the newer ChemMatrix resins. While PEGA is unique in being permeable to proteins up to 35–70 kDa, which makes it well suited for biochemical studies of peptides immobilized to the support, the ChemMatrix resins have gained in importance for standard Fmoc-SPPS.

1.9 LINKERS (HANDLES)

1.9.1 LINKER OVERVIEW

Linkers or handles are bifunctional molecules that on the one hand allow anchoring to the support and on the other hand have the characteristics of a protecting group, which enables anchoring of the growing peptide chain during SPPS and release of the final product under well-defined conditions.

Many resins are commercially available with a suitable linker attached to them. However, most resins can easily be functionalized with linkers (James, 1999). Typically, this is achieved by acylation of a resin containing a primary amino group; however, nucleophilic displacement of the chloride in Merrifield resin is also a well-established method.

Here we will focus on the chemistry for Fmoc-SPPS. Anchoring of the first amino acid normally occurs through the C-terminal, and the linker has to be chosen according to whether the peptide will have a C-terminal carboxylic acid or amide (Albericio and Giralt, 2004; Chan and White, 2004; Fields, 1997; Fields et al., 2002; Guillier et al., 2000). For peptides with a C-terminal carboxylic acid, 4-alkoxybenzyl alcohol type (Wang) linkers are an obvious choice. The first amino acid is coupled to Wang-type handles by esterification, and attention has to be paid to potential racemization in the formation of the ester bond. Release of the peptide acid is achieved by treatment with concentrated TFA. Substituted benzyl alcohol linkers, for example, dialkoxybenzyl structures, provide for higher acid-lability; hence, the release of peptides can be achieved with lower concentrations of TFA (e.g., TFA-dichloromethane). Other linkers for the synthesis of peptide acids include trityl based handles, for example, chloro-trityl chloride developed by Barlos. They have, in general, a higher acid-lability and the first amino acid is anchored racemization-free by nucleophilic displacement of the chloride.

For C-terminal amides in Fmoc-SPPS, the most common linker is a benzhydryl-type handle, the Rink amide linker; indeed the most commonly used resins are available with a Rink amide linker. The PAL (peptide amide linker) (Albericio and Barany, 1987; Albericio et al., 1990) handle, which has a trisalkoxybenzyl structure, is also suitable for Fmoc-SPPS of peptide amides.

Many biological active peptides or peptide building blocks are either C-terminal modified, meaning that they have a C-terminal functionality other than a carboxylic acid or an amide, or acyclic. Their synthesis would be difficult or impossible through C-terminal anchoring. There are specialized linkers that can provide this, for example, for the synthesis of peptide thioesters, as we will see below. However, one general strategy to obtain peptides with a C-terminus other than carboxylic acid or an amide uses anchoring of the peptide through a "vacant" backbone amide nitrogen. This so-called backbone amide linker (BAL) (Boas et al., 2009; Jensen et al., 1998) methodology is now widely used. In this strategy, the first amino acid is anchored by reductive amination, followed by acylation of the newly formed secondary amine. Thus, the growing peptide chain is anchored *not* through the C-terminal carboxyl but through a backbone amide nitrogen giving access to, in principle, any C-terminal modification (Figure 1.5). In the first implementation of this general strategy, amino acid derivatives were attached by convenient and reliable reductive amination to support-bound 5-(formyl-dimethoxyphenoxy)valeric acid, forming a trialkoxybenzylamine linkage. A defining feature of the BAL strategy is that the handle precursor is an aromatic aldehyde. BAL strategies have been used for the synthesis of peptide aldehydes (Guillaumie et al., 2000), peptide thioesters (Brask et al., 2003), cyclic peptides, and so on. The BAL strategy has mainly been used in Fmoc-SPPS, but it has also been adapted to Boc-SPPS (Bourne et al., 1999).

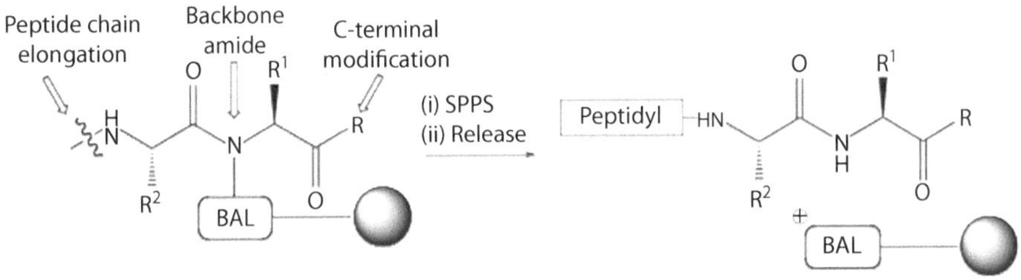

FIGURE 1.5 Backbone amide linker concept.

Another strategy for *C*-terminal modifications and cyclization is side-chain anchoring. Here, the first amino acid is anchored through the side-chain functional group. This could be Asp/Asn, Glu/Gln, or indeed most other trifunctional amino acids.

1.9.2 ACIDOLYTIC RELEASE AND DEPROTECTION OF PEPTIDES

All handles mentioned so far release the peptide from the support by acidolysis. The acid used for release of peptides in Fmoc-SPPS is normally TFA, as most resins swell well in TFA, peptides are often soluble in TFA, and TFA is volatile and easily evaporated. Concomitant to acidolytic release from the solid support, acid-labile side-chain protecting groups are also removed. This can generate a diverse set of carbocations, which are good electrophiles capable of reacting with nucleophilic side-chain functionalities. As mentioned, the cation derived from the Pbf protecting group can alkylate the indole ring in Trp. Thus, nucleophilic "scavengers" are added to the TFA. The most common is simply 5% of water, which normally is sufficient for *tert*-butyl protecting groups. Silanes such as trietheyl silane (TES) or trisisopropyl silane (TIS) are added to quench mainly trityl cations, while thiols can scavenge other cations.

1.9.3 LINKERS FOR THE SYNTHESIS OF PEPTIDE THIOESTERS

Several handles release the peptides by nucleophilic displacement, and a *C*-terminal functional group can be introduced in this step. In the 4-hydroxymethylbenzoic acid (HMBA) linker, the peptide is anchored through an ester bond. The peptide is released by nucleophiles, such as hydroxide or other competent nucleophiles. HMBA is maybe best used in combination with PEGA resins. Another handle is the so-called safety-catch linker based on a sulfonamide (Backes and Ellman, 1999; Ingenito et al., 1999). Here the first amino acid is anchored by acylation of a sulfonamide. The resultant linkage is stable during Fmoc-SPPS; however, the linkage is "activated" by *N*-alkylation, which makes it susceptible to nucleophilic displacement. This is especially useful for the synthesis of peptide thioesters, which are created by release with thiols. However, the peptide is typically released into a solution with excess of thiol in DMF or NMP, which can be difficult to remove. Recently, a new and simple approach was developed where the linker is an additional *C*-terminal glutamic acid moiety (Tofteng et al., 2009). The side-chain carboxylate is protected by a selectively

removable protecting group, such as the Ph*i*Pr, and upon its removal, the carboxylate is activated to form the five-membered pyroglutamyl imide. Treatment with a
thiolate will cleave the imide linkage, releasing the peptide thioester into solution.

1.10 INSTRUMENTS FOR AUTOMATED SPPS

SPPS is very amenable to automation, and the development of commercially available automated peptide synthesizers has come a long way allowing a high degree
of predictability and reproducibility in the assembly of peptides. These synthesizers range from semiautomated systems, with automated washing and Fmoc removal
steps, to fully automated synthesizers, which will assemble the whole sequence without intervention, to parallel synthesizers that can prepare large numbers of peptides
in parallel. Precise microwave heating has emerged as a new tool in peptide synthesis
(Pedersen et al., 2012), and this is now available in, for example, the semiautomated
Biotage Initiator⁺ SP Wave and in the fully automated Biotage Syro *Wave* (Malik
et al., 2010) instruments. The latter can either synthesize one peptide at a time with
microwave heating or up to 96 peptides in parallel at room temperature.

1.11 POST-TRANSLATIONAL MODIFICATIONS AND OTHER MODIFICATIONS INTRODUCED IN SPPS

Post-translational modifications of peptides and proteins are ubiquitous in nature.
They include phosphorylation, *O*- and *N*-glycosylation, farnesylation and palmi-
toylation, γ-carboxylation of glutamic acid, and much more (Walsh, 2006). Incorporation of some of these modified amino acids in a sequence can be achieved by
SPPS. For some of these modifications, suitably protected amino acid building
blocks are commercially available and can be incorporated into the peptide sequence
using standard methods. This is certainly the case with phosphopeptides, where
Fmoc-protected phosphorylated Ser and Thr building blocks, which carry a single
O-benzyl protecting group on the phosphate, can be used in standard Fmoc SPPS,
with longer deprotection times. Some Fmoc protected *O*- or *N*-glycosylated amino
acids, that is, Ser, Thr, and Asn, are also commercially available and can be used in
SPPS. Typically, these building blocks contain mono-, di-, or trisaccharide glycans.
The glycans are most commonly protected with *O*-acetyl or -benzoyl ester moieties,
which can be removed upon completion of the synthesis by treatment with, for example, a dilute solution of methoxide in methanol. While numerous glycopeptides have
been prepared in this manner, it adds another level of complexity to the synthesis
design and can cause new side reactions such as β-elimination of the glycan, leaving
a dehydro-alanine residue in the peptide sequence (Jansson et al., 2004). The synthesis of peptides with appended farnesyl, geranyl-geranyl, and palmitoyl groups has
been achieved (Brunsveld et al., 2006), but currently remains a task for specialists.
Other lipids, for example, simple acylation of a Lys N$^{\varepsilon}$-amine, can be achieved while
the peptide remains on the solid support or after it has been released from the support, provided there is only one Lys in the sequence.

1.12   CONCLUSIONS AND FUTURE DIRECTIONS

An understanding of the possibilities and current limitations in SPPS is a good starting point for designing peptides for biochemical, biomedical, and biophysical studies. The chemical tools available to SPPS define which synthetic peptides are available and thus which peptide-based research and development can be carried out. Many peptides can be prepared with ease and predictability by SPPS. Peptides up to 15–20 amino acids in length are normally routinely available from companies specializing in on-demand synthesis. Longer peptides and peptides carrying post-translational modifications can often be prepared by more specialized companies or, in more difficult cases, through academic collaborations. Several instrument companies supply automated peptide synthesizers, which can be used not only by the specialist but also by the nonspecialist with a chemical interest.

ABBREVIATIONS*

Acm: acetamidomethyl
BAL: Backbone Amide Linker
Boc: tert-butyloxycarbonyl
BOP: benzotriazol-1-yl-*N*-oxy-tris(dimethylamino)phosphonium hexafluorophosphate
Cbz: benzyloxycarbonyl
COMU: (1-[1-(cyano-2-ethoxy-2-oxoethylideneaminooxy)-dimethylamino-morpholino-methylene)] methanaminium hexafluorophosphate)
DIC: *N,N'*-di-isopropylcarbodiimide
DIEA: *N,N*-diisopropylethylamine
DMF: *N,N*-dimethylformamide
Fmoc: fluoren-9-ylmethyloxycarbonyl
HATU: (*N*-[(dimethylamino)-1*H*-1,2,3-triazole[4,5-b]pyridine-1-ylmethylene]-*N*-methylmethanaminium hexafluorophosphate *N*-oxide)
HBTU: (*N*-[(1*H*-benzotriazol-1-yl)(dimethylamino)methylene]-*N*-methylmethanaminium hexafluorophosphate *N*-oxide)
HOBt: 1-hydroxybenzotriazole
HOAt: 3-hydroxy-3*H*-1,2,3-triazolo[4,5-*b*]pyridine [1-hydroxy-7-azabenzotriazole]
NHS: *N*-hydroxysuccinimide
NMP: *N*-methyl-2-pyrrolidinone
Pbf: 2,2,4,6,7-pentamethyl-dihydrobenzofuran-5-sulfonyl
Pfp: pentafluorophenyl
PyBOP: (1-benzotriazolyloxy-tris-pyrrolidinophosphonium hexafluorophosphate)
TFA: Trifluoroacetic acid
TFMSA: Trifluoromethanesulfonic acid
Trt: Trityl (triphenylmethyl)

---

* See also Figure 1.4

REFERENCES

Albericio, F., and G. Barany. 1987. An acid-labile anchoring linkage for solid-phase synthesis of C-terminal peptide amides under mild conditions. *Int. J. Pept. Protein Res.* 30:206–216.

Albericio, F., and E. Giralt. 2004. Handles and Supports. In Goodman, M., Felix, A., Moroder, L., Toniolo, C. (Eds.), *Hoyben-Weyl, E22a: Synthesis of Peptides and Peptidomimetics.* Thieme: Stuttgart, 685–725.

Albericio, F., N. Kneib-Cordonier, S. Biancalana, L. Gera, R.I. Masada, D. Hudson, and G. Barany. 1990. Preparation and application of the 5-(4-(9-fluorenylmethyloxycarbonyl)aminomethyl-3,5-dimethoxyphenoxy)valeric acid (PAL) handle for the solid-phase synthesis of C-terminal peptide amides under mild conditions. *J. Org. Chem.* 55:3730–3743.

Aletras, A., K. Barlos, D. Gatos, and S. Koutsogianni. 1995. Preparation of the very acid-sensitive Fmoc-Lys(Mtt)-OH—Application in the synthesis of side-chain to side-chain cyclic peptides and oligolysine cores suitable for solid-phase assembly of MAPS and TASPs. *Int. J. Pept. Prot. Res.* 45:488–496.

Atherton, E., H. Fox, D. Harkiss, C.J. Logan, R.C. Sheppard, and B.J. Williams. 1978a. A mild procedure for solid-phase peptide synthesis—Use of the fluorenylmethoxycarbonyl amino acids. *J. Chem. Soc. Chem. Commun.* 13:537–539.

Atherton, E., H. Fox, D. Harkiss, and R.C. Sheppard. 1978b. Application of polyamide resins to polypeptide synthesis—Improved synthesis of beta-endorphin using fluorenylmethoxycarbonyl amino acids. *J. Chem. Soc. Chem. Commun.* 13:539–540.

Backes, B.J., and J.A. Ellman. 1999. An alkanesulfonamide 'safety-catch' linker for solid-phase synthesis. *J. Org. Chem.* 64:2322–2330.

Barany, G., and F. Albericio. 1985. A 3-dimensional orthogonal protection scheme for solid-phase peptide synthesis under mild conditions. *J. Am. Chem. Soc.* 107:4936–4942.

Bergmann, M., and L. Zervas. 1932. Über ein allgemeines Verfahren der Peptid-Synthese (A general method for peptide synthesis). *Berichte* 65:1192–1201.

Boas U., J. Brask, and K. J. Jensen. 2009. The backbone amide linker for solid-phase synthesis. *Chem. Rev.* 109:2092–2118.

Bourne, G.T., W.D.F. Meutermans, P.F. Alewood, R.P. McGeary, M. Scanlon, A.A. Watson, and M.L. Smythe. 1999. A backbone linker for BOC-based peptide synthesis and on-resin cyclization: Synthesis of stylostatin1. *J. Org. Chem.* 64:3095–3101.

Brask, J., F. Albericio, and K.J. Jensen. 2003. Fmoc solid-phase synthesis of peptide thioesters by masking as trithioortho esters. *Org. Lett.* 5:2951–2953.

Brunsveld, L., J. Kuhlmann, and H. Waldmann. 2006. Synthesis of palmitoylated Ras-peptides and -proteins. *Methods* 40:151–165.

Carpino, L.A., and G.Y. Han. 1970, The 9-fluorenylmethoxycarbonyl function, a new base-sensitive amino-protecting group. *J. Am. Chem. Soc.* 92:5748–5749.

Carpino, L.A., and G.Y. Han. 1972. The 9-fluorenylmethoxycarbonyl function amino-protecting group. *J. Org. Chem.* 37:3404–3409.

Carpino, L.A., H. Imazumi, A. El-Faham, F.J. Ferrer, C. Zhang, Y. Lee, B.M. Foxman et al. 2002. The uronium/guanidinium peptide coupling reagents: Finally the true uronium salts. *Angew. Chem., Int. Ed.* 41:441–445.

Carpino, L.A., H. Shroff, S.A. Triolo, E.-S.M. E. Mansour, H Wenschuh, and F. Albericio. 1993. The 2,2,4,6,7-pentamethyldihydrobenzofuran-5-sulfonyl group (Pbf) as arginine side-chain protectant. *Tetrahedron Lett.* 34:7829–7832.

Chan, W.C., and White, P.D. (Eds.). 2004. *Fmoc Solid Phase Peptide Synthesis.* Oxford University Press: Oxford.

Coste, J., D. Lenguyen, and B. Castro. 1990. PyBOP—A new peptide coupling reagent devoid of toxic by-product. *Tetrahedron Lett.* 31:205–208.

Dourtoglou, V., B. Gross, V. Lambropoulou, and C. Zioudrou. 1984. O-Benzotriazolyl-N,N,N′,N′-tetramethyluronium hexafluorophosphate as coupling reagent for the synthesis of peptides of biological interest. *Synthesis* 1984:572–574.

El-Faham, A., R.S. Funosas, R. Prohens, and F. Albericio. 2009. COMU: A safer and more effective replacement for benzotriazole-based uronium coupling reagents. *Chem. Eur. J.* 15:9404–9416.

Fields, G.B. (Ed.). 1997. *Methods in Enzymology (Solid-Phase Peptide Synthesis)*, Volume 289. Academic Press: San Diego, CA.

Fields, G.B., Z. Tian, and G. Barany. 2002. Principles and Practice of Solid-Phase Peptide Synthesis. In Grant, G.A. (Ed.), *Synthetic Peptides: A Users Guide*, 2nd Ed. W. H. Freeman & Company: New York, 93–219.

Guillaumie, F., J.C. Kappel, N.M. Kelly, G. Barany, and K.J. Jensen. 2000. Solid-phase synthesis of C-terminal peptide aldehydes from amino acetals anchored to a Backbone Amide Linker (BAL) handle. *Tetrahedron Lett.* 41:6131–6135.

Guillier, F., D. Orain, and M. Bradley. 2000. Linkers and cleavage strategies in solid-phase organic synthesis and combinatorial chemistry. *Chem. Rev.* 100:2091–2157.

Han, Y., F. Albericio, and G. Barany. 1997. Occurrence and minimization of cysteine racemization during stepwise solid-phase peptide synthesis. *J. Org. Chem.* 62:4307–4312.

Hudson, D. 1999a. Matrix assisted synthetic transformations: A mosaic of diverse contributions. I. The pattern emerges. *J. Comb. Chem.* 1:333–360.

Hudson, D. 1999b. Matrix assisted synthetic transformations: A mosaic of diverse contributions. II. The pattern is completed. *J. Comb. Chem.* 1:403–457.

Ingenito, R., E. Bianchi, D. Fattori, and A. Pessi. 1999. Solid phase synthesis of peptide C-terminal thioesters by Fmoc/t-Bu chemistry. *J. Am. Chem. Soc.* 121:11369–11374.

Isidro-Llobet, A., M. Alvarez, and F. Albericio. 2009. Amino acid protecting groups. *Chem. Rev.* 109:2455–2504.

Jansson, A.M., P.M.S. Hilaire, and M. Meldal. 2004. Synthesis of glycopeptides. In M. Goodmann et al. (Eds.), *Synthesis of Peptides and Peptidomimetics*, Volume 22b. Thieme: Stuttgart, 235–322.

James, I.W. 1999. Linkers for solid phase organic synthesis. *Tetrahedron* 55:4855–4946.

Jensen, K.J., J. Alsina, M.F. Songster, J. Vágner, F. Albericio, and G. Barany. 1998. Backbone Amide Linker (BAL) strategy for solid-phase synthesis of C-terminal modified and cyclic peptides. *J. Am. Chem. Soc.* 120:5441–5452.

Kaiser, T., G.J. Nicholson, H.J. Kohlbau, and W. Voelter. 1996. Racemization studies of Fmoc-Cys(Trt)-OH during stepwise Fmoc-solid phase peptide synthesis. *Tetrahedron Lett.* 37:1187–1190.

Kisfaludy, L., and I. Schön. 1983. Preparation and applications of pentafluorophenyl esters of 9-fluorenylmethyloxycarbonyl amino acids for peptide synthesis. *Synthesis*:325–327.

König, W., and R. Geiger. 1970. Eine neue Methode zur Synthese von Peptiden: Aktivierung der Carboxylgruppe mit Dicyclohexylcarbodiimid unter Zusatz von 1-Hydroxy-benzotriazolen (A new method for the synthesis of peptides: Activation of the carboxyl group with dicyclohexylcarbodiimide with addition of 1-hydroxybenzotriazols). *Chem. Ber.-Recl.* 103:788–798.

Malik, L., A.P. Tofteng, S.L. Pedersen, K.K. Sørensen, and K.J. Jensen. 2010. Automated 'X-Y' robot for peptide synthesis with microwave heating: Application to difficult peptide sequences and protein domains. *J. Pept. Sci.* 16:506–512.

Matysiak, S., T. Böldicke, W. Tegge, and R. Frank. 1998. Evaluation of monomethoxytrityl and dimethoxytrityl as orthogonal amino protecting groups in Fmoc solid phase peptide synthesis. *Tetrahedron Lett.* 39:1733–1734.

Merrifield, R.B. 1963. Solid phase peptide synthesis. 1. Synthesis of a tetrapeptide. *J. Am. Chem. Soc.* 85:2149–2154.

Merrifield, R.B. 1985. Solid-phase peptide synthesis (Nobel Lecture). *Science* 232:341–347.

Pedersen, S.L., A.P. Tofteng, L. Malik, and K.J. Jensen. 2012. Microwave heating in solid-phase peptide synthesis. *Chem. Soc. Rev.* 41(5):1826–1844.

Subirós-Funosas, R., R. Prohens, R. Barbas, A. El-Faham, and F. Albericio. 2009. Oxyma: An efficient additive for peptide synthesis to replace the benzotriazole-based HOBt and HOAt with a lower risk of explosion. *Chem. Eur. J.* 15:9394–9403.

Tofteng, A.P., K.K. Sørensen, K.W. Conde-Frieboes, T. Hoeg-Jensen, and K.J. Jensen. 2009. Fmoc solid-phase synthesis of C-terminal peptide thioesters via formation of a backbone pyroglutamyl imide moiety. *Angew. Chem.* 48:7411–7414.

Walsh, C.T. 2006. *Posttranslational Modifications of Proteins*. Roberts and Company Publishers: Englewood, CO, 1–490.

Yue, C.W., J. Thierry, and P. Potier. 1993. 2-Phenyl isopropyl esters as carboxyl terminus protecting groups in the fast synthesis of peptide fragments. *Tetrahedron Lett.* 34:323–326.

2 Protein Expression

Nanni Din

CONTENTS

2.1 INTRODUCTION

Protein expression occurs in all living cells, but the focus here is on tailor-made protein expression by the aid of recombinant DNA technology. Recombinant DNA (rDNA) technology, also called "gene technology" or "genetic engineering," involves taking genetic material from one source and recombining it in vitro with genetic material from another source and thereafter introducing the recombined material into a host cell. Genetic engineering, combined with cell culture technology, makes it possible to produce large quantities of virtually any protein, including therapeutically interesting proteins. The pharmaceutical industry realized the potentials of the new technology concomitant with the development of its tools in the early 1970s, and the first gene technologically produced human protein drug was approved for marketing already in 1982, that is, by industry standards after a relatively short development time; this landmark drug was insulin, a peptide hormone used to treat diabetes. At present, more than 200 peptides or proteins produced by gene technology methods have been approved as human therapeutics, and many more are in early or late development. In contrast to insulin, for which a production method based on extraction from pig or ox pancreas has existed for many years, most of the other now approved gene technology products have never been in common use as drugs before, because they were impossible to produce in sufficient quantities. Clearly, one of the great advantages of gene technology is the ability to produce proteins that previously were so difficult to isolate as to make their use as pharmaceuticals completely unrealistic.

Another advantage of gene technology is that it allows introductions of modifications into proteins at chosen positions. Although such modifications may only involve a few changes in the amino acid sequence made by introducing specific mutations in the gene encoding the protein, they may lead to advantageously altered physicochemical characteristics, for example, new solubility properties or a different physiological half-life of the active polypeptide. However, more radical changes giving rise to novel activities are also possible.

This chapter aims to describe the basic tools and concepts of gene technology as used in pharmaceutical research and production and to sum up the current status and future trends for recombinant protein therapeutics.

2.2   GENERAL METHODS IN GENE TECHNOLOGY

2.2.1   BASIC DNA CLONING TOOLS

The basic methods for recombinant DNA technology were developed in the beginning of the 1970s. Despite improvements and innovations, the principles are still based on the same tools: DNA-modifying enzymes and DNA vectors and host cells (Sambrook and Russell, 2001).

A key factor in the DNA technology breakthrough of the 1970s was the discovery of two types of enzymes, the restriction enzymes and the DNA ligases. Restriction enzymes are endonucleases that recognize and cleave DNA at specific sequences, typically 4–8 base pairs (bp) long. Cleavage generates DNA fragments of specific sizes, some of which may contain a single complete gene. Restriction endonucleases can be isolated from a variety of microorganisms, and a wide range of enzymes, each with a specific recognition sequence, is now commercially available. An example is the restriction enzyme *BamHI* (endonuclease I isolated from *Bacillus amyloliquefaciens H*), which cleaves DNA at the sequence GGATCC. In contrast to the restriction enzymes, DNA ligases have no specific sequence requirement and can join the ends of any DNA molecules. These features of the restriction enzymes and DNA ligases enable in vitro cleavage and rejoining of DNA fragments as illustrated in Figure 2.1.

In order to clone a DNA molecule, it must be introduced into a host cell in a form which allows it to be replicated. Replication requires the presence of a specific sequence (replication origin) which can be provided by circular "mini-chromosomes" called plasmids. Therefore, to generate cloneable DNA, a restriction

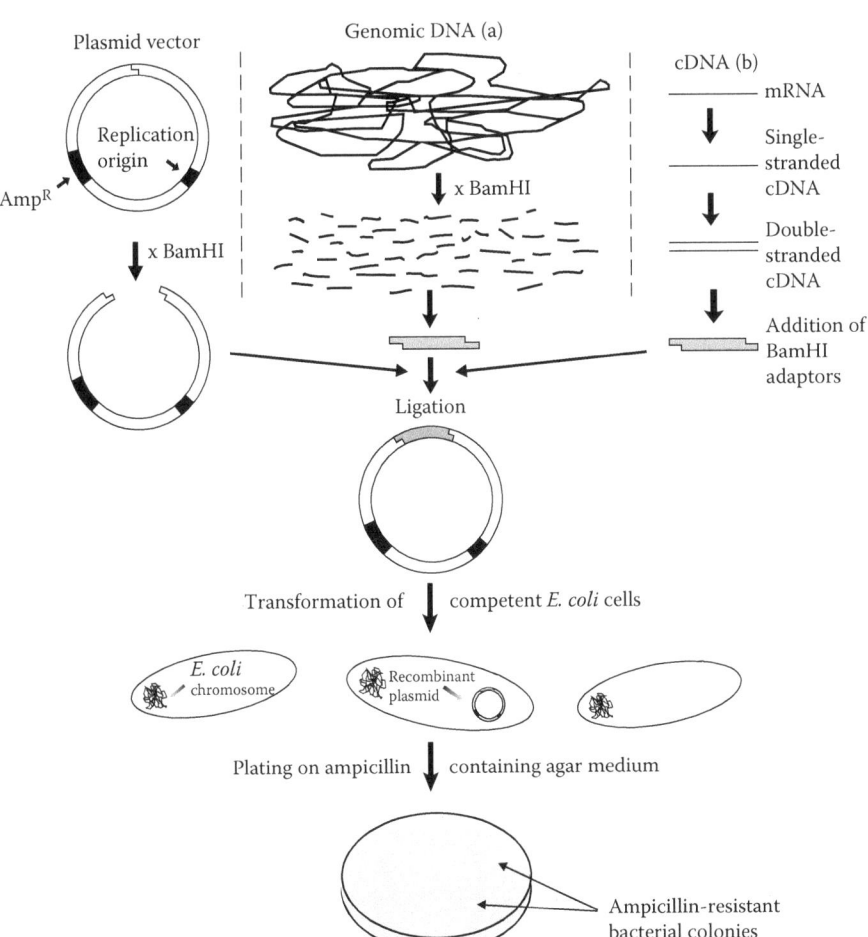

FIGURE 2.1   Overview of the steps involved in (a) the cloning of chromosomal DNA or (b) complementary DNA (cDNA) fragments into a plasmid vector. The plasmid vector carries a replication origin and a gene that confers resistance to the antibiotic ampicillin (AmpR). Outside these regions, there is a unique site for the restriction endonuclease (BamHI). In order to clone DNA fragments into this site, vector and foreign DNA are cleaved with BamHI, mixed, and joined together by the aid of the enzyme DNA ligase. The mixture of ligated DNA molecules is transformed into *E. coli* cells. Only cells that have received a plasmid will grow in the presence of ampicillin. Generation of double-stranded cDNA from mRNA requires the following steps: In the first step, a chemically synthesized oligo-dT primer is annealed to the 3′ polyadenylated tails of the mRNA molecules. Single-stranded cDNA is then synthesized by the aid of reverse transcriptase. After removal of the mRNA molecules by hydrolysis with sodium hydroxide, the population of single-stranded cDNA molecules are used as templates in a new round of DNA synthesis catalyzed by the enzyme DNA polymerase. Synthetic DNA adaptors for BamHI are subsequently ligated to the double-stranded cDNA molecules and the resulting fragments can be inserted into the BamHI site of the vector.

fragment is joined to a plasmid DNA molecule that can replicate in a suitable host cell. A very important host–plasmid system is the bacterium *Escherichia coli* and its naturally occurring small (2–10 Kbp) plasmids. Some of these plasmids have been especially modified for cloning purposes, and they are therefore called vectors. Apart from the replication origin, an important feature of the vector is the presence of a gene coding for a selectable marker, such as resistance to an antibiotic. When recombinant plasmids are introduced into *E. coli* (in a process called transformation), the resistance gene will allow the cells to grow in the presence of the antibiotic, while cells without the plasmid will die. A recombinant plasmid is able to multiply until each cell contains several copies (often in the range from 20 to 200), and as the doubling time for *E. coli* under optimal conditions is 20–30 minutes, it is possible to multiply one single copy of the recombinant plasmid to more than 10^{10} molecules within 12 hours; the plasmid with the particular gene can now be purified and further manipulated.

2.2.2   PCR

The polymerase chain reaction (PCR), invented in the early 1980s, is another useful cloning tool (Daniels, 2011). The PCR technique is based on enzymatic in vitro amplification of a specific DNA sequence. Figure 2.2 illustrates how one specific sequence, called the target or template DNA, can be amplified from a complex DNA population.

Repeated rounds of DNA synthesis are catalyzed by a thermostable DNA polymerase in a suitable buffer containing the DNA template, a specific set of two short synthetic DNA primers, and a mixture of the four deoxynucleotides. The primer set is designed so that one primer can bind near one end of the target DNA and the other can bind near the other end of the complementary strand. When the first primer anneals to its target, it creates a starting point for the DNA polymerase, and a complementary copy of the template is generated. By heating the reaction mixture, the two complementary strands are separated, and targets for both primers are thereby exposed. When the temperature is lowered, both primers can anneal to their respective targets and both strands can now be copied by DNA polymerase. Each consecutive cycle consists of the same series of strand separation, primer annealing, and DNA polymerization reactions, and after about 20 cycles the target sequence may become amplified more than a million fold. Thereafter, it is easy to clone this almost pure DNA fragment.

Some knowledge about the sequence of the template DNA is needed in order to design primers which are able to bind to the target. Initially, protein sequence data was used to design and synthesize a short DNA strand with corresponding coding capacity. However, since the completion of the Human Genome Project (see below), the sequences of prototype versions of any human protein coding gene have become available in public data bases, providing the necessary sequence information for primer design; moreover, cloned DNA encoding almost any human protein is also publicly available (cf. http://www.orfeomecollaboration.org). These resources have vastly simplified genetic engineering work aimed at expressing human proteins. Nevertheless, the experimental use of DNA still requires knowledge of the basic recombinant DNA techniques.

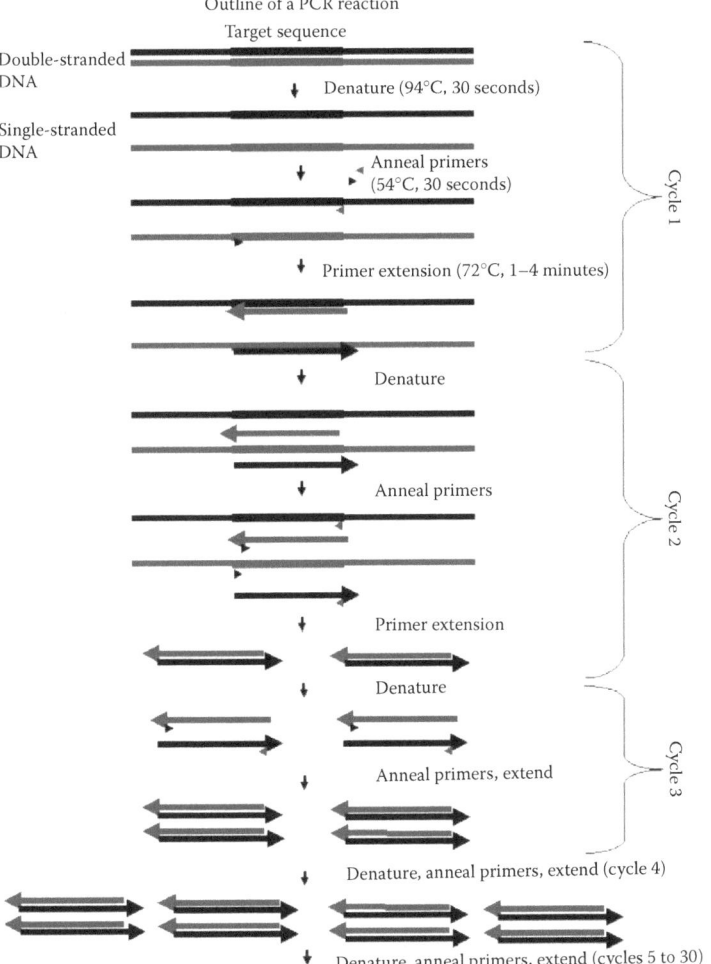

FIGURE 2.2 The polymerase chain reaction (PCR). A PCR reaction mixture contains double-stranded DNA, synthetic oligomeric single-stranded DNA (primers), a thermostable DNA polymerase, and dNTPs. There are three major steps in a PCR: denaturation of double-stranded DNA (at 94°C), annealing of the primers to their now available complementary sequences in the single-stranded DNA (at a suitable temperature for the primers, for instance, 54°C), and extension of the primers (i.e., incorporation of dNTPs, leading to new DNA synthesis) by the DNA polymerase (at 72°C, which is optimal for several thermostable polymerases). Repeated rounds of denaturation, annealing, and primer extension will produce double-stranded DNA molecules identical to the region defined by the two primers in the original target sequence. Because both strands are copied during PCR, there is an exponential increase in the number of copies of the gene. Supposing there is only one copy of the wanted gene before the cycling starts, after one cycle, there will be 2 copies, after two cycles, there will be $2^2 = 4$ copies, after three cycles there will be $2^3 = 8$ copies, and so on until the dNTPs are used up and a plateau copy level is reached. Note that the original target sequence in the complex DNA will continue to be copied, although this is omitted from cycle 2 onward for the sake of simplicity.

2.2.3   DNA Sequencing

With the availability of cloned or PCR-amplified DNA fragments that can be purified in substantial amounts, it is possible to determine the sequence of nucleotide bases (A, G, C, T) and thus the potential protein coding capability of the DNA. The first DNA sequences were obtained in the early 1970s by academic researchers using laborious methods, but nowadays there are much more efficient methods for obtaining this indispensable knowledge (Pettersson et al., 2009). The initial DNA sequencing methods relied on the ability to generate "ladders" of DNA fragments that could be separated by electrophoresis, with each step in the ladder representing one nucleotide base. This basal methodology has been ingeniously adapted to increase the speed of sequencing by orders of magnitude. High throughput sequencing methods differing significantly from the "ladder" technology have also been developed; development of sequence data analysis programs has gone hand in hand with this.

2.2.4   Genomics

In the mid-1980s, when high throughput DNA sequencing was beginning to show its impact on research, the term "genomics" was coined to describe the combination of gene mapping and sequencing with studies to uncover gene function (McKusick and Ruddle, 1987). "Genomics" then included the combined characterization of genes and their mRNA and protein products. However, nowadays the separate terms "transcriptomics" and "proteomics" are generally used to describe mRNA and protein repertoires, respectively, and genomics has become a narrower term to describe gene mapping and sequencing. Comprehensive genome sequencing has now been performed for many different organisms, including humans.

The Human Genome Project was a huge collaborative effort, combining work from both public and private laboratories, with the goal to determine the sequences of the 3 billion base pairs that make up human DNA and identify the protein coding regions (Rowen et al., 1997; also cf. Human Genome Project Information at www .ornl.gov). When formally started in 1990, the project was planned to last 15 years, but technological advances accelerated the completion date to 2003. The methods used included both cloning of human chromosomal DNA (also called genomic DNA) and of DNA representing the part of the genome transcribed into protein-coding mRNA (so-called cDNA for complementary or copy DNA). Figure 2.1 shows a schematic overview of these two cloning strategies. While genomic DNA can be isolated directly from cells, cDNA must be made from isolated mRNA by an enzymatic in vitro reaction using enzyme types aptly named "reverse transcriptase."

A surprising result of the Genome Project (cf. The Human Genome, 2001a,b) was that only a small fraction of the genome, about 2%, encodes protein; however, a quite larger fraction, around 25%, must still be denoted as protein genes (while the remaining ca. 75% of the genome is nominated as intergenic DNA and is without known genetic function). This is because most human genes have a complex structure, where protein coding regions, so-called exons, are interspersed with noncoding regions called introns. Introns can range in size from less than 100 bp to several Kbp, and most human genes contain several introns; on the average, introns contribute about 90% to human gene size. DNA regions involved in gene expression also contribute to gene size. Another surprise from the finished genome project was that the total number of genes is now estimated at 20,000 to 25,000, much lower than previous estimates of 80,000 to 140,000. These estimates were partly based on extrapolations from early analysis of gene-rich areas as opposed to the final comprehensive analysis of gene-rich and gene-poor areas, and partly on estimates of total protein complexity in all human tissues. The number of different proteins in an individual is certainly considerably higher than the number of protein-coding genes. The explanation for this apparent paradox is that many genes are subject to alternative splicing whereby one or more exons can be omitted, so different primary protein sequences originate from the same gene. Post-translational modifications of the primary protein sequence further increase the number of different proteins (see Section 2.3 for basics on transcription, splicing, translation, and post-translational modifications). The term proteome has been coined to describe the entire complement of proteins, including the modifications made to any particular set of proteins produced by an organism or cell. The proteome will therefore vary with time and different needs or stresses that a cell or organism encounters. Quite recently, a group of scientists have launched The Human Proteome Project (Rabilloud et al., 2010), which is very similar to the now completed Human Genome Project. The goal of the project is to provide a comprehensive list of proteins present in normal and diseased cells and tissues; while such knowledge would greatly help at many stages of the development of new drugs, it is currently debatable whether it is at all possible to obtain 100% comprehensive lists and whether it is worth the expense to make the attempt. However, there is no doubt that even partial proteome lists will be very useful since functions are still unknown for more than 50% of the genes discovered to date.

2.3 EXPRESSION OF RECOMBINANT PROTEINS

Production of recombinant proteins and peptides to be used as pharmaceuticals requires efficient expression systems that generate proteins with the correct structure and function. In this section, some requirements for expression will be outlined together with a discussion of the relative merits and drawbacks of different host–vector expression systems (Fernandez and Hoeffler, 1999; Hodgson, 1993).

2.3.1 BASICS: TRANSCRIPTION, TRANSLATION, AND PROTEIN MODIFICATIONS

The first step in gene expression is transcription into RNA. Transcription is initiated by the binding of the enzyme RNA polymerase to a specific DNA sequence at the beginning of the coding region, the so-called promoter. Many different promoters and other elements that influence transcription efficiency have been identified in both pro- and eukaryotes. Promoters are most often species-specific, and in higher organisms also tissue-specific. Transcription termination is also determined by specific DNA sequences, located downstream of the coding region. In prokaryotes, transcription from protein coding genes produces RNA that is used without any major modification as mRNA to direct protein synthesis. In contrast, transcription from most eukaryotic protein coding genes produces primary RNA molecules containing both exons and introns; complex enzymatic processing steps then remove the intron

TABLE 2.1

Examples of Post-Translational Protein Modifications

Modification	Occurrence
Cleavage of peptide bonds (maturation)	Pro- and eukaryotes
Formation of disulfide bonds	
N-linked glycosylation of asparagine	Eukaryotes
O-linked glycosylation of threonine or serine	
Hydroxylation of proline or lysine	
Myristylation of N-terminal glycine	
γ-carboxylation of glutamic acid	Higher eukaryotes
Amidation of C-terminal glycine	
Sulfation of tyrosine	

regions from the primary RNA in a process known as splicing, and this generates mature mRNA containing only the exons joined together, normally in the order in which they occur in the gene. The splicing process can however also give rise to mRNA where one or more exons are omitted; this is called alternative splicing. Prokaryotes do not have intron-containing genes and splicing machinery and are therefore unable to process heterologous intron-containing transcripts.

The next step in gene expression is translation of the mRNA into protein. The basic translation mechanisms are the same as in prokaryotes and eukaryotes. The codon for translation initiation is with rare exceptions universal and leads to translation products containing methionine at their N-terminus. This N-terminal methionine is, however, often removed already during protein elongation by methionyl amino peptidase associated with the ribosomes. Most proteins destined for secretion from cells are synthesized with a so-called signal peptide of 20–30 amino acids at their N-terminus. The signal peptide mediates contact between the membrane and the nascent protein, and it is usually cleaved off during protein translocation over the membrane even before the protein is fully synthesized. In comparison with prokaryotes, eukaryotes have a very complex pathway for secretion of proteins, and many essential modifications of proteins occur in the secretory pathway. Therefore, maturation of the primary translation product to a protein with correct secondary and tertiary structure cannot always be attained if the classical *E. coli* cloning host is used for expression of an eukaryotic protein. To overcome this problem, several efficient eukaryotic expression systems have been developed. Examples of post-translational protein modifications in pro- and eukaryotes are listed in Table 2.1 (Creighton, 1993).

2.3.2 CHOICE OF EXPRESSION SYSTEM

2.3.2.1 *E. coli*

Since 1977, when a human protein (somatostatin) was expressed in a functional form for the first time in *E. coli* (Itakura et al., 1977) many hundreds of eukaryotic genes have been expressed in this organism but very often not in a functional form. Because *E. coli* cannot perform many of the post-translational modifications that occur in mammalian cells, it is most suitable for production of proteins with few or no modifications. However, its ease of handling and its safety as a production organism favor the retention of *E. coli* as a host even for proteins that are not completely matured in this organism. Various ingenious methods have been devised to complete the final maturation in vitro and to increase the in vitro processing capacity (Georgiou and Valax, 1996). A number of approved recombinant pharmaceuticals are produced in *E. coli*, e.g., growth hormone, interferon-α, and interleukin 2.

E. *coli* expression can be optimized in various ways. Promoters from highly transcribed bacterial genes or synthetic promoters are used to obtain high production levels. Promoters that can be regulated by altering the growth conditions (temperature, added chemicals, etc.) are often used for production of proteins that are harmful to the cell. Examples of such promoters are the *lac* promoter from the β-galactosidase operon or the bacteriophage λ promoters P_L and P_R. The advantage of regulated promoters is that the synthesis of the harmful protein can be delayed until the cell density has reached a high level where a short protein production phase results in generation of large amounts of product before the cells are affected.

2.3.2.2 Yeast

Saccharomyces cerevisiae (baker's yeast) is a eukaryotic microorganism that performs many of the same post-translational modifications that also occur in humans. This makes it useful for production of peptides and proteins of medium complexity (Romanos et al., 1992). The long tradition of using yeast for food production contributes significantly to its acceptability as a producer of pharmaceutical proteins.

S. *cerevisiae* contains a natural plasmid (2-micron DNA) that can be used as a vector when equipped with a selectable marker. Instead of a gene conferring resistance to an antibiotic, a yeast gene that complements a mutant gene in the host genome is often used for plasmid selection. This principle (which can also be used in *E. coli* and other hosts) is preferable to antibiotic selection for large-scale work.

It can be an advantage to produce recombinant proteins from yeast in a secreted form for two reasons: because secreted proteins can undergo post-translational modifications and because they then become easier to purify. However, undesirable changes may also occur during passage through the secretory pathway, for example, incorrect *N*- and *O*-linked glycosylation. The glycosyl side chains made in yeast and humans differ, and since glycosylation of a protein may radically affect its properties (Betenbaugh et al., 2007; Li and d'Anjou, 2009), considerable efforts have been put into glycoengineering of yeast strains, that is, using genetic engineering to replace key enzymes of the yeast glycosylation machinery with those of the human machinery. In particular, the yeast *Pichia pastoris* has been extensively engineered to yield strains capable of replicating the most essential glycosylation pathways found in humans. Not only do such yeast strains produce proteins with a human glycosylation pattern, they also generate strains that, in contrast to most mammalian cells, produce homogeneous glycoproteins with a specific predetermined glycan structure (Beck et al., 2010a).

Insulin is an example of a (nonglycosylated) recombinant protein produced by secretion from *S. cerevisiae* that has been approved for human use. So far, there are no glycosylated therapeutic proteins made in yeast on the market, but several are in the R&D pipeline.

2.3.2.3 Mammalian Cells

Mammalian systems are currently the usual choice for production of glycoproteins or proteins with other complex post-translational modifications (e.g., sulfation, amidation, and γ-carboxylation). Many of these modifications are essential for the biological activity of the proteins. For example, the activity of a number of blood coagulation factors (Factors II, VII, IX, and X) are completely dependent on γ-carboxylation of glutamic acid, a modification that occurs only in cells of higher eukaryotes.

Many different mammalian host–vector systems have become available for expression of recombinant proteins (Birch and Froud, 1994; Jeffs, 2007). Examples of immortalized mammalian cell lines, which can be grown in culture and therefore are suitable as host cells, are CHO (Chinese hamster ovary), BHK (baby hamster kidney), COS (African green monkey kidney), NS0 (mouse myeloma), and HeLa (human epitheloid carcinoma) cells. Mammalian viruses that replicate autonomously in cells can be used as vectors for recombinant DNA, but it is also possible to perpetuate introduced DNA in mammalian cells without an autonomously replicating vector, because transfected DNA molecules can integrate themselves into mammalian chromosomes. Autonomously replicating vectors are preferred for short-term expression because of immediate high protein yields, whereas integration is preferable for the generation of a stable expression cell line. Viruses that have been modified to become particularly useful vectors include SV40 (*Simian virus 40*), Epstein–Barr virus, and retroviruses. The two former are self-replicating while the latter mediate cellular uptake and integration into the genome with resulting significantly greater efficiency than DNA constructs without retroviral elements. To allow easy isolation of cells that have taken up recombinant DNA, a selectable marker gene such as resistance to the cytotoxic drug neomycin is introduced into the vector. The promoters used for gene expression in mammalian cells are often strong viral promoters such as SV40 and CMV (*cytomegalovirus*) promoters.

It is important to realize that post-translational modifications may not be identical to those occurring in the authentic protein, even when mammalian cells are used for production (Walsh and Jefferis, 2006; Walsh, 2010). In the body, proteins are produced in organs composed of specialized cells, and it has become increasingly clear that different tissues have different processing capabilities. Also, immortalized cell lines may have diverged from their progenitors in their growth characteristics and in other ways. Proteins with aberrant post-translational modification may be recognized as "non-self" by the immune system and therefore provoke an immune response that will eliminate them. Therefore, careful characterization of recombinant proteins is equally important whether production occurs in mammalian cells or in microorganisms.

From an economical viewpoint, mammalian cell culture systems are less attractive than microbial systems. The slow growth rate, moderate expression levels, complex requirements to culture media, and the need for advanced cultivation equipment all contribute to the relatively high cost of producing biological drugs from mammalian cell lines. However, production costs are usually only an issue if large amounts of the protein are necessary for therapeutic purposes. The coagulation factors FVIIa and FVIII and a number of antibodies are examples of marketed recombinant proteins produced in mammalian cell cultures. While the coagulation factors are very potent proteins required in only relatively small amounts, therapeutic antibodies need to be produced in substantial amounts; alternatives to mammalian cell culture production are therefore sought after by the growing antibody market. However, since the precise glycosylation pattern of antibodies is crucial to their immunological function, a switch of production system is by no means trivial. The glycoengineered *P. pastoris* strains are however possible alternatives to mammalian cells.

2.3.2.4 Transgenic Animals

Another option for expression is the use of transgenic animals (Velander et al., 1997). Transgenic animals can be generated by injection of a foreign gene (the transgene) into fertilized eggs by in vitro micromanipulation. The injected foreign gene integrates into the chromosome of the egg, normally at random locations. The eggs are then implanted into the oviduct of a foster mother. After a normal period of pregnancy, she will give birth to transgenic progeny with the foreign gene incorporated permanently into the genome of all cells. If transgenic animals are used to produce recombinant proteins, expression of the transgene must be regulated. The reason is that the compound usually is biologically active and therefore should be produced and stored in organs or compartments where it does not affect the animal. The most attractive transgenic production systems employ regulatory elements specific to the mammary gland. These regulatory elements direct the accumulation of the protein to the milk of the animal. Transgenic rabbits, sheep, and goats have been made, which express high levels of various recombinant proteins in the milk. Human antithrombin-III, extracted from the milk of genetically engineered goats, is currently the only marketed therapeutic protein from transgenic animals.

2.3.2.5 Insect Cells

Insect cell expression systems have also been developed for both small- and large-scale production of mammalian recombinant proteins. High expression levels of active proteins can usually be attained much faster than in mammalian cells, which is a key attraction of these systems. Most vectors are based on a lytic insect virus belonging to the baculovirus family; the foreign cDNA is inserted in the viral genome without interfering with the lytic life cycle of the virus. Protein production thus occurs during a lytic infection of insect cells with recombinant viruses, and usually less than a week is needed from infection to maximal product yield (Griffiths and Page, 1997). The fact that baculoviruses are noninfectious to vertebrates and their promoters inactive in mammalian cells gives insect systems a potential advantage over other systems for expressing oncogenes or other genes that are harmful to mammals. The speed with which a new protein can be produced in large amounts could make the system attractive, for example, for vaccine production for diseases such as influenza, where new pandemics caused by mutated viruses regularly threaten. Many different insect-derived vaccines are in the pipeline around the world, and a human papillomavirus vaccine (Cervarix) produced in insect cells is already on the market for human use.

2.4 PROTEIN DESIGN

Recombinant DNA technology makes it feasible to produce not only natural proteins, but also to design and produce new types of protein molecules (Blundell, 1994). New proteins can be broadly classified in two categories: (1) natural protein variants featuring replacements, insertions, and deletions of small numbers of amino acids and (2) chimeric proteins consisting of domains originating from different proteins. The novel proteins can be generated stepwise by rational design, that is, based on prior knowledge or assumptions about the function of different domains of natural proteins, or they can be found by selection from so-called epitope display libraries. The purpose of generating such (partly) novel proteins is usually to obtain protein drugs with improved characteristics such as better stability in vivo or higher specific activity, or to combine different activities in one protein.

2.4.1 PROTEIN VARIANTS

Protein variants can be generated by a genetic engineering method called site-directed mutagenesis. With this method, it is possible to change, delete, or insert one or a few nucleotides within a coding region (Figure 2.3). Expression of a mutated gene will result in a protein variant (analogue) with a specific amino acid alteration, the function of which is then tested in various ways. One can engineer small changes in proteins for a variety of purposes, such as changing the solubility or stability properties or the affinity for substrates or receptors.

As an example, there has been extensive work to design insulin molecules with improved characteristics for diabetes treatment (Bristow, 1993; Valla, 2010). The insulin molecule contains four glutamic acid residues; exchanging one or more of these with the neutral amino acid glutamine shifts the isoelectric point toward a higher pH, resulting in an insulin variant with a lower solubility at the physiological pH of about 7.3. Such insulin variants could conceivably offer alternatives to the slow-release formulations made by complexing native insulin with protamine or zinc ions. Other amino acid substitutions that increase solubility could potentially lead to new, fast-acting insulin preparations. The actual behavior in vivo can however be quite different from those predicted from in vitro studies, and as anticipated, many variants have failed at various stages in test programs, for example, because of altered affinity for the receptors for insulin and the related hormone insulin–like growth factor. Another major problem for the clinical application of analogues is the risk of adverse reactions due to immunogenicity or altered physiological functions of degradation products. These risks are especially serious for a drug such as insulin, which is administered daily for years, and extensive test programs for adverse reactions are therefore necessary. However, several insulin analogues are now on the market: three fast-acting analogues (the first, introduced in 1996, being Insulin Lispro, so named because the alteration involves the amino acids Lys and Pro) and two slow-acting ones, all of them showing good clinical profiles.

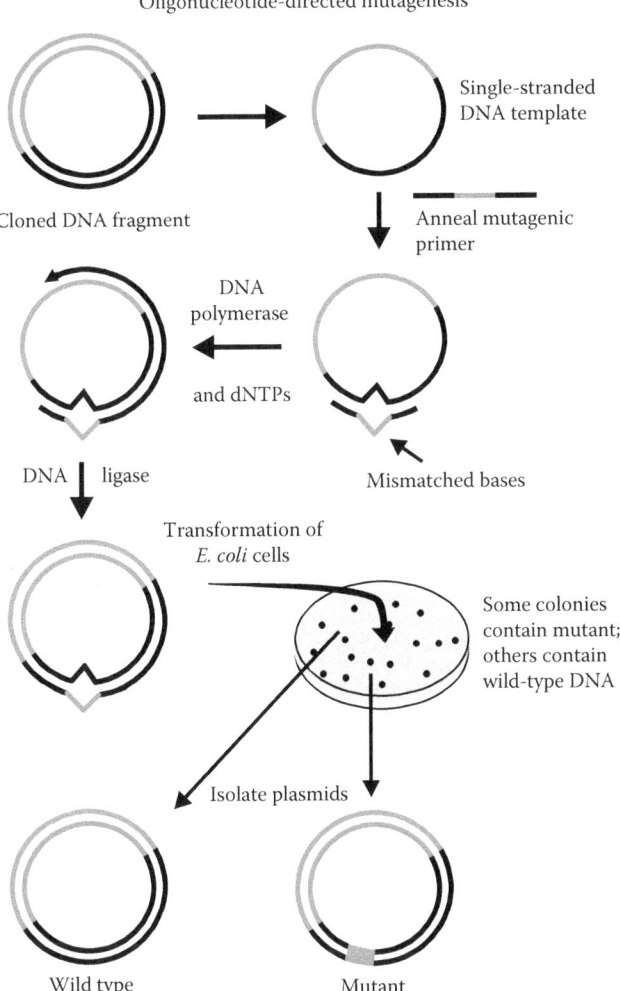

Oligonucleotide-directed mutagenesis

FIGURE 2.3 Oligonucleotide-directed mutagenesis by enzymatic primer extension. A "mutagenic" oligonucleotide encoding the desired mutation embedded in wild-type flanking sequences is annealed to a single-stranded DNA template. The sequence of the oligonucleotide is complementary to the template except for the nucleotides that define the mutation. The mutagenic oligonucleotide serves as a primer for DNA synthesis by DNA polymerase. Once the entire template has been copied, the ends of the newly synthesized strand are linked by DNA ligase. The heteroduplex DNA is transformed into *E. coli*. Replication results in segregation into separate wild-type and mutant plasmids, and further repair mechanisms generate colonies containing either one or the other plasmid. Plasmid DNA is isolated from the resulting colonies and is screened to identify mutants.

2.4.2 PROTEIN CHIMERAS

Chimeric proteins are made by fusing the coding region for one protein (or protein domain) with the coding region of another, followed by expression of the fused coding regions. The successful generation of a functional chimeric protein usually requires that the structure–function relationships of the two starting proteins are well known, so that the desired active domains will be retained within the new protein.

The chimeric protein approach has been very successful for the production of therapeutic antibodies or antibody-like structures (Dougall et al., 1994). While production of murine monoclonal antibodies based on the immortalization and proliferation of individual antibody forming murine B-cells (the so-called hybridoma technique) was established in the 1970s, it has been difficult to establish a similar production method for human monoclonal antibodies. Instead, with access to cloned antibody genes of both human and murine origin, production based on recombinant DNA expression is possible. In mice, expansion of cells producing an antibody with a desired specificity can be obtained by immunization with any chosen antigen, and the cDNA encoding the antibody can subsequently be isolated. By combining the coding regions corresponding to the constant and the framework regions from human antibodies with the variable regions from the murine antibody, a chimeric antibody can be generated, which is tolerated by the human immune system and has the desired antigenic specificity (Winter and Harris, 1993).

Humanized antibodies have numerous therapeutic applications, either alone or coupled to other biologically active substances. For example, they can be used in the elimination of toxic substances such as bacterially derived endotoxins or as regulators of the humoral immune response by binding of excess levels of cytokines (interleukin 1, tumor necrosis factor, etc.). In cancer therapy, antibodies are used to mediate selective destruction of tumor cells, both via normal immunological mechanisms and via targeted toxin delivery ("magic bullet" approach), where a tumor-cell-specific antibody fused to parts of a cellular toxin, such as Pseudomonas exotoxin A, may interact selectively with and destroy tumor cells (Vitetta et al., 1993).

Chimeric proteins, created through the combination of the extracellular domain of a receptor with the constant part of an antibody, mimic some properties of antibodies, and a number of such proteins have successfully reached the clinic. In fact, antibodies and antibody-like chimeric proteins currently constitute the most rapidly growing type of protein drugs. More than 30 different approved substances are on the market, and many more are in the pipeline (Beck et al., 2010b).

2.4.3   EPITOPE DISPLAY LIBRARIES

The use of epitope display libraries provides a useful complement to the rational design of mutated or chimeric proteins. Epitope display libraries are vast collections of peptides without any predetermined homology to known proteins or peptides, created through the cloning of complex mixtures of combinatorially synthesized oligonucleotides into specialized expression "display" vectors. The filamentous phage display system in which the expressed peptides are displayed fused to phage coat proteins has been effective in the discovery of many new ligands for various target proteins (Cortese, 1996). Affinity purification of the population of peptide phage particles on the target protein is used to recover peptides with binding activity. The identity of the selected peptide is revealed by sequencing of the appropriate segment of the DNA of each captured phage (Figure 2.4). An elegant demonstration of the power of this

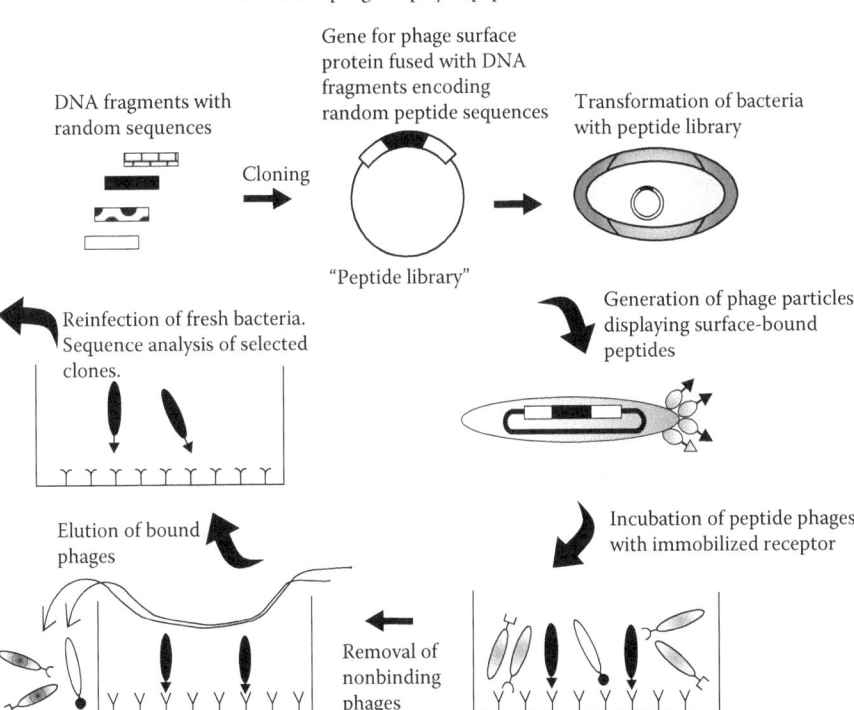

Outline of phage display of peptide libraries

DNA fragments with random sequences

Gene for phage surface protein fused with DNA fragments encoding random peptide sequences

Transformation of bacteria with peptide library

Cloning

"Peptide library"

Reinfection of fresh bacteria. Sequence analysis of selected clones.

Generation of phage particles displaying surface-bound peptides

Elution of bound phages

Incubation of peptide phages with immobilized receptor

Removal of nonbinding phages

FIGURE 2.4 Phage display of peptide epitopes. Cloning of random DNA sequences in frame with the coding region of a phage coat protein (gIIIp) results in the formation of phage particles, each displaying one unique peptide epitope on the phage surface. Phage particles displaying an epitope recognizing a specific target (e.g., a receptor or an enzyme) can be isolated by repeated rounds of affinity purification and repropagation in *E. coli* of the eluted bound phages.

technology was illustrated early on by the identification of a small peptide that binds to and activates the receptor for erythropoietin (Wrighton et al., 1996). Surprisingly, the peptide showed no sequence or structural homology to the cytokine it mimics. The sizes of phage-displayed epitopes are not limited to small peptides, and phage libraries displaying antibody fragments (single chain Fv and Fab) have been extensively used to generate high affinity human-type antibodies, some of which have reached the market.

2.5 RECOMBINANT PROTEIN THERAPEUTICS: STATUS AND FUTURE TRENDS

Recombinant protein drugs represent a wide variety of biological substances such as hormones, cytokines, enzymes, enzyme inhibitors, blood coagulation regulators, and antibodies. Many are replicates of naturally occurring proteins, but a growing number are designed or modified so that they do not have a natural counterpart. The total number of recombinant protein drugs currently on the market, including copy versions of different types, exceeds 200, of which innovative products number around 80.

Some of the drugs are available in first, second, or third generation versions, where the versions may differ by small alterations that affect stability, solubility, and/or pharmacokinetics. Some examples are the fast- and slow-acting insulin variants, in which one or more amino acids have been altered and/or acylated. Addition of inert nonprotein polymers, such as polyethylene glycol (PEG), is another common way of affecting stability, and several cytokine drugs, first made as pure proteins, are now available as more stable second generation PEGylated versions. Generally, the new generation drugs represent improved versions that can give substantial benefits to patients.

In addition to these new improved versions, copy versions with no intended changes relative to the originals have also appeared after expiry of original patents. Such copy versions are called biosimilars or follow-on biologics because although they are similar to the original product, they are not identical. A follow-on manufacturer has neither access to the originator's cell bank nor to the exact fermentation and purification process, and because no two production cell lines developed independently can be considered identical, biotech pharmaceuticals cannot be exactly copied. Small differences in the cell line, the manufacturing process, or the surrounding environment can make a major difference in clinical profile. In particular, treatment side effects may differ, for example, by provoking an immunological response. Clearly, substitution between biosimilars might have adverse clinical consequences. Therefore, these drugs are subject to an approval process that requires substantially more additional data than required for chemical generic drugs, although the requirements are not as comprehensive as for the original biotech medicine (Guideline on similar biological medicinal products, EMEA, 2005).

Protein drugs are traditionally viewed as less convenient than low-molecular drugs, both from the manufacturer's and the user's point of view. For the manufacturer, the complexity of proteins poses challenges for purification, stability, and administration. For the user, administration of a protein drug is somewhat burdensome because, in contrast to small-molecule drugs, proteins cannot usually be administered orally but have to be injected. Nevertheless, protein drugs are becoming increasingly important in the drug market. Over the last 10 years, the number of novel recombinant protein products has risen from around 30 to around 80, and this increase is expected to continue with undiminished or even accelerating speed (PipeLine, 2011). Several of the biotech drugs have reached blockbuster status (Lawrence, 2007). The main reason for this success is that they admirably fulfill unmet medical needs, but improvements in formulation and administration routes have also been a key factor for their success.

ACKNOWLEDGMENT

I wish to acknowledge the contribution of my late husband, Professor Jan Engberg, to the first edition of this book. Jan is sorely missed on many, many occasions, of which the revision of this chapter is only a minor one.

REFERENCES

Beck, A., Cochet, O., and Wurch, T. 2010a. GlycoFi's technology to control the glycosylation of recombinant therapeutic proteins. *Expert Opin. Drug Discov.* 5:95–111.

Beck, A., Wurch, T., Bailly, C., and Corvaia, N. 2010b. Strategies and challenges for the next generation of therapeutic antibodies. *Nat. Rev. Immunol.* 10:345–352.

Betenbaugh, M.J., Tomiya, N., and Narang, S. 2007. Glycoengineering: Recombinant glycoproteins. In: Johannis P. Kamerling, Editor-in-Chief, *Comprehensive Glycoscience,* Elsevier, Oxford, pp. 607–642.

Birch, J.R., and Froud, S.J. 1994. Mammalian cell culture systems for recombinant protein production. *Biologicals* 22:127–133.

Blundell, T.L. 1994. Problems and solutions in protein engineering—towards rational design. *Tibtech* 12:145–148.

Bristow, A.F. 1993. Recombinant-DNA-derived insulin analogues as potentially useful therapeutic agents. *Tibtech* 11:301–305.

Cortese, R., (Ed.). 1996. *Combinatorial Libraries: Synthesis, Screening and Application Potential.* Walter de Gruyter, Berlin, New York.

Creighton, T.E. 1993. *Proteins: Structures and Molecular Properties*, 2nd ed. W.H. Freeman & Co., New York.

Daniels, J., (Ed.). 2011. PCR protocols, 3rd ed., *Methods in Molecular Biology*, Vol. 687, Humana Press.

Dougall, W.C., Peterson, N.C., and Greene, M.I. 1994. Antibody-structure-based design of pharmacological agents. *Tibtech* 12:372–379.

Fernandez, J.M., and Hoeffler, J.P., (Eds.). 1999. *Gene Expression Systems*. Academic Press, San Diego, CA.

Georgiou, G., and Valax, P. 1996. Expression of correctly folded proteins in *Escherichia coli.* *Curr. Opin. Biotechnol.* 7:190–197.

Griffiths, C.M., and Page, M.J. 1997. Production of heterologous proteins using the baculovirus/insect expression system. *Meth. Mol. Biol.* 75:427–440.

Guideline on similar biological medicinal products, European Medicines Agency, Evaluation of Medicines for Human Use London, October 30, 2005.

Hodgson, J. 1993. Expression systems: A user's guide. *Biotechnology* 11:887–893.

http://www.orfeomecollaboration.org/. The ORFeome Collaboration was formed in late 2005 to meet the need of the research community for an unrestricted source of fully sequence-validated full-ORF human cDNA clones in a format allowing easy transfer of the ORF sequences into virtually any type of expression vector (accessed September 2012).

http://www.ornl.gov/sci/techresources/Human_Genome/home.shtml. Human Genome Project Information (accessed September 2012).

Itakura, K., Hirose, T., Crea, R., Riggs, A.D., Heyneker, H.L., Boliver, F., and Boyer, H-W. 1977. Expression in *E. coli* of a chemically synthesized gene for the hormone somatostatin. *Science* 198:1056–1063.

Jeffs, S.A. 2007. Expression of recombinant biomedical products from continuous mammalian cell lines, In: Stacey, G., and Davis, J. (Eds.), *Medicines from Animal Cell Cultures*, John Wiley & Sons, pp. 62–77.

Lawrence, S. 2007. Billion dollar babies—biotech drugs as blockbusters. *Nat. Biotechnol.* 25:380–382.

Li, H., and d'Anjou, M. 2009. Pharmacological significance of glycosylation in therapeutic proteins. *Curr. Opin. Biotechnol.* 20:678–684.

McKusick, V.A., and Ruddle, F.H. 1987. A new discipline, a new name, a new journal [editorial]. *Genomics* 1:1–2.

Pettersson, E., Lundeberg, J., and Ahmadian, A. 2009. Generations of sequencing technologies. *Genomics* 93:105–111.

PipeLine Informa Healthcare, a searchable business intelligence service monitoring global pharma and biotech R&D. Subscription only (accessed January 2011).

Rabilloud, T., Hochstrasser, D., and Simpson, R.J. 2010. Is a gene-centric human proteome project the best way for proteomics to serve biology? *Proteomics* 10:3067–3072.

Romanos, M.A., Scorer, C.A., and Clare, J.J. 1992. Foreign gene expression in yeast: A review. *Yeast* 8:423–488.

Rowen, L., Mahairas, G., and Hood, L. 1997. Sequencing the human genome. *Science* 278:605–607.

Sambrook, J., and Russell D. 2001. *Molecular Cloning: A Laboratory Manual*, 3rd ed. Cold Spring Harbor Laboratory Press, New York.

The Human Genome. 2001a. *Nature*, February 15, entire issue.

The Human Genome. 2001b. *Science*, February 16, entire issue.

Valla, V. 2010. Therapeutics of diabetes mellitus: Focus on insulin analogues and insulin pumps. *Exp. Diabetes Res.* 2010:1–14.

Velander, W.H., Lubon, H., and Drohan, W.N. 1997. Transgenic livestock as drug factories. *Scientific American* 276:70–74.

Vitetta, E.S., Thorpe, P.E., and Uhr, J.W. 1993. Immunotoxins: Magic bullets or misguided missiles? *Immunology Today* 14:252–259.

Walsh, G., and Jefferis, R. 2006. Post-translational modifications in the context of therapeutic proteins. *Nat. Biotechnol.* 24:1241–1252.

Walsh, G. 2010. Post-translational modifications of protein biopharmaceuticals. *Drug Discov. Today* 15:773–780.

Winter, G., and Harris, W.J. 1993. Humanized antibodies. *Immunol. Today* 14:243–246.

Wrighton, N. C., et al. 1996. Small peptides as potent mimetics of the protein hormone erythropoietin. *Science* 273:458–463.

3 Protein Purification

Lars Hovgaard, Lars Skriver, and Sven Frokjaer

CONTENTS

3.1 INTRODUCTION

Therapeutic proteins are available from a number of different sources, for example, animal tissue, body fluids (blood), plants, microorganisms, and cell culture systems. However, most commercially available therapeutic proteins, for example, recombinant human insulin, human growth hormone, erythropoietin (EPO), and interferons, are produced by large-scale fermentation using either microorganisms or mammalian cells as sources.

Large-scale fermentation of such recombinant organisms can produce substantial quantities of proteins. The majority of the recombinant proteins from microbial sources are produced by a limited number of microorganisms that are classified as GRAS (generally recognized as safe). GRAS microorganisms include bacteria such as *Bacillus subtilis* and *Escherichia coli*, and also yeast and other fungi.

Many proteins obtained by fermentation are secreted by the microorganisms into the culture medium. Such extracellular proteins are normally much simpler to purify in the subsequent downstream processing than intracellular proteins, as there is no requirement to disrupt the cells in order to harvest the protein. Thus, for extracellular secreted proteins, the amount of foreign protein and cellular components that needs to be separated from the product of interest is much less. Mammalian cell cultures such as Chinese hamster ovary (CHO) cells and baby hamster kidney (BHK) cells represent another very important source of therapeutic proteins. Blood coagulation factors and monoclonal antibodies are among the examples produced in such cell culture systems. Compared to microbial sources, the mammalian cell culture systems are generally a more complex protein source. Besides the basal nutritional requirements, various supplements have to be added to the culture media. Such supplements may include vitamins, growth hormone, mineral salts, amino acids, antibiotics, and serum. The added mixture of proteins and other biomolecules makes the subsequent purification protocol more complex. Addition of serum also increases the risk of contaminating the culture medium, and potentially the final product, with blood-borne pathogens; hence, much effort is devoted to developing serum-free cell culture conditions. Mammalian cells are capable of conducting important post-translational metabolic reactions such as glycosylation of proteins and routinely provide protein products that are properly folded and functional. This is therefore the preferred cell system for industrial production of large-size proteins.

The purpose of this chapter is to give a brief overview of downstream processing of protein products. Several useful books covering protein separation and purification from different points of view are available (Deutscher, 1990; Hejnaes et al., 1998; Jakoby, 1971, 1984; Jakoby and Wilchek, 1974). Another important source of information is booklets from manufacturers of separation equipment and media.

3.2 FRACTIONATION STRATEGIES

Detailed information about the characteristics of the protein and preferably also about the properties of the most important impurities is a prerequisite not only for making a successful purification scheme, but also for achieving the optimal quality of the final drug product. Important data involve approximate molecular weight, isoelectric point, solubility, presence of labile groups, and other physicochemical data. One should also define criteria with regard to the stability of the protein to be purified. The parameters affecting the chemical and physical stability of proteins are temperature, pH, organic solvents, heavy metals, oxygen, mechanical stress, and a potential risk of proteolytic degradation. Based on such information, it is possible to select fractionation techniques and conditions that effectively separate the protein of interest from other proteins and biomolecules.

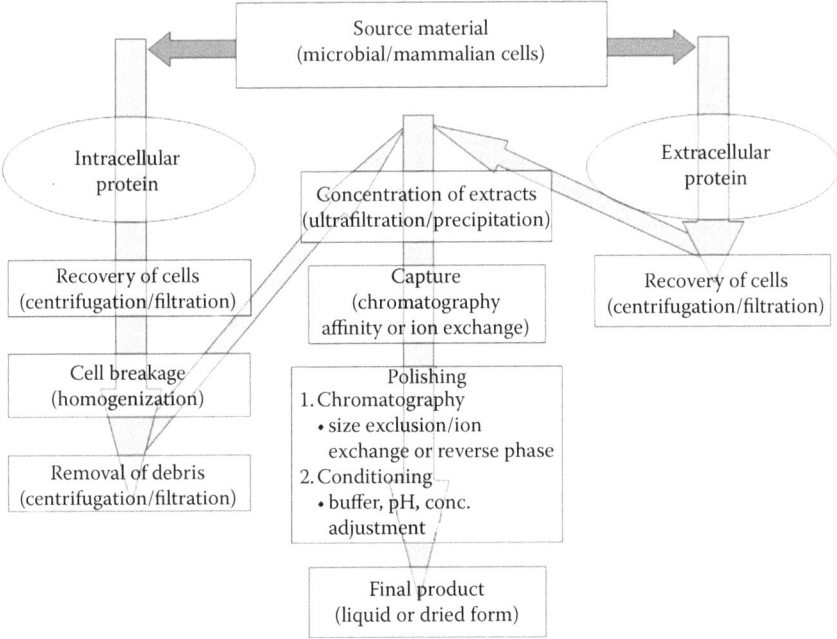

FIGURE 3.1   Outline of protein production from microbial and mammalian sources.

A typical overall purification process is outlined in Figure 3.1. More information on the various fractionation techniques is given in Chapter 4.

3.2.1   INITIAL FRACTIONATION STEP

The purpose of the initial fractionation (capture) is to obtain a solution of the protein suitable for further purification by chromatography. If the protein is intracellular, the initial step involves a cell disruption by homogenization followed by a removal of any remaining cells and cell debris, often by centrifugation or filtration. Ideally, one should use a highly selective step as early as possible in the purification process. This may reduce the number of subsequent steps that must be used to achieve the desired level of purity. In the case of extracellular proteins, a concentration and clarification step is often employed as the initial step. This can be achieved by unit operations such as precipitation and ultrafiltration. The capture is assigned to remove most of the foreign (host cell) proteins and to concentrate the product. The capture is preferably performed by chromatographic methods such as selective absorption affinity matrix or by ion exchange chromatography.

These initial steps should also result in the removal of most of the bulk foreign proteins including proteases and membrane fragments that might bind the protein of interest.

Precipitation of a protein in an extract may be achieved by changing the mother liquor solution by adding salts (e.g., ammonium sulfate and sodium sulfate), organic solvents (e.g., ethanol and acetone), polymers (e.g., polyethylene glycol) or by changing the pH or the temperature of the solution.

3.2.2 Intermediate Purification Step

In order to achieve high protein purity by removal of host cell proteins, nucleic acids, lipids, salts, and other small molecules, various chromatographic steps can be applied. The number and types of such chromatographic steps needed at this stage in the process depend primarily on the resolution achieved by each step and the economic constraints determined by the balance between the overall purity and total product recovery.

Although many different chromatographic techniques are available on an industrial scale, the most common techniques are ion exchange chromatography, affinity chromatography, hydrophobic interaction chromatography, and size-exclusion chromatography (see Chapter 4).

The order in which the different chromatographic steps are applied in a protein purification protocol is determined by factors such as capacity, process time, and cost. Monoclonal antibody-based affinity chromatography may be attractive because of its high selectivity. However, the rather viscous fermentation broth will limit the hydrodynamic properties of such a system and will reduce the column life significantly, thereby increasing the cost substantially. In such cases, more robust media, such as ion exchangers on large beads with high purity, might be advantageous.

3.2.3 Final Polishing Step

The purpose of the last polishing step in the protein purification protocol is to produce a product that fulfills the specifications (e.g., purity and limits for specific degradation products, product derivatives such as oxidized, deamidated, or degraded forms of product, and contaminants such as pyrogenic substances) required for its final processing. This means that possible aggregates, chemical degradation products, and other contaminants including ligands that may have leached from previous fraction steps must be removed to a certain limit by this procedure. At the same time, the protein product is conditioned for its use in the final pharmaceutical preparation.

Size-exclusion chromatography is often used in this final step. It may effectively remove dimeric or higher molecular weight aggregates and low molecular weight degradation products and at the same time it can be used to change the purified protein into a new buffer system suitable for further processing or storage. As a production process has strong 'artistic' elements and can be difficult to protect by patents, detailed information on protein purification schemes used in purification of therapeutic proteins is generally considered as strictly confidential for obvious commercial reasons.

3.2.4 The Finished Product

Optimization of stability of the protein is of importance not only during downstream processing but also during storage of the purified product. Loss of activity may be caused by a number of factors, as outlined in Table 3.1. This should be minimized by setting strict limits on exposure, residual content, and product handling.

TABLE 3.1

Examples of Factors That May Adversely Affect the Stability of Proteins

Chemical	Physical	Biological
Detergents	Extreme pH	Proteolytic activity
Urea	Elevated temperatures	
Organic solvents	Light	
Oxidants	Freezing and thawing	
Heavy metals	High shear forces	

Furthermore, it should be kept in mind that denatured proteins are often more susceptible to chemical degradation.

In some instances, the addition of specific stabilization agents may enhance the stability of the finished product. Such additives include substances such as substrates, cofactors, or other compounds that may interact with the protein of interest and thereby stabilize the native conformation. A variety of nonspecific additives may also be incorporated into the finished product, for example, glycerol, polyethylene glycol, sugars, and salt such as ammonium sulfate or sodium chloride, as they may have a significant stabilizing influence on some—but definitely not all—proteins. Bulk enzymes such as proteases are often marketed as solutions containing 20% sodium chloride as stabilizer.

Many processes leading to loss in biological activity of proteins in solution may be minimized by removal of water. Drying may be achieved by various techniques such as spray-drying, vacuum-drying, and drum-drying. However, these methods are relatively harsh due to high-temperature exposure and/or high mechanical stress, and are therefore unsuitable for labile proteins. Labile proteins are best dried by lyophilization, as discussed in more detail in Chapter 10.

3.3 PROTEIN STABILITY IN DOWNSTREAM PROCESSING

After the large-scale manufacture of given desired proteins in expression organisms, downstream processing poses a major challenge for the retention of protein structure and thus activity. It is of utmost importance to design or tailor-make the downstream process for the protein, as the possible unfolding of proteins caused by the expression and initial purification conditions may not be fully reversible. The unfortunate result of the unfolding is often a loss in biological activity, as this typically depends inherently on the folding of the native state protein. In the protein, hydrophobic forces and hydrogen bonding are predominantly responsible for the overall stability. If viewed in an isolated fashion, these effects are weak and can easily be broken. However, they act synergistically and are responsible for the immense activity that is seen in proteins. In processing, the proteins may undergo unfolding and, due to possible chemical modification, intermediate formation, or aggregation, the native state may not be regained after purification. Therefore, it is extremely important to maintain native structure throughout the processing, thus ensuring that the biological activity levels are comparable to the level desired for the use of the protein in the final product.

The major concerns when designing a purification scheme for a protein are the unknowns. Very often the protein chemist has very little information about chemical and physical stability. Moreover, the specific effects of solvent pH, ionic strength, redox potential, additives, cosolvents, and so on, are generally unknown. During processing, the protein is thus at risk of unintentional unfolding and destabilization, resulting in loss of activity and yield.

This chapter outlines the types of instability encountered for proteins as well as the physical and chemical factors affecting instability in downstream processing. Further details are given in the references listed at the end of this chapter (see also Chapters 5 and 6).

3.3.1 Protein Conformation Stability

The conformational stability of a protein is important in relation to its activity and use. The unfolding of globular proteins in aqueous solutions is easily induced. Only 20–60 kJ/mol separates the native/active state from the unfolded/inactive state of a protein under physiological conditions (Pace, 1975; Privalov, 1979). Proteins of very different natures exhibit similar thresholds for unfolding.

3.3.1.1 Native State

In the globular protein, specific features and folding patterns govern the native conformation. These structures are α-helix, β-sheet, and random-coil structures. The specific folding optimizes stability for a given protein due to a complex combination of weak bonds. Some of these are hydrophobic and often referred to as van der Waals' forces. They include the burial of nonpolar amino acid residues in the protein interior, α-helical dipoles and caps, and weak polar aromatic interactions (Alber, 1989). Moreover, hydrophilic and electrostatic interactions are also very abundant in the folded protein structure. Among the types are the complexes between oppositely charged amino acids (Sali et al., 1991; Serrano et al., 1992) and hydrogen bonds between hydrogen atoms on heteroatoms and neighboring heteroatoms (Stickle et al., 1992). The latter play a major role in the α-helical formations in the native state of proteins (Dill, 1990). Other more specific factors affecting protein native state stability involve certain amino acids, for example, Phe, Trp, and Tyr, as they are all hydrophobic (Burley and Petsko, 1985).

Thus, it is apparent that the native state is determined by a complex of different types of interaction and that all of this is strongly dependent on the environmental situation (Oobatake and Ooi, 1993; Seckler and Jaenicke, 1992).

3.3.1.2 Intermediate State

In the unfolding process, the protein passes through several transition conformations. One has been termed the molten globule state, and refers to a state of the protein in which it retains much of its native folding but has lost its tertiary structure (Ptitsyn and Uversky, 1994). It is believed that the protein occasionally assumes this state in downstream processing and that the state is easily unfolded to an inactive form. But if this is prevented, the native state can be assumed again (Privalov, 1979).

3.3.1.3   Unfolded State

The unfolded state of a protein defines a random coil structure whose specific nature varies from individual molecule to molecule. This inherently implies that all biological activity is absent. It is important to note that the chemical structure of the unfolded protein in almost all cases is identical to that of the native state (Creighton, 1990; Jaenicke, 1987). Thus, downstream bioactivity testing is an important measure for the folding state of the recovered protein.

3.3.2   PROTEIN INSTABILITY

The instability of proteins can be attributed to two distinctly different phenomena, physical instability and chemical instability. The former is more directly related to the conformational stability discussed above, although chemical instability may also be an integral part of a complex degradation path of a protein.

3.3.2.1   Physical Events

The physical instability of a protein involves the unfolding and subsequent undesirable interaction of several protein molecules. It does not involve chemical alterations such as covalent bond breaking or formation. Very often this kind of instability leads to the formation of higher order complexes, aggregates, or precipitates. If the physical instability leads to irreversible aggregation, the product's biological activity is greatly reduced (Brange et al., 1997). The phenomenon is difficult to predict, and the result often disastrous.

3.3.2.2   Denaturation

When a protein loses its tertiary structure, it is often said to denature. All intramolecular hydrogen bonds and hydrophobic interactions are disrupted. The organized native structure is lost, and all components are exposed (Jakoby, 1984; Tanford, 1968). The complexity of the phenomenon, of course, depends on the size and folding of the native protein (Privalov, 1979). Very often, the denaturation is an abrupt process occurring in a narrow range of environmental factors, that is, temperature, pH, ionic strength, redox potential, or denaturant concentration. It is important to note that the denaturation reactions in many cases are reversible, but that the renaturation must be performed under tightly controlled conditions. In many cases, the result of the renaturation is an insoluble gel or protein aggregate that is totally without activity (Hejnaes et al., 1992; Matthiesen et al., 1996).

3.3.2.3   Aggregation

When denatured, proteins are prone to aggregate due to favorable intermolecular hydrophobic interactions. This reaction may be very fast and may lead to an insoluble polymerized product (Figure 3.2). The aggregation is assumed to be governed by the initial dimerization process, resulting in a second-order reaction (Kiefhaber et al., 1991). However, in the case of insulin, it has been suggested that even partially unfolded conformation may lead to aggregates (Brange et al., 1997; Speed et al., 1996).

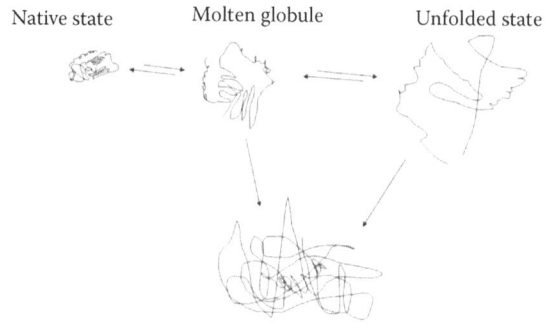

FIGURE 3.2   Unfolding and aggregation of proteins.

TABLE 3.2
The Hoffmeister Series

Cations	$NH_4^+ > K^+ > Na^+ > Li^+ > Mg^{++} > Ca^{++} > Gdn^{++}$
Anions	$SO_4^- > HPO_4^- > CH_3COO^- > Cl^- > NO_3^- > SCN^-$

3.3.2.4   Precipitation

Controlled mild precipitation is being used in the downstream purification of proteins to limit or prevent irreversible aggregation in solution. Different techniques are used. Iso-precipitation involves the phenomenon of poor solubility of proteins around the iso-electric point, merely a pH effect. Salting-in and salting-out are two opposed phenomena. The former denotes the situation where proteins' solubility is increased by low salt concentrations. The latter is a phenomenon of high-concentration salt precipitation of proteins. One technique that is very well known and widely used is addition of ammonium sulfate. The Hoffmeister series of salting-in effectiveness of cations and anions can be useful if precipitation is needed (Hejnaes et al., 1998) (Table 3.2).

3.3.2.5   Chemical Events

The chemical instability of proteins can involve the cleavage or formation of covalent bonds within the protein primary structure. This results in new chemical entities which often impose a severe separation problem for the chemist, as these derivatives are almost identical. Very specialized separation techniques are thus needed downstream. The reader is encouraged to refer to literature on the specific chemical processes, as they will be mentioned only briefly in this chapter.

3.3.2.6   Proteolytic Degradation

In the living cell, a fine balance between the degrading enzymes and the synthesized protein is maintained by subcellular compartmentation. However, all this is destroyed when the cell is lysed and disrupted as part of the protein isolation procedure. The cellular enzymes are mixed with the protein of interest and pose a potential fatal track.

Therefore, the initial steps of a downstream purification are aimed at minimizing this proteolysis. The proteolysis may be of little chemical significance and thus difficult to detect. Moreover, proteolytic removal of just a few amino acids terminally may affect the biological activity. Therefore, high-power analytical tools such as mass spectroscopy, spectroscopy, capillary electrophoresis, peptide mapping, and isoelectric focusing are used to determine the degree of proteolysis (Chapters 4 and 9). In order to minimize the enzyme activities, environmental factors such as pH, protein stabilizing compounds, and enzyme inhibitor cocktails can be used with caution (Hejnaes et al., 1998).

3.3.2.7 Nonenzymatic Degradation

In general the peptide bond is stable at alkaline pH. In dilute acid, however, where the carboxyl group of aspartic acid is not dissociated, peptide bonds of aspartic acid are cleaved very quickly. Even under normally stable conditions Asp-Pro is a major destabilizing link in a protein (Piszkiewicz et al., 1970). N-terminal degradation has been reported for human growth hormone (Battersby et al., 1994).

3.3.2.8 Deamidation

The deamidation process involves the nucleophilic attack of nitrogen in the peptide chain on the carbonyl residue in the amide side-chain of asparagine, resulting in the loss of one amino group. The chemical result is an Asp residue or an iso-Asp residue. This reaction is documented in a long list of proteins: insulin (Berson and Yalow, 1966; Brange et al., 1992), human growth hormone (Lewis et al., 1970), cytochrome c (Flatmark, 1966), and so on. Moreover, the deamidation has also been shown to occur at the glutamine amide bond. The extent of deamidation can in general be minimized by the use of buffers with low ionic strength, low temperatures, and neutral to slightly acidic pH (Scotchler and Robinson, 1974).

3.3.2.9 β-Elimination and Racemization

A very frequently reported degradation reaction is the β-elimination, which involves the abstraction of a β-hydrogen from cysteine, serine, or threonine residue in the protein backbone. A resulting carbanion can either form the unsaturated derivative or cause racemization (d-, l-amino). The abstraction of the β-hydrogen is proportional to the alkalinity; hence, pH should not be high (Hejnaes et al., 1998). Moreover, the reaction is affected at high temperatures and by divalent metals (Lee et al., 1977; Nashef et al., 1977; Sen et al., 1977).

3.3.2.10 Oxidation

Several oxidation reactions in proteins have been reported (Brot et al., 1982). In the alkaline or neutral medium, the residues of the amino acids cysteine, histidine, methionine, tryptophan, and tyrosine are especially prone to oxidation. In acidic conditions, however, methionine is sensitive. Often the oxidation reactions cause a great loss in biological activity and even immunogenicity (Morley et al., 1965). As the amino acid cysteine serves protein stability in a unique way through intramolecular disulfide bonds, its stability is especially well characterized (Gilbert, 1990; Martin and Viswanatha, 1975). The disulfide bond is degraded via a variety of chemical mechanisms. It has been shown to be very sensitive to nucleophils in neutral and alkaline media and to the presence of electrophils in strongly acidic media. However, divalent metals, for example, Cu^{++} in the presence of OH^- and O_2, are also a major reason for oxidation. Among the proteins that exhibit this cysteine instability are insulin, chymotrypsin, and ribonuclease (Gilbert, 1990; Martin and Viswanatha, 1975).

3.3.2.11   Stability of Arginine

The last type of instability reaction to be discussed in this chapter is the conversion of arginine to ornithine. The reaction involves the alkaline hydrolysis of the guanidinium group in arginine and is generally smaller than for Cys, Lys, Thr, and Ser.

3.3.3   ESSENTIAL PROCESS-RELATED PARAMETERS

Given that the crude protein product is present in an immensely complex mixture of fragmented cells and subcellular components as well as in a wealth of cellular proteins and enzymes, it is a serious challenge for purification scientists to pick the protein of interest. Moreover, to pick and purify this protein without losing bioactivity is a challenge for experts. Hence, this section will only outline the effects of parameters that are used in this separation and isolation process. Table 3.3 gives some of the effects and examples of common parameters varied in the downstream processing of proteins. As mentioned earlier, little is known about the properties of a protein at the time of isolation. Hence the pitfalls of the process are numerous. It is important to keep in mind that the described parameters can all happen unexpectedly and at the same time. Therefore, the chemist faces a tremendous challenge. Process strategy and design are prerequisites for the subsequent formulation work towards the final drug product. This involves careful assessment of choice of matrices and solutes in combination with physical factors involved in downstream purification processes.

TABLE 3.3
Parameters in the Downstream Processing of Proteins

Parameter	Effect	Example	Reference
pH	Isoelectric point stabilization	pI-precipitation	Pace (1990)
	Protonation/deprotonation	Carboxyl (pKa 3–4.7)	Hejnaes et al. (1998)
		Imidazolium (pKa 8–8.5)	
		Sulfhydryl (pKa 8–8.5)	
		Amino (pKa 7.6–10.6)	
		Q-Hydroxy (pKa 9.4–10.4)	

TABLE 3.3 (*Continued*)
Parameters in the Downstream Processing of Proteins

Parameter	Effect	Example	Reference
	Other	Deamidation (alkaline and acidic conditions)	Lee et al. (1977), Florence (1980), Porter et al. (1993)
		Racemization	
		Hydrolysis	
Temperature	High-temperature denaturation	Entropy exceeds enthalpy	Privalov et al. (1986)
	Low-temperature denaturation	Hydrogen bond breaking	
		Breakage of hydrophobic interaction	
Redox potential	Reducing	Cleavage of disulfide bonds	Brot et al. (1982)
	Oxidizing	Methionine oxidation	
Cosolvents	Sugars, some amino acids, some salts, polyols	Stabilizing through increase in surface tension, *no* protein binding	Hejnaes et al. (1998)
	Weakly interacting salts	Stabilizing through increase in surface tension, binding to charged protein residues	
	Steric exclusion and repulsion from charged groups	PEG and 2-methyl-2, 4-pentanediol	
Protein concentration	Aggregation reactions	Insulin fibrillation	Brange et al. (1997)
Pressure	Denaturation	570 MPa, 30°C, secondary structure disruption	Takeda et al. (1995)

REFERENCES

Alber, T. 1989. Mutational effects on protein stability. *Annu. Rev. Biochem.* 58:765–798.

Battersby, J.E., Hancock, W.S., Canova-Davis, E., Oeswein, J. and O'Connor, B. 1994. Diketopiperazine formation and N-terminal degradation in recombinant human growth hormone. *Int. J. Pept. Protein Res.* 44:215–222.

Berson, S.A. and Yalow, R.S. 1966. Deamidation of insulin during storage in frozen state. *Diabetes* 15:875–879.

Brange, J., Langkjaer, L., Havelund, S. and Volund, A. 1992. Chemical stability of insulin. 1. Hydrolytic degradation during storage of pharmaceutical preparations. *Pharm. Res.* 9:715–726.

Brange, J., Andersen, L., Laursen, E.D., Meyn, G. and Rasmussen, E. 1997. Toward understanding insulin fibrillation. *J. Pharm. Sci.* 86:517–525.

Brot, N., Werth, J., Koster, D. and Weissbach, H. 1982. Reduction of N-acetyl methionine sulfoxide: A simple assay for peptide methionine sulfoxide reductase. *Anal. Biochem.* 122:291–294.

Burley, S.K. and Petsko, G.A. 1985. Aromatic–aromatic interaction: A mechanism of protein structure stabilization. *Science* 229:23–28.

Creighton, T.E. 1990. Protein folding. *Biochem. J.* 270:1–16.

Deutscher, M.P. 1990. *Methods in Enzymology (Vol. 182), Guide to Protein Purification*, San Diego, CA: Academic Press.

Dill, K.A. 1990. Dominant forces in protein folding. *Biochemistry* 29:7133–7155.

Flatmark, T. 1966. On the heterogeneity of beef heart cytochrome c. 3. A kinetic study of the non-enzymic deamidation of the main subfractions (Cy I-Cy 3). *Acta Chem. Scand.* 20:1487–1496.

Florence, T.M. 1980. Degradation of protein disulphide bonds in dilute alkali. *Biochem. J.* 189:507–520.

Fortis, F., Guerrier, L., Righetti, P.G., Antonioli, P., and Boschetti, E. 2006. A new approach for the removal of protein impurities from purified biologicals using combinatorial solid-phase ligand libraries. *Electrophoresis* 27:3018–3027.

Gilbert, H.F. 1990. Molecular and cellular aspects of thiol–disulfide exchange. *Adv. Enzymol. Relat. Areas Mol. Biol.* 63:69–172.

Hejnaes, K.R., Bayne, S., Norskov, L., Sorensen, H.H., Thomsen, J., Schaffer, L., Wollmer, A. and Skriver, L. 1992. Development of an optimized refolding process for recombinant Ala-Glu-IGF-1. *Protein Eng.* 5:797–806.

Hejnaes, K., Matthiesen, F. and Skriver, L. 1998. Protein stability in downstream processing. In: Subramanian, G. (Ed.), *Bioseparation and Bioprocessing*, Vol. 3, New York: John Wiley & Sons.

Jaenicke, R. 1987. Folding and association of proteins. *Prog. Biophys. Mol. Biol.* 49:117–237.

Jakoby, W.B. 1971. *Methods in Enzymology (Vol. 22), Enzyme Purification and Related Techniques*, New York, San Francisco, London: Academic Press.

Jakoby, W.B. and Wilchek, M. 1974. *Methods in Enzymology (Vol. 34), Affinity Techniques of Enzyme Purification, Part B*, New York, London: Academic Press.

Jakoby, W.B. 1984. *Methods in Enzymology (Vol. 104), Enzyme Purification and Related Techniques, Part C*, Orlando, FL: Academic Press.

Kiefhaber, T., Rudolph, R., Kohler, H.H. and Buchner, J. 1991. Protein aggregation in vitro and in vivo: A quantitative model of the kinetic competition between folding and aggregation. *Biotechnology (NY)* 9:825–829.

Lee, H.S., Osuga, D.T., Nashef, A.S., Ahmed, A.I., Whitaker, J.R. and Feeney, R.E. 1977. Effects of alkali on glycoproteins. β-Elimination and nucleophilic addition reactions of substituted threonyl residues of antifreeze glycoprotein. *J. Agric. Food Chem.* 25:1153–1158.

Lewis, U.J., Cheever, E.V. and Hopkins, W.C. 1970. Kinetic study of the deamidation of growth hormone and prolactin. *Biochim. Biophys. Acta* 214:498–508.

Martin, B.M. and Viswanatha, T. 1975. Selective reduction of a disulfide bond in chymotrypsin A-alpha. *Biochem. Biophys. Res. Commun.* 63:247–254.

Matthiesen, F., Hejnaes, K.R. and Skriver, L. 1996. Stabilization of recombinantly expressed proteins. *Ann. NY Acad. Sci.* 782:413–421.

Milne, J.J. 2011. Scale-up of protein purification: Downstream processing issues. In: Walls, D. and Loughran, S.T. (eds), *Protein Chromatography: Methods and Protocols (Methods in Molecular Biology)*, 681, 73–85. New York, NY: Humana Press.

Morley, J.S., Tracy, H.J. and Gregory, R.A. 1965. Structure–function relationships in the active C-terminal tetrapeptide sequence of gastrin. *Nature* 207:1356–1359.

Nashef, A.S., Osuga, D.T., Lee, H.S., Ahmed, A.I., Whitaker, J.R. and Feeney, R.E. 1977. Effects of alkali on proteins. Disulfides and their products. *J. Agric. Food Chem.* 25:245–251.

Oobatake, M. and Ooi, T. 1993. Hydration and heat stability effects on protein unfolding. *Prog. Biophys. Mol. Biol.* 59:237–284.

Pace, C.N. 1975. The stability of globular proteins. *CRC Crit. Rev. Biochem.* 3:1–43.

Pace, C.N. 1990. Conformational stability of globular proteins. *Trends Biochem. Sci.* 15:14–17.

Piszkiewicz, D., Landon, M. and Smith, E.L. 1970. Anomalous cleavage of aspartyl–proline peptide bonds during amino acid sequence determinations. *Biochem. Biophys. Res. Commun.* 40:1173–1178.

Porter, W.R., Staack, H., Brandt, K. and Manning, M.C. 1993. Thermal stability of low molecular weight urokinase during heat treatment. I. Effects of protein concentration, pH and ionic strength. *Thromb. Res.* 71:265–279.

Privalov, P.L. 1979. Stability of proteins: Small globular proteins. *Adv. Protein Chem.* 33:167–241.

Privalov, P.L., Griko, Yu. V., Venyaminov, S. Yu. and Kutyshenko, V.P. 1986. Cold denaturation of myoglobin. *J. Mol Biol.* 190:487–498.

Ptitsyn, O.B. and Uversky, V.N. 1994. The molten globule is a third thermodynamical state of protein molecules. *FEBS Lett.* 341:15–18.

Rosa, P.A.J, Ferreira, I.F., Azevedo, A.M. and Aires-Barros, M.R. 2010. Aqueous two-phase systems: A viable platform in manufacturing of biopharmaceuticals. *J. Chrom. A* 1217:2296–2305.

Sali, D., Bycroft, M. and Fersht, A.R. 1991. Surface electrostatic interactions contribute little of stability of barnase. *J. Mol. Biol.* 220:779–788.

Scotchler, J.W. and Robinson, A.B. 1974. Deamidation of glutaminyl residues: Dependence on pH, temperature, and ionic strength. *Anal. Biochem.* 59:319–322.

Seckler, R. and Jaenicke, R. 1992. Protein folding and protein refolding. *FASEB J.* 6:2545–2552.

Sen, L.C., Gonzalez-Flores, E., Feeney, R.E. and Whitaker, J.R. 1977. Reactions of phosphoproteins in alkaline solutions. *J. Agric. Food Chem.* 25:632–638.

Serrano, L., Kellis, J.T. Jr, Cann, P., Matouschek, A. and Fersht, A.R. 1992. The folding of an enzyme. II. Substructure of barnase and the contribution of different interactions to protein stability. *J. Mol Biol.* 224:783–804.

Speed, M.A., Wang, D.I. and King, J. 1996. Specific aggregation of partially folded polypeptide chains: The molecular basis of inclusion body composition. *Nat. Biotechnol.* 14:1283–1287.

Stickle, D.F., Presta, L.G., Dill, K.A. and Rose, G.D. 1992. Hydrogen bonding in globular proteins. *J. Mol. Biol.* 226:1143–1159.

Takeda, N., Kato, M. and Taniguchi, Y. 1995. Pressure- and thermally-induced reversible changes in the secondary structure of ribonuclease A studied by FT-IR spectroscopy. *Biochemistry* 34:5980–5987.

Tanford, C. 1968. Protein denaturation. *Adv. Protein Chem.* 23:121–282.

4 Characterization of Therapeutic Peptides and Proteins

Marco van de Weert and Tudor Arvinte

CONTENTS

4.1 INTRODUCTION: THE CHALLENGE OF CHARACTERIZING PROTEINS

Due to the large size of proteins, as well as the importance of the three-dimensional protein fold held together by weak noncovalent forces, analytical characterization of a protein's structural integrity is more complex than that of the most conventional drug molecules. In this chapter, we will discuss the most common methods to characterize proteins, as well as outline their applicability in a pharmaceutical development setting. The methods are briefly summarized in Table 4.1. It is not possible to discuss any of the methods in detail or even all available methods, which means that the reader will have to refer to the cited literature if additional details

TABLE 4.1

List of Analytical and Physicochemical Approaches to Characterize Proteins, Their Biophysical Principles, and the Main Information They Provide about the Proteins

Methodology	Biophysical Principle	Main Information	References
X-ray crystallography	Scattering of x-rays by protein crystals	Three-dimensional atomic structure of the protein	McPherson, 1999; Rupp, 2009
NMR spectroscopy	Excitation of nuclear spin by radiofrequency radiation	Three-dimensional atomic structure of the protein; fingerprint of protein fold	Aubin et al., 2008; Bieri et al., 2011; Blommers and Cerletti, 1997; Kwan et al., 2011; Wishart, 2005
Molecular modeling	Computational methods based on empirical or ab initio calculations	Three-dimensional atomic structure	Moret and Gestel, 2005
RP-HPLC	Affinity of protein for hydrophobic stationary phase versus mobile phase	Quantitation of protein chemical degradation, in particular oxidation, hydrolysis, deamidation	Bischoff and Barroso, 2005
Amino acid analysis	Hydrolysis of amino acid chain, followed by detection of individual amino acids	Relative content of the amino acids constituting the protein	Bartolomei and Maisano, 2006; Fountoulakis and Lahm, 1998
Edman degradation[a]	Hydrolysis of N-terminal amino acid followed by subsequent identification through chromatographic methods	Amino acid sequence determination (up to ca. 30 amino acids)	Shively, 2000
Peptide mapping	Hydrolysis of protein into smaller peptides by enzymes and subsequent chromatographic separation	Protein "fingerprint"; identification of site(s) of modification; amino acid sequence determination (when coupled to MS)	Hoff and Chloupek, 1996; Lundell and Schreitmüller, 1999
SEC	Differences in transport rate of a protein through a porous particle matrix based on hydrodynamic radius	Comparative protein hydrodynamic size; quantitation of soluble protein aggregates and hydrolysis products (if significantly different in size from native protein)	Bischoff and Barroso, 2005

AC	Preferential interaction of protein with a ligand, receptor, or other compound that binds to the protein	Chemical degradation (if affecting binding site); physical degradation, e.g., partial unfolding or aggregation (if affecting binding site); potentially predictive for biological activity	Bischoff and Barroso, 2005
IEC	Interaction with oppositely charged stationary phase	Chemical degradation, in particular deamidation	Bischoff and Barroso, 2005
FFF	Differences in transport rate between small and large analytes in a laminar flow, combined with a perpendicular cross flow (or other type of perpendicular force)	Primarily protein aggregates, including very large and insoluble aggregates; hydrolysis, if the change in hydrodynamic radius is large enough to allow separation	Caldwell and Wahlund, 2005
Native PAGE	Movement of charged species through a gel, resulting in separation of molecules by their hydrodynamic radius/charge profile	Presence of soluble aggregates and hydrolysis products	Righetti et al., 2011; Westermeier, 2011
SDS-PAGE	Movement of SDS-unfolded proteins through a gel, resulting in separation based on protein molecular weight	Presence of covalent protein aggregates and hydrolysis products; identification of disulfide and nondisulfide linked aggregates	Righetti et al., 2011; Westermeier, 2011
IEF	Movement of proteins in a gel containing a pH gradient, separating the proteins based on IEP	Protein deamidation; presence of isoforms with different charge	Righetti et al., 2011
CE	Separation of proteins in an electric field; various separation principles may be applied similar to chromatography	Protein degradation, including oxidation, deamidation, and hydrolysis; protein aggregation	Catai et al., 2005; Hu and Dovichi, 2002; Rabiller-Baudry et al., 1998; Thormann, 2011
MS	Separation of charged molecules in the gas phase using an electromagnetic field	Identification of chemical degradation products; amino acid sequencing; with special methodology also rough insight into protein three-dimensional size as well as local flexibility	Bobst et al., 2008; El-Aneed, 2009; Romijn et al., 2005; Standing, 2003; van den Bremer and Heck, 2005

(Continued)

TABLE 4.1 (Continued)
List of Analytical and Physicochemical Approaches to Characterize Proteins, Their Biophysical Principles, and the Main Information They Provide about the Proteins

Methodology	Biophysical Principle	Main Information	References
AUC	Separation of molecules based on their molecular mass and size in a centrifugal field	Size distribution of the protein molecules in the sample; gold standard in characterization and quantification of protein aggregates; protein–protein interaction	Philo, 2005
UV–Vis spectroscopy	Absorption of radiation in the UV–Vis range by chromophores, in particular aromatic amino acids and the peptide backbone	Quantitation of proteins (most common detector in chromatographic analysis); changes in protein conformation around aromatic amino acids	Kueltzo and Middaugh, 2005; Mach and Middaugh, 2011
CD	Differential absorption of left and right circularly polarized light	Global secondary structure; local changes in tertiary structure around aromatic amino acids (and cystines)	Bloemendal and Jiskoot, 2005; Greenfield, 1999; Kelly et al., 2005; Li, Nguyen et al., 2011
IR spectroscopy (IR)	Absorption of IR radiation by vibrating bonds in a molecule	Global secondary structure; may identify noncovalent aggregates through intermolecular β-sheets	Byler and Susi, 1986; Goormaghtigh et al., 2009; Jackson and Mantsch, 1995; Jiang et al., 2011; van de Weert et al., 2005
Raman spectroscopy	Absorption of IR radiation by vibrating bonds in a molecule	Global secondary structure; may identify noncovalent aggregates through intermolecular β-sheets; local changes in tertiary structure, e.g., around cystines or tyrosines	Nielsen, 2005; Sane et al., 1999
Intrinsic fluorescence	Emission of radiation by intrinsic fluorophore upon excitation	Changes in tertiary structure around fluorophore	Jiskoot et al., 2005; Lakowicz, 1999

Technique	Principle	Information	References
Extrinsic fluorescence	Emission of radiation by extrinsic fluorophore interacting with a protein	Changes in tertiary structure and/or aggregation; detection of minute quantities of aggregates	Hawe et al., 2008
Fluorescence anisotropy	Level of depolarization of emitted radiation compared to polarization of excitation light	Changes in local tertiary structure; aggregation/misfolding (with extrinsic probes)	Jiskoot et al., 2005; Lakowicz, 1999
SLS	Scattering of light by particles in solution, averaged over time	Presence of particulate material; aggregation kinetics; radius of gyration; protein–protein interactions	Demeester et al., 2005; Santos and Castanho, 1996
DLS	Fluctuation over time of scattering of light by particles in solution	Presence of particular material; aggregation kinetics; radius of hydration; protein–protein interactions	Demeester et al., 2005; Santos and Castanho, 1996
SAXS/SANS	Scattering of x-rays or neutrons by particles in solution, averaged over time	Low-resolution three-dimensional structure of protein; presence of aggregates	Chacon et al., 1998; Svergun et al., 2011
Optical microscopy, including NTA	Visualization of particles using a visual light-based microscope; light scattering is used in NTA	Quantitation of particles in solution; size distribution of particles	Garidel et al., 2012
Optical imaging (e.g., microflow imaging [MFI]; fluorescence microscopy)[b]	Visualization of particles using a visual light-based microscope	Quantitation of particles in solution; size distribution of particles; shape of particles	Garidel et al., 2012
EM	Visualization of particles using electron bombardment	Low-nanometer imaging of particle size and shape	Garidel et al., 2012
AFM	Visualization of particles by measuring attractive/repulsive forces between the compound and a cantilever tip	Atomic imaging of compound	Engel et al., 2005

(Continued)

TABLE 4.1 (Continued)

List of Analytical and Physicochemical Approaches to Characterize Proteins, Their Biophysical Principles, and the Main Information They Provide about the Proteins

Methodology	Biophysical Principle	Main Information	References
DSC	Uptake or release of heat upon thermal event in the sample	Protein unfolding thermodynamics; reversibility of unfolding	Cueto et al., 2003; Privalov and Dragan, 2007; Schön and Velazquez-Campoy, 2005; Wen et al., 2012
Chemical denaturation	Unfolding of protein by increasing concentration of chemical denaturants like urea or guanidine; unfolding is usually monitored by fluorescence spectroscopy	Thermodynamic data on stability of protein fold	Pace, 1990
Temperature-induced unfolding	Unfolding of protein by increasing temperature; unfolding can be monitored using CD, fluorescence, FTIR, but also DSC	Thermodynamic data on stability of protein fold; reversibility of unfolding	Chen and Oakley, 1995
Biochemical methods	Biochemical read-out of protein function, e.g., enzyme activity, binding affinity, effect on cells or tissues, and so on	Biological activity of protein	—

[a] A single C-terminal amino acid can be cleaved off using the so-called Bergmann degradation, but further sequencing is not possible.

[b] MFI generally cannot see particles in the low-nanometer range, while shape determination requires sizes within the micrometer range.

are required. Moreover, this chapter will mainly focus on the characterization of the protein, rather than the formulation. We will end this chapter with a discussion on choosing the proper methodology.

A significant challenge in protein characterization includes the required use of numerous advanced analytical methods. That is, the complexity of the protein structure (Chapter 6) and the potential large number of degradation products (Chapters 5 and 6) cannot be captured with a single or even a small set of methods. Ideally, an analytical scientist working on therapeutic protein characterization will have a firm understanding of all available methods, allowing the selection of the most appropriate methods and proper interpretation of the results. This situation is, in practice, hardly feasible; analytical scientists are usually trained as experts in a small subset of methods and may thus develop a bias toward that subset. In contrast, most formulation scientists have generally only received superficial training in commonly used methodology, and thus by necessity have to rely on the expert's advice. Thus, continuous training in methods and methodology is necessary to assure that the protein is characterized properly and rationally throughout its whole development process.

4.2 METHODS FOR CHARACTERIZATION

4.2.1 X-ray Crystallography, NMR Spectroscopy, and Computational Methods

X-ray crystallography and nuclear magnetic resonance (NMR) spectroscopy are the only methods available that allow a full three-dimensional structural characterization of a protein. Prior knowledge about the primary sequence is, however, required. Each of the two techniques has its own unique advantages, disadvantages, and pitfalls that will be discussed below. Note that there is no regulatory requirement that the full three-dimensional structure of a therapeutic protein is characterized. However, this information can be very helpful in understanding the behavior of the protein, including degradation pathways, and is thus a common step in the development of protein therapeutics.

X-ray crystallography (discussed in depth in McPherson [1999] and Rupp [2009]) is based on the diffraction of x-rays by crystalline materials. The obtained diffraction pattern can be used to determine the relative position of the atoms within the crystal. Computer technology and increasingly better algorithms have significantly simplified the various complex mathematical steps involved in translating the diffraction pattern to a three-dimensional atomic model. However, it is still a highly specialized field and proper characterization may take weeks or months.

The most important starting point for x-ray crystallography is a high-quality crystal. Obtaining crystals is not a trivial process, and it is still not possible to predict the solution conditions, which will favor crystal formation of a new protein. The empirical process of finding those conditions is often a matter of the personal experience of the crystallographer, and it is not uncommon to find proteins that cannot be crystallized with the tested methodology. This does not mean it is impossible, but merely that the right conditions have not yet been found. Another possibly problematic issue is that the protein solution must be homogeneous. The presence of small amounts of degraded protein may be sufficient to cause poorly diffracting crystals, or prevent crystallization altogether.

The conditions that favor crystallization may also introduce a significant source of uncertainty for the end user: Can those conditions induce a structure that is not relevant for the physiological situation? Although protein crystals contain a large amount of water (typically 50%, well above monolayer coverage), one often uses salting-in salts like ammonium sulfate that may induce structure in proteins. Also, there are many protein–protein interactions due to the high protein concentration, which may induce or alter protein structure. This means that the three-dimensional structure as determined by x-ray crystallography can differ from the structure of the protein under physiological conditions in an aqueous solution (see, e.g., Morth et al., 2004).

The final outcome of an x-ray crystallographic analysis ideally shows the relative position of all atoms within the protein chain, thus yielding an accurate three-dimensional atomic picture of the protein. Moreover, increasing ambiguity in the position of certain atoms or groups of atoms is often the result of the inherent flexibility of this part of the protein, and thus provides an insight into the regions of the protein that are likely flexible.

NMR spectroscopy (reviewed in Bieri et al. [2011], Kwan et al. [2011], and Wishart [2005]) is based on the difference of spin resonance frequencies of various nuclei with spin ½ or −½ in a magnetic field. Small differences in the surroundings of a specific nucleus, for example, a proton, will slightly change the resonance frequency. This difference is measured in NMR spectroscopy and denotes the chemical shift (units: ppm). For small compounds, NMR allows one to characterize the full molecular structure. For large peptides and in particular proteins, prior knowledge about the amino acid sequence is necessary. Determining the three-dimensional structure requires multidimensional correlation spectroscopy, which also identifies interactions through space, rather than through covalent bonds. The end result of an NMR experiment is a model of the three-dimensional structure that fits the observed chemical shifts. The NMR-based models generally consist of multiple equivalent structures, that is, an ensemble of structures that all fit the NMR data. This ensemble is at least in part due to the structural flexibility of the protein in solution, rather than merely an indication of the uncertainty in the methodology. Thus, in the protein data bank (http://www.wwpdb.org/), a repository for protein structures, one will generally find a single set of coordinates for any protein structure entry file determined by x-ray crystallography, and multiple sets of coordinates in each entry file for structures determined by NMR spectroscopy.

NMR spectroscopy is in principle more widely applicable to determine protein structure, as it does not require the formation of crystals. However, the complexity of NMR spectra of large proteins may be such that they cannot be appropriately analyzed. Thus, there is a certain size limit to proteins that can be characterized fully by NMR spectroscopy. This limit is continuously being extended to larger sizes, in particular with improvement in equipment and the strength of the magnetic fields used in the instruments. An added complication of NMR spectroscopy is the frequent requirement to use labeled protein, in particular ^{13}C or ^{15}N labeling. This labeling significantly improves the capability of NMR spectroscopy to allow structure determination of larger proteins, but labeling proteins is a time-consuming and expensive process. Like x-ray crystallography, full structure determination by NMR spectroscopy also requires a homogeneous sample. In addition, relatively high concentrations may be needed (>5 mg/mL), which may be above the solubility limit of some proteins. Finally, some structures cannot easily be observed in NMR if the peptide or protein is "flip-flopping" rapidly between conformations. Thus, also the three-dimensional structure as determined by NMR is not necessarily the same as the native structure of the protein under physiological conditions.

NMR spectroscopy can also be used as a "fingerprint" technique (Aubin et al., 2008; Blommers and Cerletti, 1997). That is, the original NMR spectrum (often a correlation spectrum) of the protein is compared to that of the protein under other conditions of interest, such as in the presence of certain salts or (noninterfering) excipients. Changes in the spectrum indicate structural changes, and if the various peaks in the spectrum have been correlated properly to the various amino acids, it is even possible to determine where and which type of changes have taken place.

A third method to determine protein three-dimensional structure is the use of computer modeling (Moret and Gestel, 2005). Data from NMR spectroscopy and x-ray crystallography can be the input for such computer modeling, but it is also possible to come up with a reasonable estimate of protein structure without such data. For example, ab initio methods using quantum mechanics can in principle be used to predict the structure of a protein. This is, however, a computationally very demanding process, and thus in practice limited to peptides. The process can be simplified by using a more empirical approach, in which the protein sequence is compared to that of proteins with known structure. Using sequence homology, and empirically derived tendencies of certain amino acids to form specific secondary structures such as helices or turns, an often reasonably accurate estimate of protein structure can be obtained. However, it is highly recommended to compare the predicted structure to actual measurements with other techniques. Moreover, it is important to realize that a correctly predicted global fold does not mean that the local folding and interactions will also be accurate. The protein data bank mentioned earlier also contains some structures that are theoretical models (about 2% of the total).

4.2.2 CHROMATOGRAPHY

Liquid chromatography has long been the major workhorse in therapeutic protein analysis and also is the main method used for protein purification (see Chapter 3). The principle of liquid chromatography is the retention of compounds on a column; the magnitude of this retention depends on the physicochemical properties of the compound and column material. There are various types of chromatography, and those commonly used for therapeutic proteins are concisely summarized in Bischoff and Barroso (2005).

While most scientists are familiar with the principles of liquid chromatography, this familiarity is often limited to reversed-phase high performance liquid chromatography (RP-HPLC). In this method, protein molecules are separated based on their affinity to a hydrophobic stationary matrix. Elution often requires a gradient of organic modifier, rather than the isocratic elution used for many low molecular weight therapeutics. This organic modifier also means that the protein is denatured, which makes this technique of limited use in protein purification.

RP-HPLC is mostly used to determine the presence of and quantify chemical degradation products that result in a change in hydrophobicity, such as oxidation, covalent aggregation, deamidation, and other hydrolytic reactions. Finding the right elution conditions to separate as many degradation products as possible from the native protein is a challenging task, as some degradation products only result in very small changes in physicochemical properties. This is not just a matter of the actual change taking place, but also a matter of size: a single deamidation reaction of insulin (ca. 6 kDa) will be much easier to observe than the same reaction in a monoclonal antibody (typically around 150 kDa). Finally, when new peaks are observed in a RP-HPLC chromatogram, it is still unknown what change has taken place. Characterization of these degradation products has been simplified considerably by coupling high performance liquid chromatography to mass spectrometry (MS). The latter technique will be discussed in a subsequent section.

Special applications of RP-HPLC in the characterization of protein therapeutics are in Edman degradation (Shively, 2000), amino acid analysis (Bartolomeo and Maisano, 2006; Fountoulakis and Lahm, 1998) and peptide mapping (Hoff and Chloupek, 1996; Lundell and Schreitmüller, 1999). Edman degradation (Shively, 2000) is a special method to determine amino acid sequence from the N-terminus. In this method, single amino acids are cleaved from the terminal end of the protein, and then identified using RP-HPLC. Then the next amino acid in the sequence is cleaved and analyzed, and so on. The method is very sensitive, but if the N-terminal amino acid is blocked, for example, by acylation, problems will arise. Also, in practice the maximum sequence length that can be determined is about 30 amino acids.

Amino acid analysis (Bartolomeo and Maisano, 2006; Fountoulakis and Lahm, 1998) requires full hydrolysis of a protein into its constituent amino acids, for example, using hydrochloric acid. The amino acid mixture is then injected on a RP-HPLC column, where separation into the individual amino acids takes place. This allows quantification of the relative amounts of the amino acids. Since some amino acids are more likely to be degraded in the hydrolysis procedure, care should be taken in interpreting the results.

For peptide mapping (Hoff and Chloupek, 1996; Lundell and Schreitmüller, 1999) the protein is (usually) degraded using an enzyme, and the resulting peptides are separated by RP-HPLC. This yields a "fingerprint" of the protein, and deviations from this fingerprint may be used to identify in which part of the protein a certain chemical degradation has taken place. Repeating this analysis with different enzymes can significantly improve the accuracy of this method, as each different enzyme cuts the protein in different places. For example, whereas trypsin mainly cleaves at the carboxyl end of arginine and lysine, chymotrypsin does so at the carboxyl end of tyrosine, tryptophan, and phenylalanine residues. Coupling chromatographic peptide mapping with MS allows direct identification of the various peptides observed in the chromatogram.

Another widely used chromatographic method for protein analysis is size exclusion chromatography (SEC). In this method, molecules are separated based on their hydrodynamic radius on a column containing porous particles. Importantly, in this method, larger molecules elute faster than smaller molecules. SEC is thus a strong method to determine and quantify protein aggregation and protein truncation, provided the latter significantly alters protein size. SEC will fractionate both covalent and strongly bound noncovalent protein aggregates from the native protein; weakly bound noncovalent aggregates will be broken up in the flow and thus are not detected (Philo, 2006).

A potentially complicating factor is any interaction of the protein with the column material, introducing an additional separation factor. This is undesired, and requires modification of the eluent or the column material to minimize such interaction. Moreover, large particles (bigger than a few hundred nanometers) need to be filtered out or removed through other means to prevent the column from becoming clogged.

Affinity chromatography (AC) is mostly used in the purification of proteins, but can also be used for analytical purposes. Its separation principle is based on the affinity for a specific ligand and/or receptor. For example, certain antibodies like IgG1 and IgG2 bind with strong affinity to protein A, whereas the affinity is weaker for other antibodies, like IgM and IgG3, and absent for other proteins. Similarly, some proteins bind strongly to heparin, whereas others do not. Elution of the bound protein will depend on the nature of the interaction; in the case of heparin binding a salt gradient or a change in pH may suffice.

Changes in an AC chromatogram may be due to chemical degradation, but also physical degradation, such as unfolding, misfolding, or aggregation, as long as this degradation alters the affinity for the compound attached to the stationary phase. If the compound attached to the column material is relevant to the biological activity of the protein being studied, AC can be used as a predictive tool for changes in biological activity.

Ion-exchange chromatography (IEC) is also a form of AC, but based on the less specific electrostatic interactions between proteins and an oppositely charged column material. IEC can thus be used to separate species with a difference in net charge, which will occur on deamidation. To obtain separation, the protein sample is applied to the column and then eluted using an increasing salt gradient.

A more specialized application of IEC is chromatofocusing, in which proteins are separated based on their isoelectric point (pI) through the use of amphoteric buffers and small pH changes to achieve elution. While this method provides a much higher resolution of differently charged species than the more traditional IEC approach, having proteins in a buffer close to their pI may result in aggregation and precipitation, which can damage the column and lead to equipment failure. Thus, care is required in the selecting of proper separation conditions.

Apart from choosing the right separation principle, choosing the right instrumentation is a major decision factor. For example, in many of the separation principles used to study proteins, a gradient of some kind is required. Thus, the instrumentation must be capable of producing such a gradient. Equally important, proteins tend to adsorb to a wide variety of surfaces, and thus low-protein adsorbing surfaces should be used. In addition, some of the methods require the use of relatively high salt concentrations, which may cause corrosion of metal surfaces. Finally, the choice of detector will determine the ability to detect a wide variety of species. Most commonly a UV–Vis detector is used, but also MS is often connected to HPLCs.

Other types of detectors, for example, based on refractive index, fluorescence, conductivity, or evaporative light scattering, are less commonly used for protein samples, but may be necessary for special applications. For example, fluorescent probes added to the eluent have been used to detect and characterize very small amounts of aggregates separated by SEC (Hawe et al., 2009), while covalent modification with fluorescent probes is a common step in amino acid analysis to lower the detection limit (Bartolomeo and Maisano, 2006).

Also the column choice is an important step for protein analysis. For RP-HPLC analysis of small compounds, a scientist can make an educated guess, based on log P values of the compound, as to which column material would be suitable. This is not the case for proteins, where it is unknown how much hydrophobic area will be exposed during a RP-HPLC run. In general, however, C18 columns are best for small and hydrophilic peptides, while increasing size and hydrophobicity make C8- or C4-based columns more suitable. In SEC, electrostatic interactions are the main challenge. Thus, column materials with very little to no charge are ideal, or otherwise high salt concentrations are needed in the eluent to reduce any interaction. For AC, it is obvious the right affinity pair must be chosen, while in IEC the choice of column (cation or anion exchange) will need to be based on the pI of the protein and which type of degradation products are expected (e.g., more or less acidic).

An interesting development in the last decade has been the ability to produce very small particles of high quality to be used as column material and handling the much higher pressure required pushing a solution through columns packed with these particles. Smaller particles significantly improve separation capacity, and also push the optimal flow rate to higher values. Thus, it has become possible to obtain good separation at a faster speed. At the same time, these two factors combined narrow the chromatographic bands, resulting in improved sensitivity. This new methodology is known as ultrahigh performance liquid chromatography or rapid separation liquid chromatography. Unfortunately, it is at present only applicable to reversed phase as separation principle.

A relatively recent addition to the field of chromatography is field flow fractionation (FFF) (Caldwell and Wahlund, 2005). Although the method was already described in 1976, commercial instrumentation has not become available until quite recently. In this method, a solution is pushed through a thin rectangular chamber, creating a parabolic flow profile. When a field is applied perpendicular to this flow, certain solutes can be pushed out of the flow front (toward the chamber wall), thus resulting in a slower elution. While many different options of such perpendicular fields are available, the most common is a perpendicular flow, which pushes the solutes to the bottom of the chamber. Brownian motion and diffusion, both of which decrease with increasing size of the solute, counteract this asymmetric distribution. Thus, smaller solutes will distribute more evenly in the flow than larger particles, which means the former elute faster than the latter. This method is known as asymmetric flow field flow fractionation (AF4), and it has a wider separation range and is less affected by protein–column material interactions compared to SEC. It is an accepted orthogonal method to SEC to characterize and quantify protein aggregation and truncation.

4.2.3 ELECTROPHORESIS

Electrophoretic methods separate species in a solution using an electric field. The most important two methods in this field are gel electrophoresis (Righetti et al., 2011; Westermeier, 2011) and capillary electrophoresis (CE) (Catai et al., 2005; Hu and Dovichi, 2002; Rabiller-Baudry et al., 1998; Thormann, 2011).

Gel electrophoresis (Westermeier, 2011), and in particular polyacrylamide gel electrophoresis (PAGE), is a very common method in biochemical laboratories due to its relative simplicity. In native PAGE, charged solutes are made to move through a highly cross-linked polyacrylamide matrix, achieving separation through a sieving effect. The separation is based on a combination of charge and shape. More common than native PAGE is SDS-PAGE, where sodium dodecyl sulfate (SDS) is added to the sample (and sometimes also the gel itself) in relatively large amounts. SDS strongly binds to proteins, and for most proteins in the same mass ratio of ca. 1.4 g per gram of protein. SDS also denatures the protein, and results in a rod-like shape of the protein–SDS complex, regardless of the original structure of the protein. Thus, all protein molecules obtain approximately the same shape and charge/mass ratio. As a result, the separation in SDS-PAGE is almost solely based on the mass of the protein, and allows the detection of covalent aggregates as well as hydrolysis products. SDS-PAGE is often run both under standard and under reducing conditions. In the latter case, a reducing agent like mercaptoethanol or dithiothreitol is added that breaks cystines. By comparing the two types of SDS-PAGE runs, it is possible to determine whether the aggregates were linked through cystines, or whether other types of covalent linkages may be present.

In isoelectric focusing (IEF) (Righetti et al., 2011), a pH gradient is created in a polyacrylamide gel. When subjected to an electric field over this gel, proteins will move until they reach the part of the gel with a pH corresponding to their isoelectric point (IEP or pI). Due to the potentially high resolution power (0.01 pH units), IEF is a very powerful method to determine the presence of isoforms or degradation products with an altered charge compared to the native protein.

Increasingly popular, in particular in the field of proteomics, is the use of two-dimensional gel electrophoresis. Here the proteins are first separated in one dimension, for example by IEF, and subsequently separated by a second principle, for example, SDS-PAGE, in a perpendicular direction. Its use in protein formulation is, however, limited.

CE (Catai et al., 2005; Hu and Dovichi, 2002; Rabiller–Baudry et al., 1998; Thormann, 2011) is also based on the separation of species in an electric field. Here, however, the electric field spans a capillary column, often with the same column materials used in gas chromatography. In principle, many of the same approaches as discussed for gel electrophoresis and chromatography can be applied to CE, which makes it complementary to these two techniques, but with its own set of advantages and disadvantages. Most importantly, CE often uses less material and provides better separation, but is also more difficult to optimize. Unlike gel electrophoresis, CE allows full automation and can be coupled directly to MS, yielding increased characterization power. A further advantage is that CE can also be used, in various different configurations, to characterize the binding between proteins and ligands, the latter ranging from small molecules to very large structures like liposomes (see, e.g., Anderot et al., 2009).

4.2.4 MASS SPECTROMETRY

MS (Bobst et al., 2008; El-Aneed et al., 2009; Romijn et al., 2005; Standing, 2003; van den Bremer and Heck, 2005) has rapidly become a standard in the pharmaceutical industry to characterize various aspects of proteins, in particular the amino acid sequence of unknown proteins and chemical degradation products. However, the three-dimensional structure of proteins in solution has also been analyzed by MS.

The principle of MS is based on the separation of charged molecules in the gas phase using an electric and/or magnetic field. The latter fields separate the molecules by their mass-over-charge ratio (m/z). There are many different mass analyzers, such as time-of-flight (TOF), quadrupoles, ion traps, or ion cyclotron resonance equipment, each of which has its own set of advantages and disadvantages for the analysis of proteins. The reader is referred to literature on this topic for further details (see, e.g., El-Aneed et al., 2009). Sometimes these mass analyzers are used in tandem; in particular quadrupole–TOF and quadrupole–quadrupole combinations are popular.

Similar to the mass analyzers, there are several methods to obtain the protein molecules in the gas phase in a charged form. The latter was long a major challenge in this field, as the principles used for small molecules were too harsh for proteins (high temperature, high energy), resulting in significant fragmentation. At present, matrix-assisted laser desorption-ionization (MALDI) and electrospray ionization (ESI) are the most common ionization methods. Both methods ionize the protein by protonation or deprotonation, with the former much more common than the latter. In ESI, it is also quite common to see the addition of buffer ions present in the solution instead of a proton, such as sodium and potassium adducts. The most important difference between ESI and MALDI in terms of protein characterization is that the former generally results in the formation of multiply charged species due to addition (or removal) of multiple protons/cations, while the latter often yields only singly charged species.

Common applications of MS in the pharmaceutical industry are the accurate determination of the molecular mass and the primary structure of a protein (including post-translational modifications), and identification of chemical degradation products. In the latter case, the method is usually coupled to chromatographic or electrophoretic methods, in combination with peptide mapping (see Section 4.2.2). Thus, MS is mostly used in the early phase of the formulation development. When the identity of degradation products is established, chromatography and electrophoresis can often stand alone in formulation assessment in terms of the ability of the formulation to minimize degradation during storage and/or use.

Structural characterization of proteins using MS is based on either charge distribution of m/z species or hydrogen–deuterium exchange kinetics (van den Bremer and Heck, 2005). Hydrogen–deuterium exchange can show how fast exchangeable hydrogens are replaced by deuterium in specific regions of the protein, which provides insight into the flexibility and solvent exposure of segments of the protein. In this respect, MS can provide increased insight compared to the often more global spectroscopic methods discussed further below (see, e.g., Zhang et al., 2012). A more global picture can be obtained from the charge distribution in an ESI mass spectrum, which is different for folded and (partially) unfolded proteins as the latter will expose more ionizable groups.

4.2.5 ANALYTICAL ULTRACENTRIFUGATION

Analytical ultracentrifugation (AUC) (Philo, 2005) is one of the oldest methods to characterize the tertiary and quaternary structure of proteins, and is considered the "gold standard" in the analysis of protein aggregation. An AUC is in essence a high-speed centrifuge combined with an optical system, with the latter at set intervals scanning the sample cell from top to bottom, which allows quantification of the amount of species present at each particular height in the sample. The velocity with which a large molecule is sedimented by the centrifugal force is a function of molecular mass, diffusivity, and shape. Thus, in an AUC experiment, the various species present in the solution will be separated based on these three fundamental properties of the species. Due to the slow sedimentation velocity of proteins, AUC experiments typically take several hours, if not days, and are thus much slower than most other methods to determine protein association state, such as SEC or FFF. However, AUC has a much larger dynamic range than SEC, is a true solution method (i.e., no inter-action with column materials), is more accurate, and allows the determination of several relevant protein properties within one single experiment.

A complementary method in AUC is the use of sedimentation equilibrium, in which the time to reach an equilibrium between sedimentation and diffusion is determined. The latter is a function of molecular mass only, and also maintains the equilibria for reversible association interactions. Thus, this method allows the characterization of association constants between different species, including cases of weak interactions.

4.2.6 SPECTROSCOPIC METHODS USING RADIATION IN THE UV, VIS, AND INFRARED RANGE

In this section, UV–Vis, circular dichroism (CD), mid-infrared (and Raman) spectroscopy, and fluorescence spectroscopy will be discussed. These spectroscopic methods all follow the absorption or emission of radiation with a wavelength in the nanometer to micrometer range. UV–Vis spectroscopy and to a lesser extent fluorescence spectroscopy are also commonly used as a detection method in various chromatographic approaches. Here, we will focus on the off-line use of these methods to characterize protein structure.

UV–Vis spectroscopy (Kueltzo and Middaugh, 2005; Mach and Middaugh, 2011) is routinely used for the quantification of proteins. The most important absorbing groups are the peptide bond in the backbone and some amino acids (around ca. 190 and 210–220 nm) and the aromatic amino acids phenylalanine (ca. 255–260 nm), tyrosine (ca. 276 nm), and tryptophan (ca. 280 nm). Some proteins may also contain prosthetic groups that strongly absorb UV–Vis light, such as the heme group in hemoglobin and myoglobin. Since many compounds absorb light in the low-wavelength region (around 210 nm), including chlorine ions and water itself, it is most common to use the region around 280 nm for quantification. Analysis of a large subset of proteins has shown a relatively robust empirical relationship between the number of tyrosine (Tyr), tryptophan (Trp), and cystine (S–S bridges) residues in a protein and its molar absorption at 280 nm (Equation 4.1) (Pace et al., 1995). The typical deviation from the "true" values is only a few percent, but for some proteins it may be incorrect by more than 10%. Thus, Equation 4.1 can serve as an initial and reasonable estimation of the absorbance at 280 nm, if the amino acid sequence is known.

$$\varepsilon(280nm)(M^{-1}cm^{-1}) = (\#\,Trp)*5540 + (\#\,Tyr)*1480 + (\#\,cystine)*134 \qquad (4.1)$$

A potential confounding factor in protein quantification using the absorbance at 280 nm is the presence of other absorbing compounds. In such cases, one may first need to separate the protein from the interfering substance. Empirical correction methods are available if the interfering compound only scatters light in this spectral region (see, e.g., Mach and Middaugh, 2011). However, if uncertainty remains, or a lower quantitation limit is required, there are a variety of indirect protein quantification methods available, such as various modifications of the Biuret test (Lowry method [Lowry et al., 1951], bicinchoninic acid method [Smith et al., 1985]), the Bradford assay (Bradford, 1976), or assays based on covalent attachment of a strongly absorbing or fluorescent compound (Noble and Bailey, 2009). For a description of these methods, the reader is referred to the literature.

Although not very widely used, UV spectroscopy is also capable of tracking changes in the tertiary structure of a protein (Dasnoy et al., 2012; Kueltzo et al., 2003; Kueltzo and Middaugh, 2005; Mach and Middaugh, 1994, 2011). The absorption maxima of the three aromatic amino acids are sensitive to changes in their immediate environment, resulting in small shifts in peak position and intensity. As these changes are small, derivative spectroscopy is often used to increase resolution. Combining UV spectroscopy with diode array detectors and/or plate readers allows for a significant sample throughput, while novel technologies have significantly reduced the required amount of solution. This means that UV spectroscopy, despite its limited information content, can be an important tool in the rapid screening of a large number of formulations.

CD (Bloemendal and Jiskoot, 2005; Greenfield, 1999; Kelly et al., 2005; Li et al., 2011b) is one of the most common methods to characterize both the secondary and tertiary structure of a protein in solution. The method is based on the differential absorption of left and right circularly polarized light by optically active chiral molecules. For a protein this includes the amide bonds (far-UV CD, 170–250 nm) and the aromatic amino acids and cystine bond (near-UV CD, 250–320 nm). It has been shown that the most common secondary structural elements, such as beta-sheet, alpha helices, turns, and random coil, all yield a different CD signal. Assuming linear additivity of the signal from the different structural elements, it is then possible to calculate the secondary structure of the protein. The simplest approach only calculates the relative amount of alpha helix using the signal at 208 and/or 222 nm. More advanced methods either attempt to fit the CD signals of each structural element into the measured spectrum, or fit the spectra of proteins with known secondary structure. Each of these approaches has its own merits and issues. For example, the CD signal of less common structural elements is not well known; when the protein contains such an element without the operator being aware of this fact, the fit will unknowingly return erroneous values. Similarly, fitting the spectra of known proteins will require that these are representative for the structural elements that may be present in the protein under investigation. As a result, one should not expect the secondary structure determined by NMR or x-ray crystallography to correspond to that calculated from CD spectra. However, there may also be real structural differences between the samples analyzed by CD, NMR, and x-ray crystallography, due to the different conditions under which the sample is measured. A major advantage of CD over NMR, and in particular x-ray crystallography, is that the former can be applied to a wide variety of solution conditions, including those of relevance to a pharmaceutical formulation.

Near-UV CD spectra yield information on the microenvironment around the aromatic residues and cystine bridges. Locking these residues into the interior of a protein will give rise to a dichroic signal. This signal is very sensitive to the conformation of the protein, and thus a very useful fingerprint to assure proper folding.

The main advantages of CD spectroscopy are its ability to characterize protein structure and structural integrity of relative small amounts and low concentrations (<1 mg/mL), in a wide range of solution conditions. However, there are also several disadvantages, such as spectral interference by some commonly used excipients (e.g., chloride in the far-UV region) and the difficulty in properly characterizing proteins with high beta-sheet content. In addition, it should be remembered that CD spectroscopy, just like UV spectroscopy, only gives a "global" or "averaged" picture; that is, any observed spectral change may be due to the whole population altering structure or due to the presence of a subpopulation with a different structure. Finally, opaque samples or highly concentrated proteins cannot easily be analyzed.

In infrared (IR) spectroscopy (Byler and Susi, 1986; Goormaghtigh et al., 2009; Jackson and Mantsch, 1995; Jiang et al., 2011; van de Weert et al., 2005), the absorption of IR radiation by vibrating atoms within a molecule is determined. In the case of proteins, most attention is focused on the carbonyl stretch vibration of the backbone amide bond, observed in the region 1600–1700 cm^{-1}. The exact position of this absorption band is sensitive to the bond angles and hydrogen bonding of the amide bond, and hence sensitive to protein conformation (which is defined by bond angles and hydrogen bonding). Just like far-UV CD spectroscopy, IR spectroscopy can thus be used to characterize protein secondary structure. In contrast to CD, however, IR spectroscopy is more sensitive toward beta-sheet structure. Moreover, unlike CD spectroscopy, IR does not require information on the concentration of the protein.

Perhaps the most important advantage of IR spectroscopy is its ability to analyze protein structure regardless of the physical state of the sample. It is therefore often used to characterize protein structural changes upon lyophilization (see also Chapter 10 and Carpenter et al., 1998). In general, the long-term storage stability of a lyophilized protein is improved when the conformation of the protein in the solid state, as determined by IR spectroscopy, most closely resembles the conformation of the protein in solution prior to lyophilization.

IR spectroscopy is also very useful in the structural analysis of highly concentrated proteins (>20 mg/mL, e.g., Harn et al., 2007) and proteins entrapped in solid matrices or liposomes (e.g., van de Weert et al., 2000). Finally, intermolecular beta-sheets, often observed in noncovalent aggregates, yield a rather unique signal in the IR region, with a relatively sharp band around 1625 cm^{-1} in aqueous solution (van de Weert et al., 2001).

The main disadvantage of IR spectroscopy is the interfering influence of water on accurate analysis of the protein conformation. As a result, high protein concentrations (>10 mg/mL) are generally required. The quantitative analysis of secondary structure also generally requires more operator input than CD spectroscopy, and hence is subject to potential bias. Therefore, such quantitative analysis is not common in a pharmaceutical setting.

Raman spectroscopy (Nielsen, 2005; Sane et al., 1999) is another form of vibrational spectroscopy. Whereas in IR spectroscopy the absorption of IR radiation is measured directly, in Raman spectroscopy, the inelastic scattering of laser light is analyzed. The latter inelastic scattering results in a small shift of the scattered light, corresponding to the energy absorbed by vibrations. As it is based on a different physical principle but still results from vibrational motions, Raman spectra are both different from and similar to IR spectra. The main differences are in relative intensities of the spectral bands; in its most simplistic description, intense bands in Raman will be weak in IR, and vice versa. Raman spectra are also less affected by the water absorption band. Although Raman is not commonly used to characterize protein structure, in principle it may provide additional advantages that are as yet poorly explored. In particular, the wavelength of the laser light determines which vibrations can be excited and thus allows a more detailed insight than IR. For example, it has been shown that Raman spectroscopy can be used to characterize hydrogen-bonding patterns of cysteine sulfhydryl groups, as well as interactions involving tyrosine groups. Thus, local interactions may be explored along with the protein conformation.

Fluorescence spectroscopy (Jiskoot et al., 2005; Lakowicz, 1999) is the final spectroscopic technique that will be discussed here. In fluorescence spectroscopy, a fluorophore is excited and the resulting fluorescence measured in terms of emission maximum and intensity. These are a function of both the nature of the fluorophore itself and the microenvironment around this fluorophore. In proteins, the aromatic residues constitute the intrinsic fluorophores, with tryptophan by far the most useful for structural analysis. Its fluorescence maximum can range from about 310 nm when completely embedded inside the hydrophobic core of the protein, to 360 nm when fully exposed to the aqueous environment. The quantum yield is highly dependent on the proximity of potentially quenching side chains of other amino acids. Thus, the tryptophan fluorescence is very sensitive to small local changes.

Intrinsic fluorescence can be measured on rather small amounts of protein; depending on the content of tryptophan in the protein, concentrations in the nanogram per milliliter range can be reliably analyzed, while special cuvettes are available that require only a few tens of microliters. In contrast, high concentrations (>5 mg/mL) are more difficult to analyze, and require special cuvettes or approaches to minimize the so-called inner-filter effect.

It is also possible to covalently or noncovalently bind a fluorophore to the protein (Hawe et al., 2008). Since covalently bound fluorophores may affect the structure and stability of the protein, it is generally recommended to use a noncovalently binding fluorophore; however, those may also affect the structure and stability by acting as (usually weak) ligands. There is a wide range of such extrinsic fluorophores available, and the binding of these can be specific to the structure of the protein under investigation. For example, the dye Thioflavin T preferentially binds to amyloid fibril structures (Groenning, 2010), which subsequently results in a significant fluorescence increase of this dye. Other commonly used dyes are bis-ANS and Nile red, which bind to hydrophobic patches with concomitant changes in their fluorescence. Thus, these dyes are very useful to investigate the presence of populations of (partly) unfolded and/or aggregated species. The detection limit of such species may be very low, often much lower than that of other techniques like chromatography, and thus allows characterization of the very first events leading to large scale unfolding and/or aggregation.

Apart from intensity and wavelength maximum, it is also possible to follow the fluorescence lifetime and fluorescence anisotropy. The fluorescence lifetime of a fluorophore is very sensitive to local interactions with and around the fluorophore, and can thus give insight into the causes of any observed changes in fluorescence. Fluorescence anisotropy is a measure of the depolarization rate of the emission of a fluorophore excited by polarized light. A rapidly moving fluorophore will generally show a low anisotropy, whereas an immobile fluorophore will give a high anisotropy. Thus, insight into the rotational freedom of the fluorophore is obtained. A combination of lifetime and anisotropy measurements allows the calculation of local rotation coefficients of the fluorophore, and thus provides very detailed insight into the local viscosity around the fluorophore.

4.2.7 SCATTERING TECHNIQUES

Like all molecules proteins can scatter radiation, the magnitude of which is dependent on their size and shape, and the wavelength of that radiation. Thus, scattering can be used to obtain information about size and shape, but also about the presence of larger species in the solution, for example, aggregates (Demeester et al., 2005; Santos and Castanho, 1996). In its simplest format, the UV–Vis absorbance is monitored in regions where the protein does not normally absorb any radiation, for example, above 300 nm. An increase in absorbance then indicates the formation of larger species that scatter light, which usually suggests aggregate formation.

Scattering can be determined more sensitively by measuring the intensity of the scattered light at an angle compared to the incident light. This can be done on specially adapted UV spectrometers and also fluorescence spectrometers. However, it is more common to use dedicated instruments that use laser light, which may also allow measurements of scatter intensity at different angles. This method is generally referred to as static light scattering (SLS) and is especially useful to accurately determine the so-called radius of gyration, R_g (root mean square distance of all parts of the species compared to its center of gravity). The latter analysis is difficult when species with different sizes are present, as the R_g will then be an average of all species. Also, light scattering scales nonlinear to species size, which means that small amounts of very large particles will contribute most of the observed scattering to a sample consisting mainly of small particles. This makes quantification difficult, if not impossible.

Scattering of neutrons and x-rays by protein solutions may also be determined, known as small angle neutron and small angle x-ray scattering (SANS and SAXS, respectively) (Chacon et al., 1998; Svergun et al., 2011). Although still uncommon in analysis of protein pharmaceuticals, these two techniques can provide more detailed structural information than SLS. They generally require much larger and more expensive instrumentation than SLS, as well as significantly more expertise, and these methods are thus most appropriate for the early characterization of the protein. However, efforts are being made to adapt the method to very small volumes and microfluidics chips (Hura et al., 2009; Toft et al., 2008), which would allow rapid formulation screening.

A different approach to utilizing scattering is to measure the fluctuation of the intensity of the scattered light on very short time scales. This is known as dynamic light scattering (DLS; also sometimes called photon correlation spectroscopy or quasielastic light scattering). Due to the Brownian motion of molecules in a solution, the scattering of light by such a solution changes over the timescales related to that Brownian motion. These fluctuations can then be used to determine the hydrodynamic radius, R_h (the particle moves as a sphere, including its hydration shell, of this size) of particles in a solution through the Stokes–Einstein equation. Advanced algorithms also allow the analysis of polymodal samples, that is, samples containing multiple species with different sizes. However, care should be taken not to overinterpret the results of such analyses. That is, there are numerous assumptions in the algorithms that may be appropriate for some samples, but less so for others. The obtained sizes and distributions may thus deviate significantly from reality.

Combining data from SLS and DLS further increases the insight into the sample; instruments are available that can do both analyses in one experiment. In particular, the R_g/R_h ratio is a useful parameter, as it provides an insight into the shape of the particle. For spheres this ratio is close to 0.77, while for a rod it is about 1.7. It is also common to combine these light scattering techniques online with chromatographic techniques, in particular FFF.

4.2.8 Microscopy

A wide variety of microscopic techniques are available (Garidel et al., 2012), ranging from simple optical microscopy via various electron microscopic (EM) techniques to advanced atomic force microscopy (AFM) (the latter is discussed in Engel [2005]). The main differences between these microscopic techniques are the resolution and the required sample preparation.

Optical microscopy can be used to determine the presence of relatively large particles, generally in the micrometer range. The advantages in image analysis software and automation have significantly increased the throughput of these analyses. For example, a particle-containing solution flowing through a narrow channel allows imaging of the individual particles in that solution, and gives information on the number of particles per volume unit, size of these particles, and also their shape. This usually works best for particles well above 1 μm in size.

Combining an ultramicroscope with a digital camera allows the determination of both number and size of (spherical) particles in the low nanometer to micrometer range. The so-called nanoparticle tracking analysis (NTA) instruments "track" the Brownian motion of individual particles, allowing a combination of particle size determination (through the Stokes–Einstein equation) and the number of particles per volume unit (average number of scatterers within the frame). This overcomes the inherent uncertainty of the algorithms used in DLS for the analysis of polymodal samples.

Fluorescence microscopy can also be used to image protein samples, and is used in particular in combination with extrinsic fluorescent probes. When these probes bind to specific protein populations, for example to an aggregated protein only, the formation and relative size of these aggregates may be followed over time. Fluorescence microscopy can be very sensitive to even minute amounts of such species, and is thus ideally suited for analyzing samples that appear "clean" by other imaging methods.

Electron microscopic techniques include scanning electron microscopy (SEM) and transmission electron microscopy, and allow low-nanometer resolution. AFM even generates a three-dimensional picture. However, the latter method has a rather limited scanning area, and it may thus be difficult to obtain a representative sample without having to analyze the same sample multiple times. In all of these microscopic methods, the sample is deposited on a solid substrate; thus, the potential impact of the sample preparation method should be taken into account.

4.2.9 MEASURING THE THERMODYNAMICS OF THE PROTEIN FOLD

Most proteins fold into specific three-dimensional structures, held together by non-covalent forces. Various degradation reactions involve an initial (partial) unfolding of this structure (see Chapter 6). The strength of the forces keeping this structure together can thus be an interesting parameter to study in connection with formulation development. For example, if the protein fold is thermodynamically more stable in a certain formulation, it is generally assumed that its physical stability is also higher. Another important observation in these thermodynamic analyses is the reversibility of the unfolding process; that is, how much of the protein refolds into its native state when the stress factor is removed. Larger reversibility indicates less physical degradation, in particular when temperature is used as the stress factor.

Protein thermodynamic stability can be determined as a function of temperature as well as a function of denaturant concentration. The protein fold can be monitored by a variety of methods discussed above, for example, CD, Fourier transform infrared (FTIR), or fluorescence. In the case of temperature-induced unfolding, it is also possible to use differential scanning calorimetry (DSC) (Chen and Oakley, 1995; Cueto et al., 2003; Schön and Velazquez-Campoy, 2005; Privalov and Dragan, 2007; Wen et al., 2012). In the latter method, the energy required to heat a sample at a certain rate is measured versus the energy required to measure the same sample in the absence of protein. This method yields the melting temperature (T_m), the enthalpy of unfolding, and heat capacity change upon unfolding, allowing a full thermodynamic characterization. Also the reversibility of the unfolding process can be determined. Protein formulations with higher T_m and larger reversibility are very likely to be much more stable during storage, but in formulations where T_m increases and reversibility decreases this is more difficult to predict. In the latter case, the decisive factor is which of the two processes is rate-limiting; if unfolding is rate-limiting, the protein will be more stable, whereas when the physical degradation (usually aggregation) is rate-limiting, the formulation may be less stable.

At high concentration known protein denaturants, such as urea and guanidine hydrochloride, can also unfold a protein (Pace, 1990). Various spectroscopic techniques can be used to follow this unfolding, although fluorescence spectrometry is most compatible with the added denaturants. Performing the analysis at multiple temperatures will help obtain more detailed thermodynamic information. However, very high concentrations of denaturant are usually needed (several molar), which may not be compatible with the formulations the investigator wants to study.

4.2.10 High-Throughput Methodology

A relatively recent development in protein formulation development is the use of high-throughput analytical methodology. That is, methods that can rapidly and/ or simultaneously evaluate many formulations (see, e.g., Bhambhani et al., 2012; Capelle and Arvinte, 2008; Capelle et al., 2009; Dasnoy et al., 2011, 2012; Goldberg et al., 2011; He et al., 2011; Li et al., 2011a). Most of these methods aim at characterizing the protein's physical stability and are often based on 96 or 384 well plates. In general, some sort of stress factor, such as elevated temperatures or shaking, is used to increase the degradation rate. The readout methods include UV absorbance, intrinsic or more commonly extrinsic fluorescence, or light scattering. Ideally, several different readouts are used to obtain a more complete picture.

A special aspect of these high-throughput methods is the data analysis. Due to the large amount of data that is generated, it is often not useful to investigate each sample manually. Thus, thresholds can be assigned, ranking the sample from good to moderate to poor in terms of the stability factor that is being analyzed. Multidimensional plots can then be prepared with color codes, which allow an appealing visualization of the ranking of the samples as a function of the formulation parameter(s). The most promising formulations can then be characterized in a more elaborate stability study.

4.3   CHOOSING THE PROPER METHODOLOGY

The protein drug development scientist has a significant challenge in deciding which methods are to be used and when. The regulatory agencies generally do not require a specific set of methods for a protein-based drug product. For example, there is no requirement that the detailed three-dimensional structure is determined. At the same time, this information can be highly useful for a better understanding of both function and behavior of the protein. Moreover, there are still only a limited number of monographs in the various Pharmacopoeias, and thus for many proteins there are no clear guidelines. In an ideal situation, all existing analytical methods are used to characterize the protein and its degradation products. However, the observations in those analytical assays will then also need to be explained, including their potential impact on the clinical behavior of the protein drug. In addition, it is not feasible to use the large number of potential methodologies, both from a time and financial perspective. As a result, the protein scientist has a significant challenge in determining which methods to use and when, above and beyond the challenge in understanding the outcome of the assays. In this section, we provide some guidance on choosing the proper methodology, which is summarized in Table 4.2. We stress again that we focus solely on the methods that characterize the protein and its degradation products, and

TABLE 4.2
Protein Drug Development Phases and Common Methodology

Main Aims	Standard Methodology	Comments
Fundamental description of protein structure	X-ray crystallography; NMR spectroscopy; MS; CD; FTIR; AUC; thermodynamic analysis; biochemical methods	Characterization of the primary, secondary, tertiary, quaternary structure, as well as folding thermodynamics and general biological activity can be a great asset to understand the behavior of the protein in terms of (long-term) stability and potential formulation approaches
Forced degradation study; evaluation of analytical methodology	Chromatography; electrophoresis; DLS; SLS; fluorescence; FTIR; CD; AUC; UV–Vis; MS; MFI; NTA; FFF	The deliberate degradation of the protein, both by chemical and physical stresses, is used to evaluate the various analytical approaches in their ability to detect and characterize the degradation products
Stress testing	Same as in forced degradation study	Common stresses (such as shaking, pumping, heat stress, freeze-thawing, etc.) provide insight into the relative instability of the protein and the potentially required formulation approaches
Testing stability of protein in a formulation space; formulation screening	Same as in forced degradation study, but including biochemical methods	This phase is aimed at identifying promising formulations in which the protein is most stable. The panel of techniques used is often smaller than in the forced degradation studies, focusing on methods that are found most suitable to detect the most important degradation products
Testing stability of protein in most promising formulations	Same as in stress testing	A subset of the most stable formulations is subjected to real-time storage stability testing, allowing the selection of the optimal formulation(s)
Lot release; product release	UV–Vis; chromatographic methods; biochemical methods	Simple methods that provide a good indication of identity and the level of impurities are used in both lot release and product release analysis. Lot release still commonly involves some type of biochemical test to assure identity through its biological function

that further methods will be required to characterize other aspects of the formulation, such as degradation of the excipients or the presence of impurities. Moreover, the description below should not be considered a timeline as many of the different "phases" in the analytical development occur simultaneously and can be iterative.

When the bioactive protein to be developed into a protein drug has been chosen, it may be useful to obtain a detailed three-dimensional structure of the protein, if not already available. This requires methods like NMR or x-ray crystallography. At the same time, less detailed methods like CD, IR, light scattering, AUC, and thermodynamic characterization can be used to establish a more global insight into the native protein's structure. The results from these latter methods can be used as the control in the next phase of the development. Typically, potency tests are also developed during this phase, to establish a predictive read-out for the biological activity. These are often biochemical methods, ranging from animal models to cell/tissue models and ELISA assays.

The next phase consists of a forced degradation study of the protein, in which the protein is exposed to a number of stresses that are known to result in certain types of degradation products (see also Hawe et al., 2012). For example, the addition of hydrogen peroxide to a protein will result in oxidation of several amino acids, whereas heat stress generally results in aggregation. This type of testing allows the formulation scientist to determine the ability of various analytical methods to detect and characterize potential degradation products. Typically, a variety of chromatographic methods, MS, and some of the spectrometric methods are used in this phase of the development, as well as various spectroscopic methods. However, there is always a danger that the best stability-indicating analytical methodology is not amongst these chosen methods. The forced degradation study also provides valuable insight into important degradation pathways of the protein; this insight can be used to guide stabilization approaches for the protein.

Based in part on the forced degradation studies, the potentially suitable protein formulation space is determined. These formulations are then exposed to stress testing. Here, the protein is exposed to a number of stresses that are typically encountered during production, formulation, and storage of the intended product. These stresses include exposure to pH extremes and high intensity light, high and low temperatures (including repeated freezing and thawing), shaking, and various surfaces. The same analytical methodology developed during the forced degradation study is generally used in stress testing. Importantly, mechanistic information may not be necessary at this stage. Rather, the focus is on changes of the original profile (e.g., the spectrum or chromatogram) upon applying the stresses. Those formulations that show the least changes compared to the starting protein are likely also the most stable formulations. Several of these most stable formulations can then be tested in a real-time stability study to select the final optimal formulation(s). Importantly, this latter testing should be performed in connection with preclinical and/or clinical studies, to assure that the formulations also perform as intended in vivo.

The final phase is the lot release of the final product. Here the analytical methods to evaluate the protein itself are generally limited. Identity will need to be established, which may be performed by one (or more) of the methods described above, but also by biochemical methods that are indicative of biological function. In fact, most monographs of proteins in the various pharmacopeias contain a reference method to establish biological function as a part of identity confirmation. Quantity will also need to be established, which may be a simple UV spectroscopic or chromatographic measurement. Finally, purity and impurities may need to be determined, again often by chromatographic methods.

Throughout the various phases of development, the potential impact of the methodology on the outcome should always be taken into account (Arvinte, 2005). That is, the analytical method may require manipulation of the sample, which in turn may lead to changes that result in incorrect ranking of the various formulations or incorrect conclusions on the formed degradation products. For example, high protein concentration formulations, which can be well above 100 mg/mL, may require significant dilution before certain analytical methods can be applied. The dilution itself will alter the various interactions (protein–protein, protein–excipient) in the solution, while the process of dilution may introduce variation between samples. These potential confounding factors make it more difficult to draw firm conclusions. Likewise, many methods may be affected by the presence of particulate material, for example, due to scattering or because of the danger of clogging a column. Any sample pretreatment that removes these particles will essentially change the formulation, with unknown effects on how this will change the comparison between different formulations. Thus, ideally methods should be chosen that can be used on the formulation without any significant manipulation. This may require innovative or creative use of the available analytical methodology, and thereby creates an even greater challenge to the formulation scientist to understand the analytical methods.

Another important aspect of protein analysis is that the potential aggregation of the protein should be studied by at least two orthogonal methods, that is, methods that are based on different principles. Protein aggregates can consist of many different types of associates, which can be classified by their interaction strength, sizes, bonding type (covalent or noncovalent), and protein structure (native or unfolded) (Joubert et al., 2011). Many individual techniques will only probe a subset of these potential classifications, and thus provide very limited insight into the nature of the aggregates. In some cases, the aggregate may also not be observed; for example, large insoluble aggregates cannot enter a SEC column, while weakly interacting aggregates may be broken up in a chromatographic flow. Similarly, SDS-PAGE is very well suited to characterize covalent aggregates, but generally results in dissociation of noncovalent aggregates.

A final aspect to consider is that there may always be degradation routes that are missed by the, by necessity limited, analytical toolbox that is used, or where the impact of observed differences is insufficiently appreciated. A good example is shown by Taschner et al. (2001), who studied the substantial loss of antigenicity of an anti-idiotype monoclonal antibody upon lyophilization. Using a combination of SEM, various fluorescence methodologies, and CD, they could identify subtle differences between the lyophilized formulation and the starting material, which could be linked to the in vivo failure of this particular formulation. This also shows the danger of only selecting one lead formulation to use in clinical studies, in particular if there is no suitable biochemical read-out method.

4.4 CONCLUDING REMARKS

In this chapter, we briefly reviewed the main analytical techniques used in the analysis of protein pharmaceuticals. The list provided here is by no means complete, nor can the chapter do justice to the various subtleties of the methods; moreover, new methods are continuously being developed. This creates opportunities but also significant challenges for the analytical and formulation scientist in finding the most appropriate methods. Understanding the various available methods is an important starting point for this selection, and requires that formulation scientists are properly trained in the available analytical methodology.

REFERENCES

Anderot, M., Nilsson, M., Végvári, A., Moeller, E.H., van de Weert, M., and Isaksson, R., 2009. Determination of dissociation constants between polyelectrolytes and proteins by affinity capillary electrophoresis. *J. Chromatogr. B* 877:892–896.

Arvinte, T., 2005. Concluding remarks: Analytical methods for protein formulations. In: *Methods for Structural Analysis of Protein Pharmaceuticals*, Biotechnology: Pharmaceutical Aspects, Jiskoot, W., and Crommelin, D.J.A. (Eds.), Arlington, TX, AAPS Press, pp. 661–666.

Aubin, Y., Gingras, G., and Sauvé, S., 2008. Assessment of the three-dimensional structure of recombinant protein therapeutics by NMR fingerprinting: Demonstration on recombinant human granulocyte macrophage-colony stimulation factor. *Anal. Chem.* 80:2623–2627.

Bartolomeo, M.P., and Maisano, F., 2006. Validation of a reversed-phase HPLC method for quantitative amino acid analysis. *J. Biomol. Techn.* 17:131–137.

Bhambhani, A., Kissmann, J.M., Joshi, S.B., Volkin, D.B., Kashi, R.S., and Middaugh, C.R., 2012. Formulation design and high-throughput excipient selection based on structural integrity and conformational stability of dilute and highly concentrated IgG1 monoclonal antibody solutions. *J. Pharm. Sci.* 101:1120–1135.

Bieri, M., Kwan, A.H., Mobli, M., King, G.F., Mackay, J.P., and Gooley, P.R., 2011. Macromolecular NMR spectroscopy for the non-spectroscopist: Beyond macromolecular solution structure determination. *FEBS J.* 278:704–715.

Bischoff, R., and Barroso, B., 2005. Liquid chromatography. In: *Methods for Structural Analysis of Protein Pharmaceuticals*, Biotechnology: Pharmaceutical Aspects, Jiskoot, W., and Crommelin, D.J.A. (Eds.), Arlington, TX, AAPS Press, pp. 277–330.

Bloemendal, M., and Jiskoot, W., 2005. Circular dichroism spectroscopy. In: *Methods for Structural Analysis of Protein Pharmaceuticals*, Biotechnology: Pharmaceutical Aspects, Jiskoot, W., and Crommelin, D.J.A. (Eds.), Arlington, TX, AAPS Press, pp. 83–130.

Blommers, M.J.J., and Cerletti, N., 1997. High resolution NMR, a useful tool to control batch-to-batch consistency of disulphide bonds in biopharmaceuticals: Application to transforming growth factor-B$_3$. *Pharm. Sci.* 3:29–36.

Bobst, C.E., Abzalimov, R.R., Houde, D., Kloczewiak, M., Mhatre, R., Berkowitz, S.A., and Khaltashov, I.A., 2008. Detection and characterization of altered conformations of protein pharmaceuticals using complementary mass spectrometry-based approaches. *Anal. Chem.* 80:7473–7481.

Bradford, M.M., 1976. A rapid and sensitive method for the quantitation of microgram quantities of protein utilizing the principle of protein-dye binding. *Anal. Biochem.* 72:248–254.

Byler, D.M., and Susi, H., 1986. Examination of the secondary structure of proteins by deconvolved FTIR spectra. *Biopolymers* 25:469–487.

Caldwell, K.D., and Wahlund, K-G., 2005. Field-flow fractionation. In: *Methods for Structural Analysis of Protein Pharmaceuticals*, Biotechnology: Pharmaceutical Aspects, Jiskoot, W., and Crommelin, D.J.A. (Eds.), Arlington, TX, AAPS Press, pp. 413–434.

Capelle, M.A.H., and Arvinte, T., 2008. High-throughput formulation screening of therapeutic proteins. *Drug Discov. Today Technol.* 5:e71–e79.

Capelle, M.A.H., Gurny, R., and Arvinte, T., 2009. A high throughput protein formulation platform: Case study of salmon calcitonin. *Pharm. Res.* 26:118–128.

Carpenter, J.F., Prestrelski, S.J., and Dong, A., 1998. Application of infrared spectroscopy to development of stable lyophilized protein formulations. *Eur. J. Pharm. Biopharm.* 45:231–238.

Catai, J.R., de Jong, G.J., and Somsen, G.W., 2005. Capillary electrophoresis. In: *Methods for Structural Analysis of Protein Pharmaceuticals*, Biotechnology: Pharmaceutical Aspects, Jiskoot, W., and Crommelin, D.J.A. (Eds.), Arlington, TX, AAPS Press, pp. 331–378.

Chacon, P., Moran, F., Diaz, J.F., Pantos, E., and Andreu, J.M., 1998. Low-resolution structures of proteins in solution retrieved from x-ray scattering with a genetic algorithm. *Biophys. J.* 74:2760–2775.

Chen, T., and Oakley, D.M., 1995. Thermal analysis of proteins of pharmaceutical interest. *Thermochim. Acta* 248:229–244.

Cueto, M., Dorta, M.J., Munguía, O., and Llabrés, M., 2003. New approach to stability assessment of protein solution formulations by differential scanning calorimetry. *Int. J. Pharm.* 252:159–166.

Dasnoy, S., Dezutter, N., Lemoine, D., Le Bras, V., and Préat, V., 2011. High-throughput screening of excipients intended to prevent antigen aggregation at air-liquid interface. *Pharm. Res.* 28:1591–1605.

Dasnoy, S., Le Bras, V., Préat, V., and Lemoine, D., 2012. High-throughput assessment of antigen conformational stability by ultraviolet absorption spectroscopy and its application to excipient screening. *Biotechnol. Bioeng.* 109:502–516.

Demeester, J., De Smedt, S.S., Sanders, N.N., and Haustraete, J., 2005. Light scattering. In: *Methods for Structural Analysis of Protein Pharmaceuticals*, Biotechnology: Pharmaceutical Aspects, Jiskoot, W., and Crommelin, D.J.A. (Eds.), Arlington, TX, AAPS Press, pp. 245–276.

El-Aneed, A., Cohen, A., and Banoub, J., 2009. Mass spectrometry, review of the basics: Electrospray, MALDI, and commonly used mass analyzers. *Appl. Spectrosc. Rev.* 44: 210–230.

Engel, A., 2005. Atomic force microscopy. In: *Methods for Structural Analysis of Protein Pharmaceuticals*, Biotechnology: Pharmaceutical Aspects, Jiskoot, W., and Crommelin, D.J.A. (Eds.), Arlington, TX, AAPS Press, pp. 591–614.

Fountoulakis, M. and Lahm, H-W., 1998. Hydrolysis and amino acid composition analysis of proteins. *J. Chromatogr. A* 826:109–134.

Garidel, P., Herre, A., and Kliche, W., 2012. Microscopic methods for particle characterization in protein pharmaceuticals. In: *Analysis of Aggregates and Particles in Protein Pharmaceuticals*, Mahler, H-C., and Jiskoot, W. (Eds.), Hoboken, NJ, John Wiley & Sons, pp. 269–302.

Goldberg, D.S., Bishop, S.M., Shah, A.U., and Satish, H.A., 2011. Formulation development of therapeutic monoclonal antibodies using high-throughput fluorescence and static light scattering techniques: Role of conformational and colloidal stability. *J. Pharm. Sci.* 100:1306–1315.

Goormaghtigh, E., Gasper, R., Bénard, A., Goldsztein, A., and Raussens, V., 2009. Protein secondary structure content in solution, films and tissues: Redunacy and complementarity of the information content in circular dichroism, transmission and ATR FTIR spectra. *Biochim. Biophys. Acta* 1794:1332–1343.

Greenfield, M.J., 1999. Applications of circular dichroism in protein and peptide analysis. *Trends Anal. Chem.* 18:236–244.

Groenning, M., 2010. Binding mode of thioflavin T and other molecular probes in the context of amyloid fibrils—current status. *J. Chem. Biol.* 3:1–18.

Harn, N., Allan, C., Oliver, C., and Middaugh, C.R., 2007. Highly concentrated monoclonal antibody solutions: Direct analysis of physical structure and thermal stability. *J. Pharm. Sci.* 96:532–546.

Hawe, A., Kasper, J.C., Friess, W., and Jiskoot, W., 2009. Structural properties of monoclonal antibody aggregates induced by freeze-thawing and thermal stress. *Eur. J. Pharm. Sci.* 38:79–87.

Hawe, A., Sutter, M., and Jiskoot, W., 2008. Extrinsic fluorescent dyes as tools for protein characterization. *Pharm. Res.* 25:1487–1499.

Hawe, A., Wiggenhorn, M., van de Weert, M., Garbe, J.H.O., Mahler, H.-C., and Jiskoot, W., 2012. Forced degradation of therapeutic proteins. *J. Pharm. Sci.* 101:895–913.

He, F., Woods, C.E., Trilisky, E., Bower, K.M., Litowski, J.R., Kerwin, B.A., Becker, G.W., Narhi, L.O., and Razinkov, V.I., 2011. Screening of monoclonal antibody formulations based on high-throughput thermostability and viscosity measurements: Design of experiment and statistical analysis. *J. Pharm. Sci.* 100:1330–1340.

Hoff, E.R., and Chloupek, R.C., 1996. Analytical peptide mapping of recombinant DNA-derived proteins by reversed-phase high-performance liquid chromatography. *Methods Enzymol.* 271:51–68.

Hu, S., and Dovichi, N.J., 2002. Capillary electrophoresis for the analysis of biopolymers. *Anal. Chem.* 74:2833–2850.

Hura, G.L., Menon, A.L., Hammel, M., Rambo, R.P., Poole II, F.L., Tsutakawa, S.E., Jenney Jr, F.E., Classen, S., Frankel, K.A., Hopkins, R.C., Yang, S., Scott, J.W., Dillard, B.D., Adams, M.W.W., and Tainer, J.A., 2009. Robust, high-throughput solution structural analyses by small angle x-ray scattering (SAXS). *Nat. Methods* 6:606–612.

Jackson, M., and Mantsch, H.H., 1995. The use and misuse of FTIR spectroscopy in the determination of protein structure. *Crit. Rev. Biochem. Mol. Biol.* 30:95–120.

Jiang, Y., Li, C., Nguyen, X., Muzammil, S., Towers, E., Gabrielson, J., and Narhi, L., 2011. Qualification of FTIR spectroscopic method for protein secondary structural analysis. *J. Pharm. Sci.* 100:4631–4641.

Jiskoot, W., Visser, A.J.W.G., Herron, J.N., and Sutter, M., 2005. Fluorescence spectroscopy. In: *Methods for Structural Analysis of Protein Pharmaceuticals*, Biotechnology: Pharmaceutical Aspects, Jiskoot, W., and Crommelin, D.J.A. (Eds.), Arlington, TX, AAPS Press, pp. 27–82.

Joubert, M.K., Luo, Q., Nashed-Samuel, Y., Wypych, J., and Narhi, L.O., 2011. Classification and characterization of therapeutic antibody aggregates. *J. Biol. Chem.* 286:25118–25133.

Kelly, S.M., Jess, T.J., and Price, N.C., 2005. How to study proteins by circular dichroism. *Biochim. Biophys. Acta* 1751:119–139.

Kueltzo, L.A., Ersoy, B., Ralston, J.P., and Middaugh, C.R., 2003. Derivative absorbance spectroscopy and protein phase diagrams as tools for comprehensive protein characterization: A bGCSF case study. *J. Pharm. Sci.* 92:1805–1820.

Kueltzo, L.A., and Middaugh, C.R., 2005. Ultraviolet absorption spectroscopy. In: *Methods for Structural Analysis of Protein Pharmaceuticals*, Biotechnology: Pharmaceutical Aspects, Jiskoot, W., and Crommelin, D.J.A. (Eds.), Arlington, TX, AAPS Press, pp. 1–26.

Kwan, A.H., Mobli, M., Gooley, P.R., King, G.F., and Mackay, J.P., 2011. Macromolecular NMR spectroscopy for the non-spectroscopist. *FEBS J.* 278:687–703.

Lakowicz, J.R., 1999. *Principles of Fluorescence Spectroscopy*, New York, Kluwer Academic/Plenum.

Li, Y., Mach, H., and Blue, J.T., 2011a. High throughput formulation screening for global aggregation behaviors of three monoclonal antibodies. *J. Pharm. Sci.* 100:2120–2135.

Li, C.H., Nguyen, X., Narhi, L., Chemmalil, L., Towers, E., Muzammil, S., Gabrielson, J., and Jiang, Y., 2011b. Applications of circular dichroism (CD) for structural analysis of proteins: Qualification of near- and far-UV CD for protein higher order structural analysis. *J. Pharm. Sci.* 100:4642–4654.

Lowry, O.H., Rosebrough, N.J., Farr, A.L., and Randall, R.J., 1951. Protein measurement with the Folin phenol reagent. *J. Biol. Chem.* 193:265–275.

Lundell, N., and Schreitmüller, T., 1999. Sample preparation for peptide mapping—a pharmaceutical quality-control perspective. *Anal. Biochem.* 266:31–47.

Mach, H., and Middaugh, C.R., 1994. Simultaneous monitoring of the environment of tryptophan, tyrosine, and phenylalanine residues in proteins by near-ultraviolet second-derivative spectroscopy. *Anal. Biochem.* 222:323–331.

Mach, H., and Middaugh, C.R., 2011. Ultraviolet spectroscopy as a tool in therapeutic protein development. *J. Pharm. Sci.* 100:1214–1227.

McPherson, A., 1999. *Crystallization of Biological Macromolecules*, Cold Spring Harbor, NY, Cold Spring Harbor Laboratory Press.

Moret, E.E., and Gestel, D., 2005. Molecular modeling. In: *Methods for Structural Analysis of Protein Pharmaceuticals*, Biotechnology: Pharmaceutical Aspects, Jiskoot, W., and Crommelin, D.J.A. (Eds.), Arlington, TX, AAPS Press, pp. 615–660.

Morth, J.P., Feng, V., Perry, L.J., Svergun, D.I., and Tucker, P.A., 2004. The crystal and solution structure of a putative transcriptional antiterminator from Mycobacterium tuberculosis. *Structure* 12:1595–1605.

Nielsen, O.F., 2005. Raman spectroscopy. In: *Methods for Structural Analysis of Protein Pharmaceuticals*, Biotechnology: Pharmaceutical Aspects, Jiskoot, W., and Crommelin, D.J.A. (Eds.), Arlington, TX, AAPS Press, pp. 167–198.

Noble, J.E., and Bailey, M.J.A., 2009. Quantitation of protein. *Methods Enzymol.* 463:73–95.

Pace, C.N., 1990. Measuring and increasing protein stability. *TIBTECH* 8:93–98.

Pace, C.N., Vajdos, F., Fee, L., Grimsley, G., and Gray, T., 1995. How to measure and predict the molar absorption coefficient of a protein. *Protein Sci.* 4:2411–2423.

Philo, J.S., 2005. Analytical ultracentrifugation. In: *Methods for Structural Analysis of Protein Pharmaceuticals*, Biotechnology: Pharmaceutical Aspects, Jiskoot, W., and Crommelin, D.J.A. (Eds.), Arlington, TX, AAPS Press, pp. 379–412.

Philo, J.S., 2006. Is any measurement method optimal for all aggregate sizes and types? *AAPS J.* 8(3):65.

Privalov, P.L., and Dragan, A.I., 2007. Microcalorimetry of biological macromolecules. *Biophys. Chem.* 126:16–24.

Rabiller-Baudry, M., Bouguen, A., Lucas, D., and Chaufer, B., 1998. Physico-chemical characterization of proteins by capillary electrophoresis. *J. Chromatogr. B* 706:23–32.

Righetti, P.G., Fasoli, E., and Righetti, S.C., 2011. Conventional isoelectric focusing in gel slabs and capillaries and immobilized pH gradients. In: *Protein Purification: Principles, High Resolution Methods, and Applications*, Janson, J-C. (Ed.), Hoboken, NJ, John Wiley & Sons, pp. 379–410.

Romijn, E.P., Krijgsveld, J., and Heck, A.J.R., 2005. Mass spectrometry: Proteomics. In: *Methods for Structural Analysis of Protein Pharmaceuticals*, Biotechnology: Pharmaceutical Aspects, Jiskoot, W., and Crommelin, D.J.A. (Eds.), Arlington, TX, AAPS Press, pp. 465–500.

Rupp, B., 2009. *Biomolecular Crystallography: Principles, Practice, and Application to Structural Biology*, New York, Garland Science.

Sane, S.U., Cramer, S.M., and Przybycien, T.M., 1999. A holistic approach to protein secondary structure characterization using amide I band Raman spectroscopy. *Anal. Biochem.* 269:255–272.

Santos, N.C., and Castanho, M.A.R.B., 1996. Teaching light scattering spectroscopy: The dimension and shape of tobacco mosaic virus. *Biophys. J.* 71:1641–1650.

Schön, A., and Velazquez-Campoy, A., 2005. Calorimetry. In: *Methods for Structural Analysis of Protein Pharmaceuticals*, Biotechnology: Pharmaceutical Aspects, Jiskoot, W., and Crommelin, D.J.A. (Eds.), Arlington, TX, AAPS Press, pp. 573–590.

Shively, J.E., 2000. The chemistry of protein sequence analysis. *EXS* 88:99–117.

Smith, P.K., Krohn, R.I., Hermanson, G.T., Mallia, A.K., Gartner, F.H., Provenzano, M.D., Fujimoto, E.K., Goeke, N.M., Olson, B.J., and Klenk, D.C., 1985. Measurement of protein using bicinchoninic acid. *Anal. Biochem.* 150:76–85.

Standing, K.G., 2003. Peptide and protein de novo sequencing by mass spectrometry. *Curr. Opin. Struct. Biol.* 13:595–601.

Svergun, D.I., Shtykova, E.V., Volkov, V.V., and Feigin, L.A., 2011. Small-angle x-ray scattering, synchroton radiation, and the structure of bio- and nanosystems. *Crystallogr. Rep.* 56:725–750.

Taschner, N., Müller, S.A., Alumella, V.R., Goldie, K.N., Drake, A.F., Aebi, U., and Arvinte, T., 2001. Modulation of antigenicity related to changes in antibody flexibility upon lyophilization. *J. Mol. Biol.* 310:169–179.

Thormann, W., 2011. Capillary electrophoretic separations. In: *Protein Purification: Principles, High Resolution Methods, and Applications*, Janson, J-C. (Ed.), Hoboken, NJ, John Wiley & Sons, pp. 451–486.

Toft, K.N., Vestergaard, B., Nielsen, S.S., Snakenborg, D., Jeppesen, M.G., Jacobsen, J.K., Arleth, L., and Kutter, J.P., 2008. High-throughput small angle x-ray scattering from proteins in solution using a microfluidic front-end. *Anal. Chem.* 80:3648–3654.

van de Weert, M., Haris, P.I., Hennink, W.E., and Crommelin, D.J.A., 2001. Fourier transform infrared spectrometric analysis of protein conformation: Effect of sampling method and stress factors. *Anal. Biochem.* 297:160–169.

van de Weert, M., Hering, J.A., and Haris, P.I., 2005. Fourier transform infrared spectroscopy. In: *Methods for Structural Analysis of Protein Pharmaceuticals*, Biotechnology: Pharmaceutical Aspects, Jiskoot, W., and Crommelin, D.J.A. (Eds.), Arlington, TX, AAPS Press, pp. 131–166.

van de Weert, M., van't Hof, R., van der Weerd, J., Heeren, R.M.A., Posthuma, G., Hennink, W.E., and Crommelin, D.J.A., 2000. Lysozyme distribution and conformation in a biodegradable polymer matrix as determined by FTIR techniques. *J. Control. Release* 68:31–40.

van den Bremer, E.T.J., and Heck, A.J.R., 2005. Mass spectrometry: Protein conformational analysis and molecular recognition. In: *Methods for Structural Analysis of Protein Pharmaceuticals*, Biotechnology: Pharmaceutical Aspects, Jiskoot, W., and Crommelin, D.J.A. (Eds.), Arlington, TX, AAPS Press, pp. 435–464.

Wen, J., Arthur, K., Chemmalil, L., Muzammil, S., Gabrielson, J., and Jiang, Y., 2012. Applications of differential scanning calorimetry for thermal stability analysis of proteins: Qualification of DSC. *J. Pharm. Sci.* 101:955–964.

Westermeier, R., 2011. Electrophoresis in gels. In: *Protein Purification: Principles, High Resolution Methods, and Applications*, Janson, J-C. (Ed.), Hoboken, NJ, John Wiley & Sons, pp. 365–378.

Wishart, D., 2005. Nuclear magnetic resonance spectroscopy. In: *Methods for Structural Analysis of Protein Pharmaceuticals*, Biotechnology: Pharmaceutical Aspects, Jiskoot, W., and Crommelin, D.J.A. (Eds.), Arlington, TX, AAPS Press, pp. 199–244.

Zhang, A., Singh, S.K., Shirts, M.R., Kumar, S., and Fernandez, E.J., 2012. Distinct aggregation mechanisms of monoclonal antibody under thermal and freeze-thaw stresses revealed by hydrogen exchange. *Pharm. Res.* 29:236–250.

5 Chemical Pathways of Peptide and Protein Degradation

Teruna J. Siahaan and Christian Schöneich

CONTENTS

5.1 INTRODUCTION

The number of approved protein and peptide drugs increased after the approval of recombinant human insulin in 1982. Since that time, the number of protein drugs approved has grown four to five times, with sales reaching around $71 billion in 2008 (Trivedi et al., 2009b). However, development of these types of drugs has been hampered due to difficulties in formulating and delivering them. This is because peptide and protein therapeutics have stability, delivery, analysis, and metabolism problems (Manning et al., 1989, 2010). Because the stability of a drug substance is essential to its shelf life and delivery to the patients, it is important to elucidate factors that influence the stability of peptide and protein therapeutics. Some of the challenges involve maintaining chemical and physical stabilities of these molecules in formulation since chemical instability of a protein can contribute to its physical instability. Some amino acids have reactive functional groups, which can catalyze degradation reactions that compromise the chemical stability of peptide and protein therapeutics. Often, many known side reactions found during peptide synthesis also transpire as degradation reactions in peptides and proteins. Therefore, this chapter is designed to describe known reactions in peptide and protein degradations because knowledge about the kinetics and mechanisms of these reactions could help formulators to solve problems of instability during formulation development of peptides and proteins.

Intrinsic and extrinsic factors influence the chemical and physical degradation of peptides and proteins in solution and solid state. The intrinsic factors include the primary, secondary, tertiary, and quaternary structures of proteins and the primary and secondary structures of peptides. The extrinsic factors include pH, buffers, concentration, excipients, moisture, and temperature. The different structures (i.e., primary, secondary, and tertiary) of proteins can influence their chemical and physical instability. These structural types are governed by forces such as covalent bonds (i.e., disulfide bond and cross-linking), electrostatic interactions (i.e., dipole–dipole, charge–dipole, charge–charge, and hydrogen bonds), weakly polar and apolar (hydrophobic) interactions, solvation or hydration, and post-translational modifications (i.e., glycosylation and phosphorylation). For example, the thermal stability of thermolysin is influenced by electrostatic interactions such as salt bridges. Ion pairs can enhance the secondary and tertiary structure stability of peptides and proteins with the stabilization energy around 3.0–4.0 kcal/mol. Hydrogen bonds (a combination of dipole–dipole and charge transfer interactions) have a major contribution to the structural stability of peptides and proteins, and they are involved in secondary structure formation (i.e., alpha-helix, beta-sheet, beta-turn, and gamma-turn). The effects of intrinsic and extrinsic factors on peptide and protein chemical stability will be elucidated here. Many known degradation reactions such as deamidation, oxidation, hydrolysis, elimination, and bond rearrangement will be described.

5.2   DEAMIDATION REACTION

Deamidation, which occurs in the Asn residue, has been well studied in peptides and proteins (Patel and Borchardt, 1990a, 1990b, 1990c). Physical instability can be induced upon deamidation reaction in proteins such as antibodies. The extrinsic factors (e.g., pH and temperature) and intrinsic factors (primary, secondary, and tertiary structures of peptides and proteins) influence the rate and mechanism of the deamidation reaction. In addition to conversion of the Asn to Asp residue, the deamidation reaction also causes peptide bond cleavage, backbone-to-side chain rearrangement (i.e., iso-Asp), and racemization of the chiral center (Li et al., 2003).

5.2.1 EFFECT OF pH

The deamidation reaction is due to the release of ammonia from an Asn or Gln group via acid or base catalysis (Figure 5.1). The Asn residue is more prone to deamidation than is the Gln residue. Below pH 4, the deamidation of Asn residue is via direct hydrolysis of the side chain to form an Asp residue (Patel and Borchardt, 1990a, 1990b, 1990c). At pH higher than 6.0, the deamidation reaction causes cyclic imide (Asu) formation upon attack of the amide group of the Asn side chain by the backbone nitrogen of the $(n + 1)$ residue to release ammonia. A water molecule can attack two different carbonyl groups on the cyclic imide to open the five-membered ring in an unsymmetrical manner, forming the Asp and iso-Asp residues. The formation of Asp and iso-Asp changes the charge in the peptide or protein, and the formation of iso-Asp extends the backbone length by adding one bond length. The presence of Asp or iso-Asp can further lead to peptide and protein fragmentation (see Asp degradation).

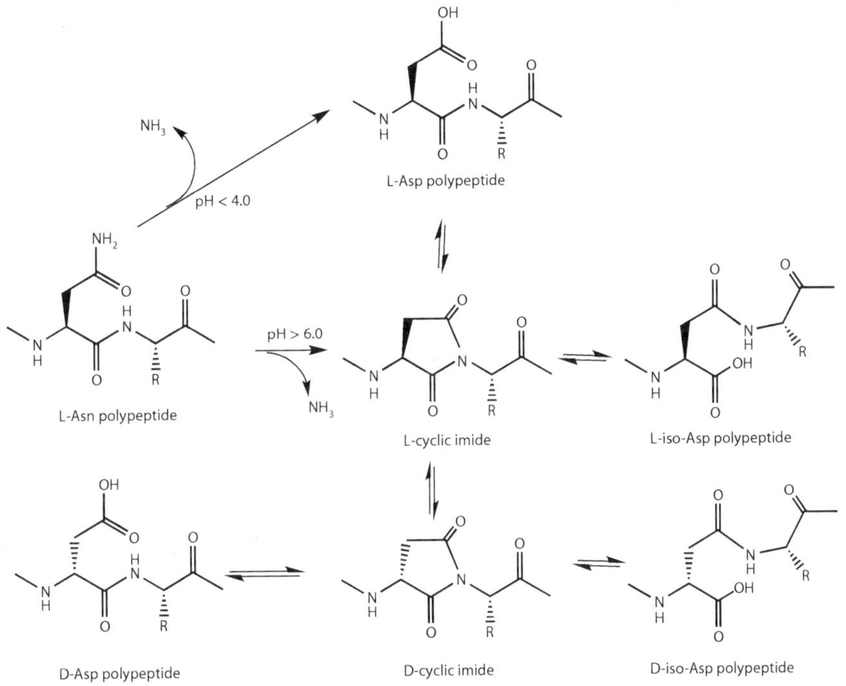

FIGURE 5.1 Mechanism of deamidation of Asn residue at different pH values. Below pH 4.0, the deamidation reaction proceeds via direct hydrolysis of the amide side chain group. At pH higher than 6.0, the deamidation reaction is via a cyclic imide intermediate and generates Asp and iso-Asp. The cyclic imide can undergo racemization at the chiral center, which ultimately produces D-Asp and D-iso-Asp residues.

5.2.2 Primary Structure Effect

The primary sequence influences the degradation rate of Asn-containing peptides and proteins. To study the effect of primary structure on the deamidation rate in VYPNGA, the $(n + 1)$ Gly residue to the Asn was mutated to Ser and Val to give VYPNSA and VYPNVA peptides, respectively (Patel and Borchardt, 1990a, 1990b, 1990c). Then, the $(n - 1)$ Pro residue in VYPNGA was mutated to the Gly residue to give VYGNGA peptide. The deamidation rates and the ratio between iso-Asp/Asp were determined at pH 1.0 and 7.5. At pH 1.0, the deamidation reaction product was exclusively the Asp residue and the rate of deamidation reaction was not influenced by mutation of the residues flanking the Asn residue (Figure 5.1). The results indicated that the mechanism of the reaction was via direct hydrolysis of the side chain amide group without forming the cyclic imide intermediate. Thus, the primary sequence has no influence on the deamidation reaction at pH 1.0. In contrast, the deamidation reaction at pH 7.5 was influenced by the primary sequence of the peptide. The rate of degradation of the parent peptide (VYPNGA) was faster than those of mutated peptides (VYPNSA and VYPNVA) with the slowest rate of degradation for VYPNVA. The degradation products at pH 7.5 contained iso-Asp and Asp with higher amounts of iso-Asp than Asp (i.e., ratio around 4:1, depending on peptide sequence). The higher amount of iso-Asp than Asp indicates that the more accessible carbonyl backbone in the cyclic imide has greater potential for hydrolysis than the side-chain carbonyl. Finally, VYGNGA had the fastest degradation rate; this could be due to the presence of a more flexible Gly residue at the $(n - 1)$ position to the Asn residue compared to a more rigid $(n - 1)$ Pro residue.

The effects of $(n + 1)$ and $(n + 2)$ residues on the rate of deamidation reactions in solution and solid state have been evaluated further (Goolcharran et al., 2000a; Wakankar and Borchardt, 2006). The $(n - 1)$ residue has less influence on the rate of deamidation than the $(n + 1)$ residue (Robinson and Robinson, 2001). To evaluate the effect of $(n + 1)$ and $(n + 2)$ residues on the rate of deamidation, several different peptides were studied, including GQNGH, GQNHH, GQNVH, and GQNGG. The GQNHH, GQNVH, and GQNGH peptides display His, Val, and Gly residues, respectively, at the $(n + 1)$ position, and these peptides also contain a His residue at the $(n + 2)$ position. The GQNGG peptide lacks a His residue at both the $(n + 1)$ and $(n + 2)$ positions and serves as a control peptide.

At pH < 6.5, the rate of deamidation of GQNHH was faster than that of GQNVH. This result suggested that deamidation of GQNHH was affected by acid catalysis where the protonated side chain of the $(n + 1)$ His residue influenced the deamidation reaction. The rate of deamidation of GQNVH was slower than that of GQNGH, indicating that the $(n + 1)$ Val residue imposed steric hindrance in the deamidation compared to a more flexible $(n + 1)$ Gly residue in GQNGH. The steric factor of the $(n + 1)$ residue also influences the deamidation of GQNHH, GQNVH, and GQNGH peptides at pH 6.5. The rates of deamidation of GQNHH and GQNVH were slower than that of GQNGH. The rate of deamidation of GQNHH was also faster than that of GQNVH; this was due to base catalysis by the unprotonated side chain of the His residue to extract a proton from the backbone NH for the formation of the cyclic imide. In summary, the nature of the $(n + 1)$ residue (i.e., steric and electrostatic inductive effects) can influence the rate of deamidation, and the flexibility and ionization of the $(n + 1)$ side chain residue affects the rate of deamidation reaction (Goolcharran et al., 2000b).

At pH < 6.5, the rate of deamidation of GQNGH was similar to that of GQNGG, indicating that the effect of the $(n + 2)$ His residue on deamidation was negligible (Wakankar and Borchardt, 2006). The effect of the $(n + 2)$ His residue was pronounced at pH 6.0, but not at pH 10; the rate of deamidation of GQNHH was twofold faster than the rate of deamidation of GQNHG, indicating that there was an acid catalysis effect of the $(n + 2)$ His residue on the deamidation reaction (Wakankar and Borchardt, 2006).

The deamidation reaction in the rubbery solid of GQNHH was significantly slower than in solution (Wakankar and Borchardt, 2006). This is attributed to the lower mobility of the peptide backbone in the rubbery solid compared to solution, and it leads to an increased energy barrier for the deamidation reaction. It is interesting to find that the change in rates of deamidation of GQNGH in the solid was less pronounced than in solution. This was due to the backbone flexibility of the Gly residue in GQNGH compared to the His residue in GQNHH. In the solid rubbery state at pH 6.0, the rate of deamidation of GQNGH was fivefold faster than that of GQNHH, while in solution the rates of deamidation of these peptides were comparable. This suggested that the effect of acid catalysis by the $(n + 1)$ His residue was attenuated in the solid state compared to the solution state, and the steric effects were more predominant in the solid state (Wakankar and Borchardt, 2006). In the solid state at pH 10, the rate of deamidation of GQNGH was twelvefold faster than that of GQNHH, implying the important role of the steric effect of the $(n + 1)$ His residue. In addition, the rate of deamidation of GQNHH was faster than that of GQNVH, indicating the effect of base catalysis of the $(n + 1)$ His residue to accelerate the reaction.

5.2.3 Secondary Structure Effect

The secondary structure (α-helix and β-turn) has an impact on the rate of deamidation (Figures 5.2 and 5.3). For helical peptides, a 32-mer peptide from bovine growth hormone–releasing factor (bGRF), referred to as Leu[27]bGRF, and its derivatives (Ala[15]Leu[27]bGRF and Pro[15]Leu[27]bGRF) were used to evaluate the effect of helical structure on the rate of deamidation (Stevenson et al., 1993b). The rate of the deamidation reaction at the Asn8 residue was evaluated as a function of helical content on each peptide as determined by circular dichroism (CD) spectroscopy and nuclear magnetic resonance spectroscopy (Stevenson et al., 1993a).

The helical content of the peptide was increased upon increase in methanol content in the solvent as determined by CD spectroscopy. Although the parent peptide (Leu[27]bGRF) with Gly15 has a high helical content, due to the presence of Gly15 (a helical breaker) the peptide has two helical segments; it has a helical structure from Phe6 to Gly15 and from Arg20 to Leu27 with a random structure in between (Stevenson et al., 1993a, 1993b). In Ala[15]Leu[27]bGRF, Gly15 was mutated to Ala15 (a helix inducer); this peptide has a continuous helical segment from Asn8 to Gln30. Finally, the Pro[15]Leu[27]bGRF contains a strong helix breaker at the Pro15 residue, and this peptide has a random coil structure with no segment of helical structure.

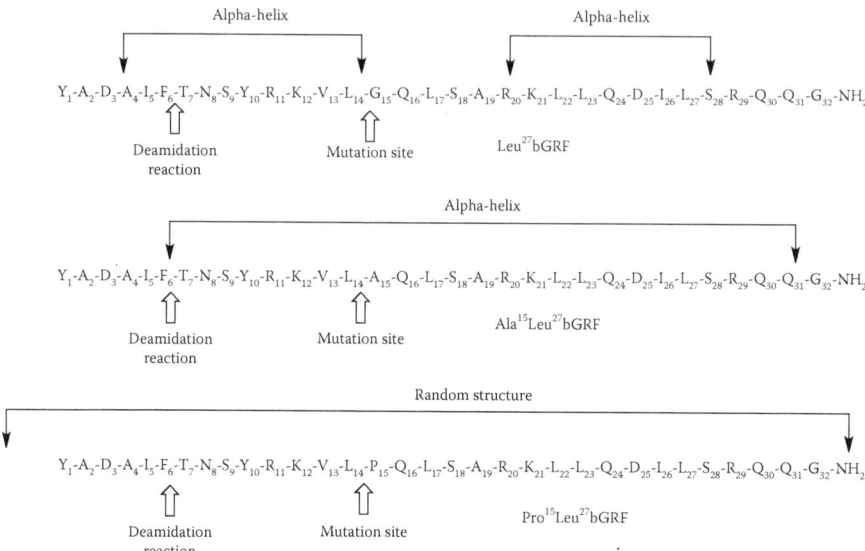

FIGURE 5.2 The role of secondary structure on the rate of deamidation of the Asn8 residue in bGRF derivatives. The parent Leu[27]bGRF contains two segments of alpha-helix at Phe6-Gly15 and Arg20-Leu27. Mutation at Gly15 to Ala15 to give Ala[15]Leu[27]bGRF generates a molecule with a stable continuous alpha-helix from Asn8 to Gln30. Pro[15]Leu[27]bGRF mutant has a random coil structure with no observable helix structure.

Ala[15]Leu[27]bGRF, which has the highest helical content, has the longest half-life (21.53 ± 2.83 hours) compared to the parent peptide Gly[15]Leu[27]bGRF (15.74 ± 2.45 hours), and Pro[15]Leu[27]bGRF has the shortest half-life (10.78 ± 2.95 hours) (Stevenson et al., 1993b). This result indicates that the stability of the Asn residue can be correlated with the increase in helical content or secondary structure of the peptide around the Asn residue (Figure 5.2).

The β-turn structure has also been shown to affect the rate of deamidation, and the role of $(n + 1)$ and $(n + 2)$ residues of β-turn on the deamidation reaction was also explored (Figure 5.3). Deamidation in two linear peptides called AcNG and AcGN as well as two cyclic peptides called cNG and cGN was investigated (Xie et al., 2000). All four peptides have a β-turn structure at Asn-Gly and Gly-Asn with the Asn residue at the $(n + 1)$ or $(n + 2)$ of the β-turn. In general, the rates of deamidation of the cyclic peptides (cGN = $< 2.2 \times 10^4 \text{ s}^{-1}$ and cNG = $9.36 \times 10^7 \text{ s}^{-1}$) were slower than those of the respective linear peptides (AcGN = $20.1 \times 10^7 \text{ s}^{-1}$ and Ac-NG = $42.2 \times 10^7 \text{ s}^{-1}$) at pH 8.8 (Xie et al., 2000). The propensity to stabilize the β-turn structure is higher in the cyclic peptides than in the linear peptides because the backbone is more rigid in the cyclic peptides than in the linear peptides. The position of the Asn residue in the β-turn affected its rate of deamidation in peptides; the deamidation rate of Asn at the $(n + 2)$ position of the β-turn was slower than that of the Asn residue at the $n + 1$ position. This is reflected by the higher stability of cGN than cNG as well as the higher stability of AcGN than AcNG peptide. The deamidation preference of the $(n + 1)$ Asn residue is due to the ability

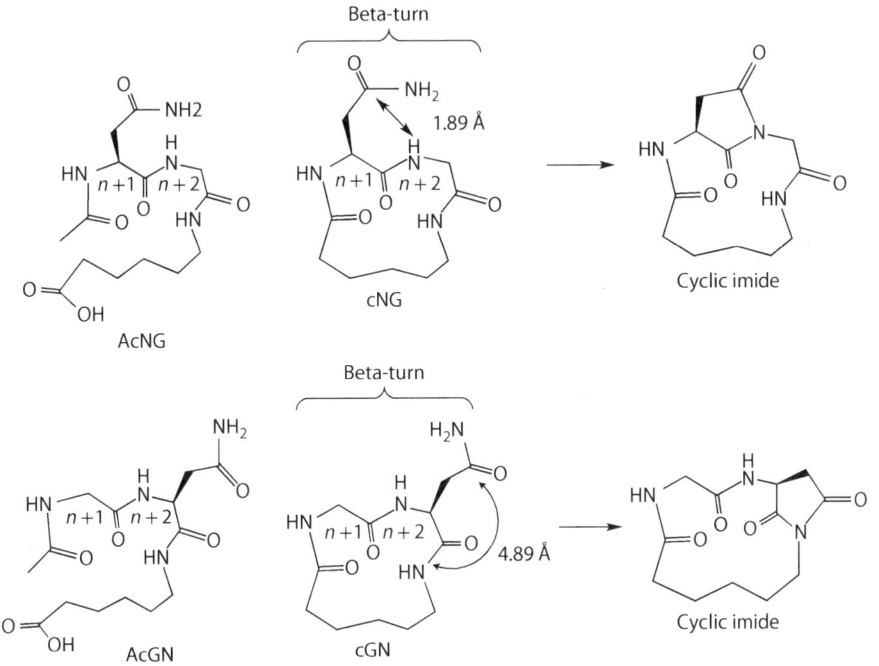

FIGURE 5.3 Structures of linear (AcNG and AcGN) and cyclic (cNG and cNG) peptides that contain a beta-turn conformation where the Gly-Asn and Asn-Gly sequences are at the $(n + 1)$ and $(n + 2)$ positions. The rate of deamidation of the Asn residue depends on its position on a beta-turn and the flexibility of the backbone.

of the backbone nitrogen of the $n + 2$ residue of the β-turn to attack the carbonyl side chain of Asn $(n + 1)$ to form a cyclic imide. This is because the shortest possible distance between the backbone nitrogen of the $(n + 2)$ residue and the side chain carbonyl of the $(n + 1)$ Asn is 1.89 Å. In contrast, when the Asn is in the $(n + 2)$ position, formation of the cyclic imide is less favorable because the distance between the nitrogen of the $(n + 3)$ residue and the carbonyl side chain of the Asn $(n + 2)$ residue is 4.8 Å. It can be concluded that β-turn structure and the position of the position of the Asn residue in the β-turn influence the rate of deamidation.

5.3 ASPARTIC ACID DEGRADATION

The Asp residue degradation is similar to Asn degradation via cyclic imide formation and is followed by the formation of iso-Asp and racemic products (Oliyai and Borchardt, 1993, 1994; Oliyai et al., 1994). In addition, the Asp residue undergoes backbone hydrolysis at the *N*- and *C*-termini. There are two different pathways of the Asp-mediated peptide bond cleavage (Figure 5.4) (Bogdanowich-Knipp et al., 1999b). First, the oxygen of the Asp carboxyl side chain attacks the backbone carbonyl carbon of the $(n - 1)$ residue to form a tetrahedral intermediate with a six-membered ring. The rearrangement of the tetrahedral intermediate breaks the C–N backbone between the $(n - 1)$ residue and the Asp residue to generate the acid

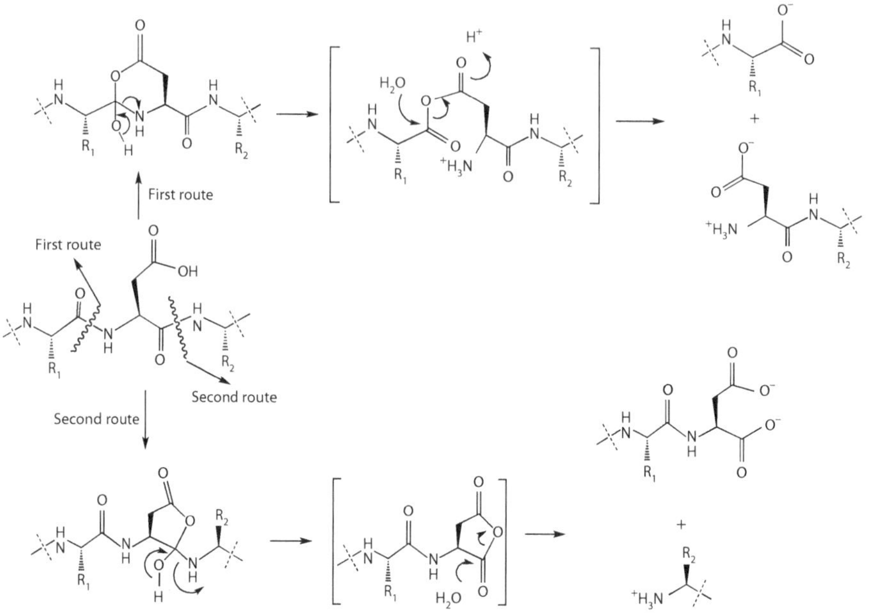

FIGURE 5.4   The peptide bond cleavage mediated by the Asp acid residue proceeds via two different routes.

anhydride intermediate. The hydrolysis of the acid anhydride produces two peptide fragments upon cleavage of the peptide bond between $(n - 1)$ and the Asp residues. In the second Asp-mediated backbone fragmentation, the oxygen of the carboxylic side chain of the Asp residue reacts with the electropositive backbone carbonyl carbon of the Asp reaction to generate a five-membered ring with a tetrahedral intermediate at the C-terminal of Asp residue. Rearrangement of the tetrahedral intermediate results in two peptide fragments upon the cleavage of the peptide bond between the Asp residue and the $(n + 1)$ residue. Similar to that of the Asn residue, the rate of Asp-mediated peptide degradation is related to peptide conformation and the rigidity of the peptide backbone. The rate of degradation of the Asp-containing cyclic peptide was slower than that of the Asp-containing linear peptide (Bogdanowich-Knipp et al., 1999b). Molecular dynamics simulation was used to explain the slow degradation of the cyclic peptide compared to linear peptides. This study indicated that, on average, the reactive atoms in the cyclic peptide are much farther away from each other than are those in the linear peptide (Bogdanowich-Knipp et al., 1999a).

5.4   DIKETOPIPERAZINE REACTION

The diketopiperazine (DKP) reaction occurs via an intramolecular reaction at the N-terminus of peptides and proteins that contain the Xaa-Pro sequence (Goolcharran and Borchardt, 1998; Goolcharran et al., 2000a; Kertscher et al., 1993; Okada et al., 1991). As well as in peptide and protein formulation, this

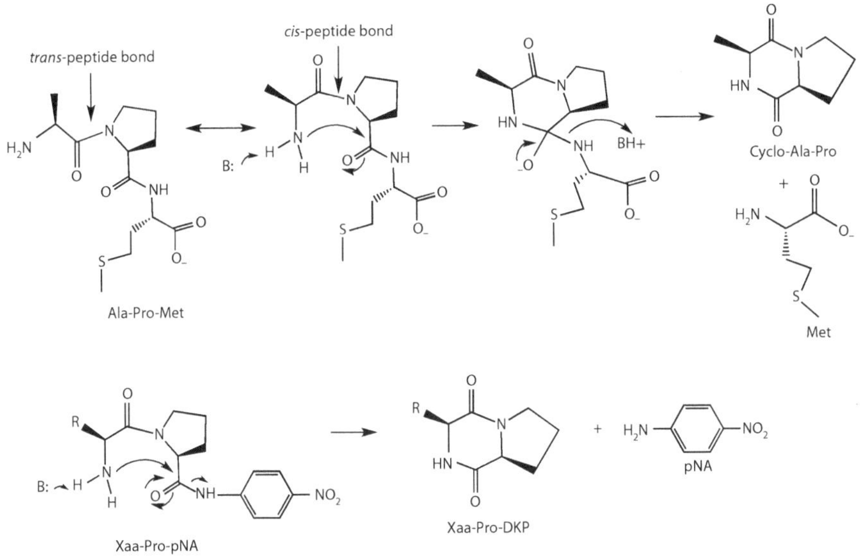

FIGURE 5.5 Diketopiperazine formation from Ala-Pro-Met and Xaa-Pro-pNA.

reaction occurs in biosynthetic pathways and in tissues and body fluids; it also occurs as a side reaction during solution- and solid-state peptide synthesis. The formation of DKP is due to nucleophilic attack on the peptide bond carbonyl carbon of the Pro residue by the *N*-terminus nitrogen to form a cyclic six-membered ring dipeptide of Xaa-Pro followed by the release of a deleted protein fragment. Thus, the p*K*a of the amino group of the *N*-terminus affects the rate of the DKP formation. To form the six-membered ring, the peptide bond on Xaa-Pro must form a *cis*-peptide bond (Figure 5.5) (Goolcharran et al., 2000a). In this case, the Xaa-Pro sequence accommodates the reaction because the peptide Xaa-Pro bond normally contains both *cis*- and *trans*-conformation due to the structural nature of the Pro residue. In other words, the energy barrier to form the *cis*-peptide bond in Xaa-Pro sequence is lower than in the other sequence without the Pro residue (i.e., Xaa-Xaa). The nucleophilic attack on the Pro carbonyl of Xaa-Pro by the nitrogen of the *N*-terminus residue involves the formation of a *cis*-Xaa-Pro peptide bond to accommodate the formation of a six-membered ring of DKP. The stability of the *cis*-peptide bond of the Xaa-Pro sequence depends on the flanking residues to Xaa-Pro and the electrostatic environment around the peptide bond. In peptide and protein formulation, the formation of DKP is p-sensitive and is catalyzed by acid or base.

The formation of DKP has been shown in recombinant human vascular endothelial growth factor (rhVEGF), where the *N*-terminal Ala-Pro forms the DKP (i.e., cyclo-Ala-Pro) and releases the remaining rhVEGF fragment that lacks Ala-Pro dipeptide (Goolcharran et al., 2000a). During the stability study of rhVEGF, the first tryptic digest fragment of this protein (T1 fragment = APMAEGGGQNHHEVVK) indicated that this *N*-terminus fragment had several degradation products, include DKP formation, Met-3 oxidation, and Asn deamidation (Goolcharran et al., 2000a). In rhVEGF, the formation of DKP was found at neutral pH. To further elucidate the nature of DKP reaction and Met oxidation in rhVEGF, the stability of Ala-Pro-Met (APM) tripeptide was evaluated at different pH values (Goolcharran et al., 2000a). The DKP formation and oxidation of Met-3 were observed simultaneously in APM peptide. It is interesting to find that the majority of the degradation in APM was due to DKP formation and not to the oxidation of the Met-3 residue. The overall rate of degradation of APM was faster at pH 8.0 (14.85×10^4 h^{-1}) than at pH 5.0 (2.125×10^4 h^{-1}). At both pH 8.0 and 5.0, the rate of DKP formation was faster than the rate of Met-3 oxidation, and the majority of DKP was produced from the starting material APM, not from the Met-oxidized APM (or APM-O). For example, at pH 8.0, the rate of formation of DKP from APM-O was very slow ($k_{dkp}^{ox} = 0.252 \times 10^4$ h^{-1}) compared to the rate of DKP formation from the parent APM ($k_{dkp} = 7.633 \times 10^4$ h^{-1}).

The kinetics of DKP formation was studied in more detail using Phe-Pro-pNA peptide at pH 3–8 (Figure 5.5). Between pH 3.0 and 8.0, the Phe-Pro-DKP degradation product was observed in the absence of Phe-Pro-OH, indicating that the primary reaction was DKP formation (Goolcharran and Borchardt, 1998). The direct hydrolysis of Phe-Pro-pNA to Phe-Pro-OH was observed only below pH 3.0 or above pH 8.0. In addition, Phe-Pro-DKP was stable between pH 3.0 and 8.0. Furthermore, the Phe-Pro-DKP could be hydrolyzed to Phe-Pro-OH only below pH 3.0 and above pH 8.0, suggesting that the presence of Phe-Pro-OH was due to the hydrolysis of Phe-Pro-DKP and not to direct hydrolysis of Phe-Pro-pNA.

The primary sequence of the peptide affecting the rate of DKP formation and the role of the Xaa residue in Xaa-Pro-pNA was investigated at pH 7.0 (Goolcharran and Borchardt, 1998). The rates of DKP formation in several Xaa-Pro-pNA peptides were determined at pH 7.0 and 37°C, where the Xaa amino acids were Phe, Arg, beta-cyclo-hexylamine, Ala, and Pro. The rates of DKP formation were in the following order: Phe-Pro (5.82×10^2 h^{-1}) > Arg-Pro (3.73×10^2 h^{-1}) > b-cyclo-hexylamine-Pro (3.5×10^2 h^{-1}) > Ala-Pro (2.5×10^2 h^{-1}) > Val-Pro (1.16×10^2 h^{-1}) > Gly-Pro (0.57×10^2 h^{-1}). It was found that the rank of rates of DKP formation could not be explained by simply using the bulkiness of the side chain of Xaa amino acid because the rate of DKP formation of Ala-Pro-pNA was faster that Val-Pro-pNA. The lack of effect from the side chain bulkiness can be explained by the structure of the intermediate prior to the formation of DKP. The structure of the DKP intermediate has a *cis*-Xaa-Pro peptide bond prior to cyclization with the side chains of the two amino acids (Xaa and Pro) in the opposite sites; thus, the side chains have little effect on the rate of DKP formation. It was proposed that the fast rate of DKP formation of Phe-Pro was due to ring–ring interaction between the aromatic ring of Phe side chain and the pNA aromatic ring to form an energetically stable boat conformation of the six-membered ring of the DKP backbone. The ring–ring interaction is caused by intramolecular dipole-induced dipole interaction to stabilize the tetrahedral intermediate of the boat form. Replacing the aromatic ring of Phe with cyclohexyl reduced the rate of DKP formation, indicating the importance of aromatic–aromatic interaction to the rate of DKP formation. In summary, the formation of DKP is influenced by several factors, including the pKa of the N-terminus, the nature of the Xaa side chain in Xaa-Pro, the stability of the *cis*-peptide bond conformation of Xaa-Pro, and the stability of the DKP backbone ring.

5.5 N,O-ACYL MIGRATION AND DEHYDRO-ALANINE FORMATION IN SERINE

The N,O-acyl migration reaction often occurs in the Ser residue in acidic conditions where there is a rearrangement between the backbone and side chain of the Ser residue as found in the Asn and Asp residues (Figure 5.6) (Sohma et al., 2005). The rearrangement is often reversible from acidic to basic conditions. In acidic conditions, the rearrangement reaction can take place via two different routes. The first route is in acidic condition and proceeds via the formation of an oxazoline intermediate. The oxazoline intermediate is generated upon protonation of the Ser side chain hydroxyl group followed by dehydration reaction. The attack by water molecule at the carbon of the $C = N$ bond of oxazoline forms the five-membered ring hemiacetal intermediate. Bond rearrangement to open the ring results in the final ester product. As mentioned previously, treatment of the ester with mild basic conditions can reverse the reaction to form the parent peptide. The second route is by protonation of the carbonyl oxygen of the $(n - 1)$ residue to the Ser residue followed by attack of the carbon of the same carbonyl by the side chain oxygen of the Ser residue to form a five-membered ring hemiacetal intermediate. The ring opening of the hemiacetal intermediate forms the N,O-acyl migration product of the Ser-containing peptide.

Under acidic conditions, serine amino acid can degrade to dehydroalanine (Dha). In this case, the hydroxyl group of the side chain is protonated followed by abstraction of the proton of the alpha carbon by solvent (i.e., water). The release of a water molecule produces Dha residue.

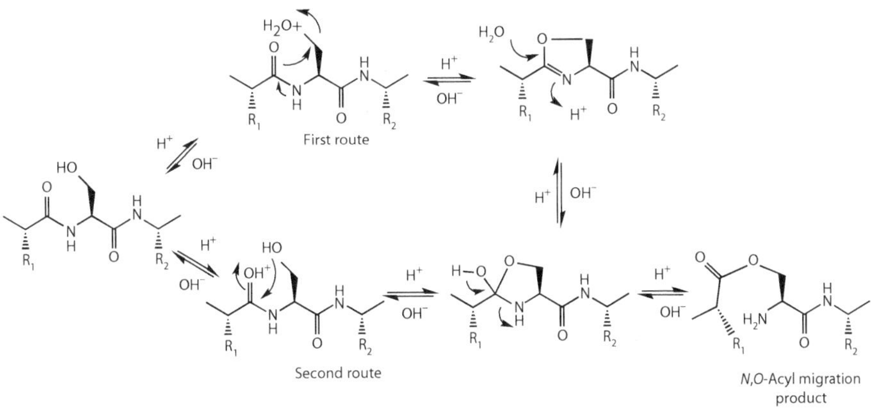

FIGURE 5.6 The N,O-acyl migration reaction occurs at the Ser and Thr residues. This reaction produces a rearrangement of the peptide backbone at the N-terminus of the side chain of the Ser or Thr residue to form an ester bond.

5.6   ALPHA- AND BETA-ELIMINATION REACTIONS

Many proteins contain one or more disulfide bonds, and often the disulfide bond imposes structural rigidity on the protein. Disulfide bond degradation or reduction often changes the biological efficacy as well as the stability of the protein in formulation (Trivedi et al., 2009b). Proteins with multiple disulfide bonds can undergo intramolecular and intermolecular disulfide bond exchange in basic conditions (Figure 5.7) (Volkin and Klibanov, 1987). The intramolecular disulfide bond exchange causes a change in protein conformation that can render the protein inactive. The intermolecular disulfide bond exchange can form in a high protein concentration to produce dimers and oligomers. This polymerization reaction could lead to protein precipitation (Trivedi et al., 2009a). The disulfide exchange normally initiated by thiolate anion on the free Cys residue in the protein attacks one of the sulfur atoms on the disulfide bond to form a new disulfide bond and a new free thiol group at a different site within the protein.

The disulfide exchange reaction can be initiated by hydroxyl anion upon the attack of a sulfur atom of the disulfide bond to generate thiolate anion and sulfenic acid on the Cys residues (Bogdanowich-Knipp et al., 1999b; He et al., 2006; Trivedi

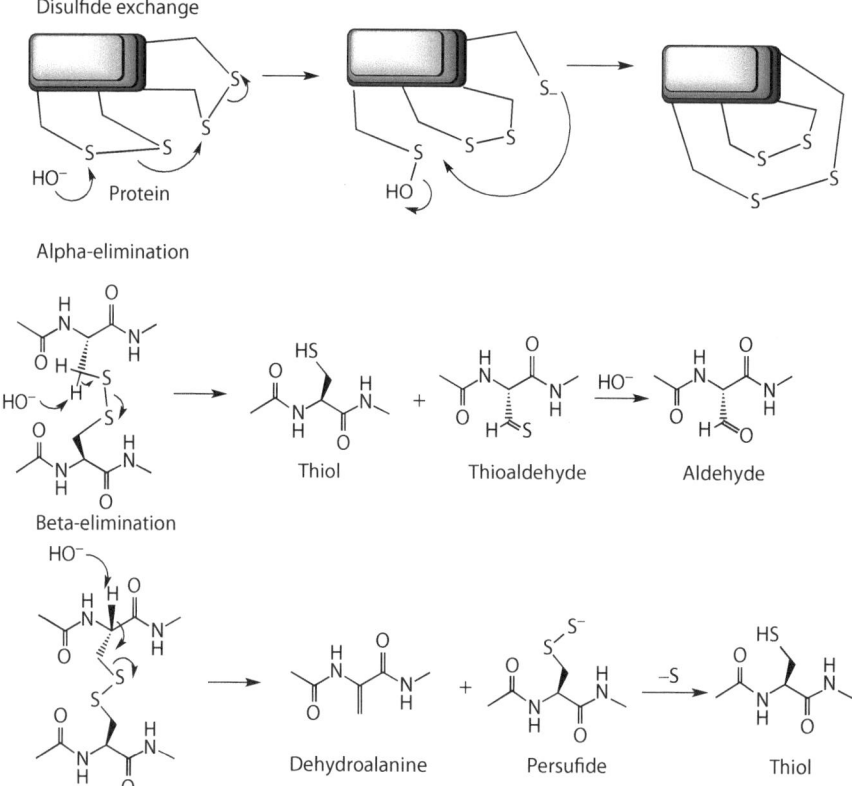

FIGURE 5.7   Degradation reactions of a disulfide bond via disulfide exchange, alpha-elimination, and beta-elimination.

et al., 2009b). Then, the thiolate anion reacts further with a neighboring disulfide bond to produce a new disulfide bond and a thiolate anion. Further reaction of the thiolate anion with the sulfenic acid sulfur produces a new disulfide bond by releasing the hydroxyl anion.

The disulfide bond also undergoes alpha-elimination, where the beta-proton of the Cys residue is abstracted by base to generate carbanion at the beta-carbon of the Cys residue. Carbanion-initiated bond rearrangement produces thioaldehyde and thiolate anion. Further reaction of thioaldehyde with hydroxyl anion generates aldehyde functional group. In addition to the alpha-elimination reaction, the beta-elimination reaction occurs in disulfide bonds started via base extraction of the alpha proton of the Cys residue to form a carbanion at the C-alpha of Cys residue; further bond rearrangement generates Dha and persulfide anion. Sulfur extrusion from the persulfide anion yields a thiolate anion on the Cys residue.

5.7 OXIDATION

The term "protein oxidation" encompasses a large manifold of potential protein degradation pathways. These are initiated through the reactions of reactive oxygen species with amino acid side chains and/or the protein backbone, that is, with the C_α-H bonds. The ensuing reaction products can give rise to additional oxidation reactions, and potentially involve in chain oxidation processes (Neuzil et al., 1993). The selectivity of an oxidation process is controlled by various parameters, depending on the nature of the oxidant and the reaction mechanism. (1) Oxidation can proceed via the direct reaction of an oxidant with a protein site. Such direct oxidation reactions are frequently controlled by the access of the oxidant to a specific target site on the protein, that is, through the solvent-accessible surface area (SASA) of a specific amino acid residue within a given protein domain. (2) Oxidation reactions can be catalyzed through metals, and the selectivity of such reactions is frequently controlled by the geometry of specific protein domains which ensure binding of the metal. (3) Oxidation reactions can be assisted through neighboring groups, which may stabilize transition states and influence reduction potentials. For such reactions, through-sequence and through-space interactions of target amino acids with other amino acids are important. For all these mechanisms, examples can be found in the pharmaceutical, chemical, and biochemical literature. The following sections will analyze these mechanisms in detail, and provide specific references for selected peptides and proteins, for which these mechanisms have been characterized. In addition, we will provide details on the reactions of individual amino acid residues with specific reactive oxygen species.

5.7.1 AUTOXIDATION

The term "autoxidation" denotes "the spontaneous oxidation in air of a substance not requiring catalysts" (Miller et al., 1990). For most protein pharmaceuticals such "autoxidation" reactions must be considered very slow, rationalized by the ground state electronic configurations of molecular oxygen and organic molecules (Taube, 1965). The Marcus equation (Equation 5.1(I)) provides a quantitative relationship between the rate constant of the forward reaction, k_f, of the oxidation–reduction equilibrium (Equilibrium 5.1) between O_2 and an electron donor, X^{n+}, and the equilibrium constant K (Merényi et al., 1993).

$$O_2 + X^{n+} \rightleftharpoons O_2^- + X^{(n+1)+} \qquad (5.1)$$

$$ln(k_f) = ln(Z) - \frac{\lambda^0}{4RT}\left(1 - \frac{RT}{\lambda^0}ln(K)\right)^2 \qquad (I)$$

For a series of phenolates, indophenolates, and some additional electron donors Merényi et al. (1993) showed a very good fit of a plot of $log(k_f)$ vs. $log(K)$, consistent with an outer-sphere electron transfer. Hence, by comparison to the reduction potentials of the model phenolates some approximations can be made for proteins: the reduction potentials for 4-cyanophenolate and 2, 4, 6-trichlorophenolate are given as 1.12 and 0.88 V, respectively, and the measured rate constants for one-electron oxidation by O_2 are $k_f = 9.7 \times 10^{-13}$ $M^{-1} \cdot s^{-1}$ and 7.0×10^{-8} $M^{-1} \cdot s^{-1}$, respectively (Merényi et al., 1993). We can now compare the reduction potentials of these phenolates to that of Trp, where $E^0(Trp\bullet, H^+/Trp, pH\ 7.0) = 1.0$ V (Harriman, 1987). Hence, by comparison to the phenolates we expect a rate constant for the oxidation of Trp by oxygen of between $k_f = 9.7 \times 10^{-13}$ $M^{-1} \cdot s^{-1}$ and 7.0×10^{-8} $M^{-1} \cdot s^{-1}$, and for an estimation we take $k_f < 7.0 \times 10^{-8}$ $M^{-1} \cdot s^{-1}$. Under pseudo-first-order conditions with $[O_2] \gg [Trp]$, and $[O_2] = 2.5 \times 10^{-4}$ M (i.e., an air-saturated solution), the oxidation of Trp by O_2 would then proceed with a half-life $t_{1/2} > 1255$ years, or a shelf-life of $t_{90} > 190$ years. Trp represents one of the most oxidation-sensitive amino acid residues in proteins, but if "autoxidation" were responsible for Trp degradation we would not expect to observe Trp oxidation over a reasonable period of time. Similar estimates can be made for Tyr and Met (Hovorka and Schöneich, 2001; Schöneich et al., 1997): the one-electron reduction potential of Met is even more positive compared to that of Trp, suggesting that "autoxidation" would proceed with an even slower rate. Therefore, we conclude that "autoxidation" of Trp (as well as His, Tyr, Met, and Phe) cannot be responsible for the degradation of such residue(s) in proteins within time frames, which are normally observed during the production and storage of protein pharmaceuticals. In other words, the oxidation of Trp, Tyr, Phe, and His proceeds most likely via reaction mechanisms other than "autoxidation." Similar calculations can be made for the "autoxidation" of Cys. However, Cys oxidation can proceed via a chain reaction (Lal et al., 1997), and in such a case the "autoxidation" step would be the chain *initiation*. It is important to note, however, that chain processes of thiol oxidation have been experimentally defined for low molecular weight thiols. With superoxide as an intermediate, the propagation of the chain relies on the reaction of a sulfinyl radical (RSO•) with a second thiol (RSH) (Winterbourn and Metodiewa, 1999). Such reactions may not proceed well in proteins if the initial oxidation targets a Cys residue, which does not have access to a nonoxidized Cys residue either within the protein sequence or on a second protein molecule. If such conditions would prevail, that is, the oxidation would *not* occur via a chain reaction, we can estimate the "autoxidation" kinetics of a protein Cys residue by comparison to the data of Merényi et al. (1993). Under pseudo-first-order conditions with $[O_2] \gg$ [protein-Cys], pH $> pK_{a,Cys}$, that is, the Cys residue present in its anionic form, CysS$^-$, and taking E^0(CysS• + e$^-$ → CysS$^-$) = 0.75 V (Surdhar and Armstrong, 1987), we estimate that for "autoxidation" $t_{1/2} \approx 32{,}000$ days, which is unrealistic in view of the high experimental oxidation sensitivity especially of deprotonated Cys residues (Schöneich et al., 1997).

5.7.2 PHARMACEUTICALLY RELEVANT OXIDANTS AND THEIR FORMATION IN FORMULATIONS

In pharmaceutical formulations, reactive oxygen species can be present as a result of chemical sterilization processes or form through interaction of the formulation with adventitious impurities or with light. When hydrogen peroxide (H_2O_2) is used for sterilization (Corveleyn et al., 1997), residual levels of this oxidant can remain adsorbed on container walls, available for the oxidation of protein and/or excipients added to the container. When γ-radiation is used for sterilization, reactive oxygen species are generated through radiation-induced chemical processes (von Sonntag, 1987). A few articles have recommended the use of ultraviolet (UV) light at λ = 253.7 nm for the decontamination of protein preparations from viruses (Wigginton et al., 2010). UV light of such wavelength is absorbed by the aromatic amino acids and cystine (Dose, 1967), and can lead not only to direct bond cleavages (e.g., of cystine) but also the generation of reactive oxygen species (see below).

For the most part, oxidation reactions by molecular oxygen will require the conversion of O_2 into reactive oxygen species. The one-electron reduction of O_2 yields superoxide radical anion (O_2^-). In a pharmaceutical formulation, such reduction of molecular oxygen involves most likely a transition metal-catalyzed reaction. Superoxide reacts with a measurable rate constant with Cys (Jones et al., 2002), but not with the other 19 essential amino acids present in human proteins. However, superoxide reacts efficiently with amino acid *radicals* or *radical cations* generated through the reaction of other reactive oxygen species with amino acids. Superoxide can be further reduced to H_2O_2, an oxidant which in proteins targets predominantly Met and Cys. Further one-electron reduction of H_2O_2 through a transition metal-catalyzed process (the "Fenton reaction") (Reaction 5.2) leads to the hydroxyl radical (HO•), a very strong oxidant, which reacts with all 20 essential amino acids, albeit the rate constants vary with the structures of the amino acids (Davies, 2005).

$$Fe^{2+} + H_2O_2 \rightarrow Fe^{3+} + HO^- + HO \bullet \qquad (5.2)$$

In pharmaceutical formulations, the reactions of HO• with protein amino acids and/or excipients will yield carbon-centered radical (R•; Reaction 5.3); these carbon-centered radicals add oxygen to yield peroxyl radicals (ROO•; Reaction 5.4) (Davies, 2005). Peroxyl radicals can accept an electron or hydrogen atom from suitable electron or hydrogen donors to yield hydroperoxides (ROOH; Reaction 5.5).

$$R\text{-}H + HO\bullet \rightarrow R\bullet + H_2O \qquad (5.3)$$

$$R\bullet + O_2 \rightarrow ROO\bullet \qquad (5.4)$$

$$ROO\bullet + R'\text{-}H \rightarrow ROOH + R'\bullet \qquad (5.5)$$

$$ROOH + e^- \rightarrow RO\bullet + HO^- \qquad (5.6)$$

$$ROOH + Fe^{3+} \rightarrow Fe^{2+} + H^+ + ROO\bullet \qquad (5.7)$$

Hydroperoxides can oxidize Met to Met sulfoxide (MetSO). In addition, the one-electron reduction of hydroperoxides leads to very reactive alkoxyl radicals (RO•; Reaction 5.6). Frequently pharmaceutical excipients are contaminated by peroxides, hydroperoxides, and/or H_2O_2 (Harmon et al., 2006; Wasylaschuk et al., 2007), so that excipients must be considered a prominent source for reactive oxygen species, which subsequently target the pharmaceutical protein(s) (Harmon et al., 2006). Polysorbates, and especially polysorbate 80 (PS80), are prominent targets for peroxide accumulation. Baseline levels of peroxides are frequently present prior to the preparation of the formulation but can continue to form within the formulation during manufacturing and storage. Polysorbates contain unsaturated and polyunsaturated fatty acids, which are known for their propensity to undergo oxidation (Yao et al., 2009). Classic pathways for the oxidation of un- and polyunsaturated fatty acids include the ene reaction (with singlet oxygen), or allyl/pentadienyl radical formation, followed by oxygen addition and hydrogen abstraction. The latter reaction generates hydroperoxide and propagates the chain reaction. An important mechanistic detail of polysorbate chain oxidation is the initiation reaction. A reasonable mechanism would be the reaction of a hydroperoxide, initially present, with a transition metal, for example, ferric iron (Fe^{III}), yielding ferrous iron (Fe^{II}) and a peroxyl radical (Reaction 5.7). However, based on model experiments with linoleic acid esters, it was concluded that hydroperoxides were not significantly involved in the initiation process of fatty acid chain oxidation (Morita and Tokita, 2006). Instead, the authors proposed polymeric peroxides as the source of the initiation process, analogous to polymeric peroxides formed during the oxidation of styrene (Mayo and Miller, 1956). Polymeric peroxides would form via addition of a peroxyl radical to the double bond of an unsaturated/polyunsaturated fatty acid, and the ensuing carbon-centered radical would add oxygen to form another peroxyl radical. The latter could add to the double bond of another unsaturated/polyunsaturated fatty acid, increasing the polymer chain and the number of peroxide functions in the polymer. These polymeric peroxides can undergo a concerted fragmentation process, which yields aldehydes and alkoxyl radicals (Morita and Tokita, 2006). This process is exothermic because of the formation of saturated and unsaturated aldehydes. The ensuing alkoxyl radicals can then initiate radical chain oxidation processes. Hence, if such polymeric peroxides would be present, even in trace amounts, in bulk polysorbate, they could be responsible for the initiation of chain oxidation processes under appropriate conditions in pharmaceutical formulations. While the initiation of fatty acid chain oxidation may not rely on the reactions of hydroperoxides with Fe^{III} (Reaction 5.7), the oxidation of the active pharmaceutical ingredient may depend on such a mechanism. In model experiments, Harmon et al. (2006) subjected various organic compounds to oxidation by PS80 in the presence of Fe^{III} and, in parallel, by peroxyl radicals generated through the thermal decomposition of 2,2′-azobis(2-amidinopropane) dihydrochloride (AAPH). The reactivity of these organic molecules toward oxidation by peroxyl radicals from AAPH was similar to the reactivity toward oxidation by the PS80/Fe^{III} system, suggesting that peroxyl radicals were also responsible for the oxidation in the PS80/Fe^{III} system. Peroxyl radicals from AAPH were also demonstrated to mimic the oxidation sensitivity of parathyroid hormone 1–34, a polypeptide containing Met, His, and Trp (Ji et al., 2009), suggesting that AAPH should be considered as a source for reactive oxygen species for accelerated stability studies and for the development of stability-indicating assays. However, AAPH-depending reactions are not without complication, as described below.

5.7.3 Mechanistic Studies with Pharmaceutically Relevant Oxidants

The reactive oxygen species described above will react with selected amino acids within specific protein domains. In accelerated stability studies, pharmaceutical proteins and their formulations will be exposed to oxidation by these reactive oxygen species, and the experimental results utilized for the design of more oxidation-resistant formulations. For a correct interpretation of the experimental results, some mechanistic details of the formation and reactions of these reactive oxygen species must be known in addition to the reaction mechanisms with specific amino acids. This shall be briefly illustrated for two examples: (1) the Fenton reaction and (2) oxidations initiated by AAPH.

5.7.3.1 The Fenton Reaction

In the traditional Fenton reaction, H_2O_2 is reduced by Fe^{II} to yield ultimately Fe^{III}, HO^-, and $HO\bullet$ (Koppenol, 1993). The kinetics of this reaction will depend on the ligand spheres of Fe^{II} and Fe^{III}, which define the reduction potential of the metal and the accessibility of H_2O_2 to Fe^{II}. The $HO\bullet$ radical will be available for reaction with the protein, but will also have the opportunity to react with the ligands of the metal complex. If the amino acids of the protein provide the ligand sphere for the metal, then the metal-binding amino acids may be oxidized preferentially by the $HO\bullet$ radical generated in their vicinity, a process referred to as "site-specific" oxidation (Stadtman, 1993). Many examples for the site-specific metal-catalyzed oxidation of proteins exist in the biochemical and pharmaceutical literature. However, if the metal is complexed with a chelator present in the formulation, that is, ethylenediaminetetraacetic acid (EDTA) or citrate, the generated $HO\bullet$ radicals may, in part, react with the chelator, potentially generating carbon-centered radicals and subsequently peroxyl radicals (analogous to Reactions 5.3 and 5.4). Hence, the product selectivity observed in a "Fenton system" may contain contributions from radicals other than the $HO\bullet$ radical. In addition, the metal may play a role in product formation from the amino acid. When the $HO\bullet$ radical is generated in the vicinity of an amino acid, potentially a metal-binding amino acid, the intermediate radical generated by the reaction of an amino acid with HO• will have the opportunity to react with the metal. Consider the reaction of HO• with Met: radiation chemical studies have shown that the HO• radical adds to the sulfur of Met, and subsequently eliminates either HO^- or H_2O (depending on pH) generating a Met sulfide radical cation (Schöneich et al., 2003). The Met sulfide radical cation will deprotonate either at the C_γ-H or the C_ε-H bond, forming an α-(alkylthio)alkyl radical (Schöneich et al., 2003). These radicals may reduce the oxidized metal and/or react with O_2 ultimately leading to carbon-sulfur cleavage generating thiol and aldehyde (Barata-Vallejo et al., 2010). Importantly, organic sulfide radical cations generate only low yields of sulfoxide, consistent with the predominant formation of α-(alkylthio)alkyl radicals (Schöneich et al., 1993). On the other hand, when a Met-containing peptide was oxidized with Fe^{II}/H_2O_2 (no chelator added, reaction conducted with $(NH_4)_2Fe(SO_4)_2$ in $NaHCO_3/Na_2CO_3$ buffer), MetSO represented the main reaction product (ca. 70% yield) while specific reaction products characteristic for the reaction of HO• were only observed at low yields (<3%) (Schöneich and Yang, 1996). The product selectivity changed when EDTA was present during the oxidation reaction, that is, the Fenton system consisted of $[Fe^{II}(EDTA)]^{2-}/H_2O_2$, reducing the yield of MetSO to ca. 19% and increasing the reaction products characteristic for HO• radicals to ca. 18%. Hence, the oxidation of Met by Fe^{II}/H_2O_2 gives products which, at first glance, are not consistent with an oxidation reaction by HO•. These can be rationalized via at least two mechanisms. First, the initial step of the Fenton reaction is an addition of H_2O_2 to Fe^{II} and it may be possible that this complex oxidizes Met to MetSO in competition to the formation of HO• radicals. Second, the HO• radical adds to the Met sulfur, and the ensuing hydroxysulfuranyl radical reduces Fe^{III} to Fe^{II}, concomitantly forming MetSO. Importantly, the presence of EDTA changes product selectivity toward product patterns which are more consistent with the formation of "free" HO•. This example illustrates the complexity of Fenton reactions and the potential effect of metal and chelator on product formation.

5.7.3.2 Oxidations Initiated by Exposure to AAPH

The reactions of AAPH were recently revisited (Werber et al., 2011). Thermolysis of AAPH leads to the generation of two carbon-centered radicals within a solvent cage. These carbon-centered radicals can add oxygen to yield the corresponding peroxyl radicals. However, within the solvent cage, prior to oxygen addition, the carbon-centered radicals can recombine or disproportionate, reducing the yield of peroxyl radicals derived from AAPH thermolysis. Importantly, spin-trapping experiments in conjunction with EPR detection revealed that the employed spin traps only showed reaction with alkoxyl radicals. Alkoxyl radicals can form through the bimolecular reaction of peroxyl radicals (Reaction 5.8); in addition, the possibility exists that alkoxyl radicals are formed through a reaction sequence involving the addition of one oxygen molecule to one carbon-centered radical formed in the solvent cage, followed by recombination of the ensuing peroxyl radical with another carbon-centered radical, followed by homolytic cleavage of the intermediary peroxide (Reactions 5.9 through 5.11).

$$2 \, ROO\bullet \rightarrow O_2 + 2 \, RO\bullet \qquad (5.8)$$

$$[R\bullet\bullet R]_{cage} + O_2 \rightarrow [ROO\bullet\bullet R] \qquad (5.9)$$

$$[ROO\bullet\bullet R] \rightarrow [ROOR] \qquad (5.10)$$

$$ROOR \rightarrow 2 \, RO\bullet \qquad (5.11)$$

Both the initial peroxyl radical and the alkoxyl radical will oxidize constituents of pharmaceutical formulations, with the alkoxyl radical being more reactive compared to the peroxyl radical in hydrogen transfer and electron transfer reactions. In these oxidation processes, the alkoxyl radical is converted into an alcohol while the peroxyl radical is converted into a hydroperoxide. The latter can directly oxidize substrates such as sulfides. Hence, when oxidation reactions are initiated by the thermal decomposition of AAPH, one must take into account the possibility that oxidation is carried out by peroxyl radicals, alkoxyl radicals, and/or hydroperoxides. Another important mechanistic detail is the pH-dependent hydrolysis of the amidino group of AAPH. The hydrolysis of both amidino groups of AAPH yields 2, 2′-azobis (2-carbamoylpropane), a molecule which is thermally stable at 40°C. The hydrolysis of one side of AAPH yields azo-2-carbamoyl-2′-amidinopropane, a compound which decomposes thermally (elimination of N_2) with rate constants of 7.5×10^{-7} s^{-1} and 1.3×10^{-6} s^{-1} at pH 3.9 and 6.9, respectively. Instead, the thermal decomposition of AAPH at 40°C proceeds with a rate constant of 2.1×10^{-6} s^{-1}. Hence, the hydrolysis of the amidino groups slows the thermal generation of radicals, and experimental oxidation yields may need to be corrected for the amount of radicals/ oxidants generated through thermal decomposition if AAPH hydrolysis can occur.

5.7.4 PARAMETERS CONTROLLING PROTEIN OXIDATION: THE OXIDATION OF MET

The oxidation of Met presents a stability problem for a large number of proteins, and more recently Met oxidation within the Fc fragments of IgG1 was reported to affect Fc receptor binding and serum half-life (Bertolotti-Ciarlet et al., 2009; Wang et al., 2011). The most prominent oxidation product identified in a large number of accelerated stability studies is MetSO, although we note that other oxidation products of Met should not be neglected such as Met sulfone and carbon–sulfur bond cleavage products (see above). Examples for the control of Met oxidation by SASA, site-specific reactions, and neighboring group effects can be found among the large quantity of available data; therefore, these parameters will be discussed in more detail with emphasis on the oxidation of Met.

To characterize oxidation-sensitive proteins in accelerated stability studies, proteins have been traditionally exposed to oxidation by either H_2O_2 or tert-butyl hydroperoxide. These oxidants react directly with the thioether side chain of Met, and differences in the oxidizability of specific Met residues have been rationalized by differences in SASA. However, more recently the mechanism of Met oxidation, especially by

H_2O_2, was revisited leading to the conclusion that particularly the availability of water molecules within two solvation shells around the sulfur atom (i.e., within 5.5 Å) controls the kinetics of Met oxidation by stabilization of a transition state (Chu et al., 2004). The sensitivity of four Met residues of granulocyte colony-stimulating factor (G-CSF) toward H_2O_2 correlated practically linearly with the two-shell water coordination number (Chu et al., 2004). The presence of common excipients such as sorbitol, sucrose, and trehalose did not significantly protect G-CSF against oxidation by H_2O_2 (actually, even a slight increase in sensitivity was observed for some Met residues), whereas EDTA did protect especially the buried Met[122] from oxidation (Yin et al., 2005). In fact, the order of Met protection in G-CSF by EDTA was Met[122] > Met[127] > Met[138] > Met[1], that is, an order that was the reverse of the order of reactivity of the specific Met residues toward H_2O_2. This protective effect of EDTA could be rationalized, in part, through the chelation of redox-active transition metals, that is, provide a hint that oxidation of Met[122] may, in part, occur through site-specific transition metal-catalyzed oxidation. In fact, frequently metal-catalyzed oxidation does not target Met residues that display the highest SASA, so observed for the metal-catalyzed oxidation of a buried Met[92] in brain-derived neurotrophic factor (BDNF) (Jensen et al., 2000a, 2000b) and for the less solvent-exposed Met B4 in human relaxin (Li et al., 1995).

These product selectivities for the metal-catalyzed oxidation processes can be explained with a higher frequency of metal-binding residues located in the vicinity of the target Met residues. We note that metal-catalyzed oxidation may involve a series of different reactive oxygen species (as also pointed out in the description of the Fenton reaction, see above). Hence, the oxidation of the buried Met residues in G-CSF, BDNF, and relaxin does not necessarily need to involve a reaction with H_2O_2, but possibly proceeds through reactive oxygen species derived from H_2O_2. The oxidation of Met to MetSO constitutes a two-electron oxidation of the sulfur of Met. Formally, such two-electron oxidation may proceed via two consecutive one-electron transfer processes (Hong and Schöneich, 2001), such as described above for one possible pathway of the Fenton oxidation of Met, where an initial one-electron oxidation by the HO• radical was followed by an electron-transfer to Fe[III]. Here, the first one-electron oxidation (to the hydroxysulfuranyl radical, which exists in equilibrium with HO− and the sulfide radical cation) and the second one-electron oxidation should depend on the oxidation potentials of Met and the hydroxysulfuranyl radical/sulfide radical cation, respectively. A mass spectrometry study has provided evidence that the second one-electron oxidation step, that is, the oxidation of the sulfide radical cation, is actually more facile compared to the first one-electron oxidation, that is, that of the nonoxidized sulfide (Drewello et al., 1987). Therefore, the one-electron oxidation potential of the sulfide should control the ease by which Met residues in proteins may be oxidized by pathways involving consecutive one-electron transfer processes. Mechanistic studies with Met-containing peptides and model compounds have shown that Met sulfide radical cations can be stabilized through interaction with heteroatoms, and in particular through interaction with the oxygen and nitrogen atoms of the peptide bond (Schöneich et al., 2003). Such interaction can actually lower the oxidation potential of Met significantly, as demonstrated through electrochemical studies with organic model compounds in which the Met residue was incorporated into a rigid norbornane framework (Glass et al., 2009,

2011). Further evidence for the interaction of Met sulfide radical cations with peptide bonds in a protein was obtained during time-resolved studies on the one-electron oxidation of calmodulin by HO• radicals (Nauser et al., 2005); moreover, the oxidation of Met[35] in β-amyloid may, in part, benefit from the interaction of the Met sulfur with the peptide bond C-terminal of Ile[31] (Pogocki and Schöneich, 2002).

5.7.5   OXIDATION OF ALIPHATIC AMINO ACIDS

The HO• radical will not only react with aromatic or sulfur-containing amino acids, but efficiently oxidizes aliphatic amino acid residues (Davies, 2005). For example, the rate constant for the reaction of HO• with Leu is k(HO• + Leu) = 1.7×10^9 $M^{-1}\cdot s^{-1}$, that is, only a factor of ca. 5 lower compared to the rate constant for the reaction of HO• radicals with Cys and Met. Considering the importance of site-specific metal-catalyzed oxidation processes, which may proceed via HO• radical formation, we need to consider also the aliphatic amino acids as potential targets for protein oxidation. The reaction of HO• with aliphatic amino acids will target both the C_α-H bond and side chain C-H bonds, leading to carbon-centered radicals, which subsequently convert into peroxyl radicals by addition of oxygen. Further hydrogen transfer can generate amino acid hydroperoxides. Both amino acid peroxyl radicals and hydroperoxides are reactive intermediates that can trigger additional oxidation reactions within a given protein.

5.7.6   OXIDATION REACTIONS OF THIOLS AND DISULFIDES

Oxidation processes of thiols and disulfides have been studied in substantial detail with individual amino acids (i.e., Cys and cysteine). The mercapto group of Cys can be oxidized via two-electron oxidation, that is, by H_2O_2, leading to sulfenic acid (RSOH), which is chemically rather unstable and will react with a second non-oxidized Cys to form water and cystine (Roos and Messens, 2011). Only if the initially formed sulfenic acid is inaccessible, that is, kinetically "stabilized," can it be isolated. Sulfenic acid can also be further oxidized to sulfinic and sulfonic acid. The one-electron oxidation of the Cys mercapto group, for example, by a carbon- or oxygen-centered radical, results in formation of a Cys thiyl radical, CysS• (Wardman and von Sonntag, 1995). These CysS• thiyl radicals exist in equilibria with carbon-centered radicals through hydrogen transfer reactions with C-H bonds on peptide/protein backbone and side chains (Mozziconacci et al., 2010), for example, in insulin (Mozziconacci et al., 2008). A proposed 1,3-hydrogen shift within CysS• itself leads to the Cys $•C_\alpha$ radical, which can subsequently eliminate HS• to yield Dha (Mozziconacci et al., 2011). Dha can react via Michael addition with thiols and amines potentially leading to chemical cross-links between protein chains.

Disulfides can react with various reactive oxygen species where mechanisms and products depend on the nature of the oxidant. For example, HO• radicals add to one of the sulfur atoms of the disulfide. This hydroxyl radical adduct can react via two pathways: homolytic cleavage of the sulfur–sulfur bond to give thiyl radical and sulfenic acid and elimination of HO^- to give a disulfide radical cation. In contrast, singlet oxygen oxidizes disulfides to thiolsulfinate.

5.7.7   Oxidation Reactions of Aromatic Amino Acids

Aromatic amino acids are prominent targets for oxidation, mainly via metal-catalyzed and photochemical processes. These reactions have been summarized before (Kerwin and Remmele, 2007; Maskos et al., 1992; Ronsein et al., 2008) and will, therefore, only be covered briefly here. The oxidation of Trp may proceed via reaction with HO• radicals, peroxyl radicals, and singlet oxygen, resulting in the formation of Trp peroxide(s), hydroxytryptophan, kynurenin, and/or N-formylkynurenin. Especially kynurenin and N-formylkynurenin can be monitored by fluorescence detection. Tyr can be hydroxylated to 3,4-dihydroxyphenylalanine or oxidized via one-electron oxidation to yield a tyrosyl radical. Tyrosyl radicals can recombine to yield dityrosine, a fluorescent product, or react with superoxide to give hydroperoxides. The oxidative conversion of His into 2-oxo-histidine benefits from the metal-binding properties of His, which promotes a site-specific oxidation reaction.

5.8   PHOTOCHEMICAL DEGRADATION

Proteins are exposed to light during purification and storage, and light-induced pathways of degradation are increasingly recognized by the pharmaceutical industry (Kerwin and Remmele, 2007). In proteins, light is absorbed predominantly by Trp, Tyr, Phe, and Cys, with Trp showing the largest absorption coefficient between 250 and 310 nm (Dose, 1967). However, absorption of light is only one prerequisite for photodegradation, the efficiency of which depends on the quantum yields for the respective processes. The quantum yield for the photodegradation of a specific amino acid within a protein can depend on its location. For example, the quenching of an excited triplet state of Trp, ^{3}Trp*, by triplet molecular oxygen, $^3O_2$, depends on the location of the respective Trp residue with regard to the surface of the protein, and the rate constant of quenching varies by a factor of 10 between ^{3}Trp* in liver aldehyde dehydrogenase and alkaline phosphatase (Strambini and Cioni, 1999). In a first approximation, the quantum yield for the photolytic degradation of an entire protein depends on the sum of the quantum yields for the degradation of the individual amino acids weighed by their individual absorption coefficients relative to the absorption coefficient of the entire protein (Equation II), where the symbols represent the following parameters: n_i = number of amino acids of type i, ε_i = absorption coefficient of amino acid of type i, $\varepsilon_{Protein}$ = absorption coefficient of the protein, φ_i = quantum yield for the photolytic degradation of amino acid i, f_i = fraction of light absorbed by the amino acid i (Dose, 1967).

$$\Phi_{Protein} = \sum_i n_i \times \frac{\varepsilon_i}{\varepsilon_{Protein}} \times \varphi_i = \sum_i f_i \times \varphi_i \tag{II}$$

Taking into account the possibility that an amino acid may show different quantum yields for different photochemical processes depending on its location within a protein, the parameter φ_i represents an average quantum yield for a specific type of amino acid i. Corrections to Equation II must also be made for the photolytic decomposition of cystine, which proceeds through a combination of (1) the direct absorption of light and (2) the photo-induced electron transfer from Trp, Tyr, and/or Phe, where the efficiency of these electron-transfer processes depends on the proximity of the aromatic amino acids to the cystine moiety (Dose, 1967, 1968). The latter processes have been investigated in more detail for the photolytic decomposition of goat α-lactalbumin (Vanhooren et al., 2006), which suffers fairly selective reductive disulfide cleavage following the excitation of Trp residues. The photolytic sensitivity of pharmaceutical proteins varies with the spectrum (or wavelength) of the incident light. For example, when stored in glass vials under laboratory light, Trp-mediated photoprocesses will likely dominate photodegradation processes due to the absorbance spectrum of Trp, which extends to wavelengths >300 nm. The photosensitivity of a protein may be enhanced through the accumulation of oxidation and/or photoproducts. For example, the Trp oxidation product N-formylkynurenine displays an absorption maximum at $\lambda_{max} = 318$ nm, and can function as a photosensitizer (Walrant and Santus, 1974). If the lipid peroxidation product malondialdehyde (MDA) can form in a pharmaceutical formulation, for example, during the oxidative degradation of polyunsaturated fatty acids present in polysorbates, MDA can react with protein Lys residues to form dihydropyridine-lysine (Lamore et al., 2010). This product represents a potent photosensitizer with an absorption maximum of $\lambda_{max} = 390$ nm.

More recently, the photolytic formation of reactive oxygen species during the exposure of antibodies to light has received significant attention. Wentworth and coworkers provided evidence that the exposure of antibodies to light resulted in the formation of a series of reactive intermediates including H_2O_2, $^1O_2$, ozone (O_3), and H_2O_3 (Wentworth et al., 2001, 2002, 2003; Zhu et al., 2004). These reactive intermediates were able to kill cells, but also resulted in the oxidative degradation of selected amino acid residues (Trp, Glu) within these antibodies. While the action spectrum of photolytic formation of reactive oxygen species displayed $\lambda_{max} = 280$ nm, that is, is consistent with the involvement of Trp, most of the reported experimental data were obtained upon exposure of these antibodies with light of 312 nm, that is, a wavelength transmitted by glass, suggesting that the observed phenomena are relevant to pharmaceutical formulations of antibodies.

5.9 CONCLUSION

The chemistry of peptide and protein degradation is complex and an increasing number of chemical reactions and products are being discovered. To date, most of the mechanistic studies have been performed in solution. It will be important to extend such mechanistic studies to proteins formulated as solids and in drug delivery matrices in order to obtain a complete picture of degradation pathways relevant for pharmaceutical formulations.

REFERENCES

Barata-Vallejo S., Ferreri C., Postigo A., and Chatgilialoglu C., 2010. Radiation chemical studies of methionine in aqueous solution: Understanding the role of molecular oxygen. *Chem. Res. Toxicol.* 23:258–263.

Bertolotti-Ciarlet A., Wang W., Lownes R., Pristatsky P., Fang Y., Mckelvey T., Li Y., et al., 2009. Impact of methionine oxidation on the binding of human IgG1 to FcRn and Fcγ receptors. *Mol. Immunol.* 46:1878–1882.

Bogdanowich-Knipp S.J., Jois S.D., and Siahaan T.J., 1999a. Effect of conformation on the conversion of cyclo-(1,7)-Gly-Arg-Gly-Asp-Ser-Pro-Asp-Gly-OH to its cyclic imide degradation product. *J. Pept. Res.* 54:43–53.

Bogdanowich-Knipp S.J., Jois D.S., and Siahaan T.J., 1999b. The effect of conformation on the solution stability of linear vs. cyclic RGD peptides. *J. Pept. Res.* 53:523–529.

Chu J., Yion J., Brooks B.R., Wang D.I.C., Speed Ricci M., Brems D.N., and Trout B.L., 2004. A comprehensive picture of non-site specific oxidation of methionine residues by peroxides in protein pharmaceuticals. *J. Pharm. Sci.* 93:3096–3102.

Corveleyn S., Vandenbosche G.M.R., and Remon J.P., 1997. Near-infrared (NIR) monitoring of H_2O_2 vapor concentration during vapor hydrogen peroxide (VHP) sterilization. *Pharm. Res.* 14:294–298.

Davies M., 2005. The oxidative environment and protein damage. *Biochim. Biophys. Acta.* 1703:93–109.

Dose K., 1967. Theoretical aspects of the ultraviolet inactivation of proteins containing disulfide bonds. *Photochem. Photobiol.* 6:437–443.

Dose K., 1968. The photolysis of free cysteine in the presence of aromatic amino acids. *Photochem. Photobiol.* 8:331–335.

Drewello T., Lebrilla C.B., Schwarz H., De Koning L.J., Fokkens R.H., Nibbering N.M.M., Ansklam E., and Asmus K-D., 1987. Formation of a two-center, three-electron, sulfur-sulfur bond in the gas phase. *J. Chem. Soc. Chem. Commun.* 18:1381–1383.

Glass R.S., Hug G.L., Schöneich C., Wilson G.S., Kuznetsova L., Lee T., Ammam M., et al., 2009. Neighboring amide participation in thioether oxidation: Relevance to biological oxidation. *J. Am. Chem. Soc.* 131:13791–13805.

Glass R.S., Schöneich C., Wilson G.S., Nauser T., Yamamoto T., Lorance E., Nichol G.S., and Ammam M., 2011. Neighboring pyrrolidine amide participation in thioether oxidation. Methionine as a hopping site. *Org. Lett.* 13:2837–2839.

Goolcharran C., and Borchardt R.T., 1998. Kinetics of diketopiperazine formation using model peptides. *J. Pharm. Sci.* 87:283–288.

Goolcharran C., Cleland J.L., Keck R., Jones A.J., and Borchardt R.T., 2000a. Comparison of the rates of deamidation, diketopiperazine formation and oxidation in recombinant human vascular endothelial growth factor and model peptides. *AAPS PharmSci* 2:E5.

Goolcharran C., Stauffer L.L., Cleland J.L., and Borchardt R.T., 2000b. The effects of a histidine residue on the C-terminal side of an asparaginyl residue on the rate of deamidation using model pentapeptides. *J. Pharm. Sci.* 89:818–825.

Harmon P.A., Kosuda K., Neslon E., Mowery M., and Reed R.A., 2006. A novel peroxy radical based oxidative stressing system for ranking the oxidizability of drug substances. *J. Pharm. Sci.* 95:2014–2028.

Harriman A., 1987. Further comments on the redox potentials of tryptophan and tyrosine. *J. Phys. Chem.* 91:6102–6104.

He H.T., Gursoy R.N., Kupczyk-Subotkowska L., Tian J., Williams T., and Siahaan T.J., 2006. Synthesis and chemical stability of a disulfide bond in a model cyclic pentapeptide: Cyclo(1,4)-Cys-Gly-Phe-Cys-Gly-OH. *J. Pharm. Sci.* 95:2222–2234.

Hong J., and Schöneich C., 2001. The metal-catalyzed oxidation of methionine in peptides by Fenton systems involves two consecutive one-electron transfer processes. *Free Rad. Biol. Med.* 31:1432–1441.

Hovorka S.W., and Schöneich C., 2001. Oxidative degradation of pharmaceuticals: Theory, mechanisms and inhibition. *J. Pharm. Sci.* 90:253–269.

Jensen J.L., Kolvenbach C., Roy S., and Schöneich C., 2000a. Metal-catalyzed oxidation of brain-derived neurotrophic factor (BDNF): Analytical challenges for the identification of modified sites. *Pharm. Res.* 17:190–196.

Jensen J.L., Kuczera K., Roy S., and Schöneich C., 2000b. Metal-catalyzed oxidation of brain-derived neurotrophic factor (BDNF): Selectivity and conformational consequences of histidine modification. *Cell. Mol. Biochem.* 46:685–696.

Ji A.J., Zhang B., Cheng W., and Wang Y.J., 2009. Methionine, tryptophan, and histidine oxidation in a model protein, PTH: Mechanisms and stabilization. *J. Pharm. Sci.* 98:4485–4500.

Jones C.M., Lawrence A., Wardman P., and Burkitt M.J., 2002. Electron paramagnetic resonance spin trapping investigation into the kinetics of glutathione oxidation by the superoxide radical: Re-evaluation of the rate constant. *Free Rad. Biol. Med.* 32:982–990.

Kertscher U., Bienert M., Krause E., Sepetov N.F., and Mehlis B., 1993. Spontaneous chemical degradation of substance P in the solid phase and in solution. *Int. J. Pept. Prot. Res.* 41:207–211.

Kerwin B.A, and Remmele R.L., 2007. Protect from light: Photodegradation and protein biologics. *J. Pharm. Sci.* 96:1468–1479.

Koppenol W.H., 1993. The centennial of the Fenton reaction. *Free Rad. Biol. Med.* 15:645–651.

Lal M., Rao R., Fang X., Schuchmann H-P., and Von Sonntag C., 1997. Radical-induced oxidation of dithiothreitol in acidic aqueous solution. *J. Am. Chem. Soc.* 119:5735–5739.

Lamore S.D., Azimian S., Horn D., Anglin B.L., Uchida K., Cabello C.M., and Wondrak G.T., 2010. The malonaldehyde-derived fluorophore DHP-lysine is a potent sensitizer of UVA-induced photooxidative stress in human skin cells. *J. Photochem. Photobiol. B* 101:251–264.

Li B., Borchardt R.T., Topp E.M., Vandervelde D., and Schowen R.L., 2003. Racemization of an asparagine residue during peptide deamidation. *J. Am. Chem. Soc.* 125:11486–11487.

Li S., Nguyen T.H., Schöneich C.H., and Borchardt R.T., 1995. Aggregation and precipitation of human relaxin induced by metal-catalyzed oxidation. *Biochemistry* 34:5762–5772.

Manning M.C., Chou D.K., Murphy B.M., Payne R.W., and Katayama D.S., 2010. Stability of protein pharmaceuticals: An update. *Pharm. Res.* 27:544–575.

Manning M.C., Patel K., and Borchardt R.T., 1989. Stability of protein pharmaceuticals. *Pharm. Res.* 6:903–918.

Maskos Z., Rush J.D., and Koppenol W.H., 1992. The hydroxylation of phenylalanine and tyrosine: A comparison with salicylate and tryptophan. *Arch. Biochem. Biophys.* 296:521–529.

Mayo F.R., and Miller A.A., 1956. Oxidation of unsaturated compounds. II. Reactions of styrene peroxide. *J. Am. Chem. Soc.* 78:1023–1034.

Merényi G., Lind J., and Jonsson M., 1993. Autoxidation of closed shell organics: An outer-sphere electron transfer. *J. Am. Chem. Soc.* 115:4945–4946.

Miller D.M., Buettner G., and Aust S.D., 1990. Transition metals as catalysts of "autoxidation" reactions. *Free Rad. Biol. Med.* 8:95–108.

Morita M., and Tokita M., 2006. The real radical generator other than main-product hydroperoxide in lipid autoxidation. *Lipids* 41:91–95.

Mozziconacci O., Kerwin B.A., and Schöneich C., 2008. Reversible *intra*molecular hydrogen transfer between protein cysteine thiyl radicals and $^{\alpha}$C-H bonds in insulin: Control of selectivity by secondary structure. *J. Phys. Chem. B* 112:15921–15932.

Mozziconacci O., Kerwin B.A., and Schöneich C., 2010. Reversible hydrogen transfer between cysteine thiyl radical and glycine and alanine in model peptides: Evidence for carbon-centered radical through radical-radical reactions and L- to D-Ala conversion. *J. Phys. Chem. B* 114:6751–6762.

Mozziconacci O., Kerwin B.A., and Schöneich C., 2011. Reversible hydrogen transfer reactions of cysteine thiyl radicals in peptides: The conversion of cysteine into dehydroalanine and alanine, and of alanine into dehydroalanine. *J. Phys. Chem. B* 115:12287–12305.

Nauser T., Jacoby M., Koppenol W.H., Squier T.C., and Schöneich C., 2005. Calmodulin methionine residues are targets for one-electron oxidation by hydroxyl radicals: Formation of S∴N three-electron bonded radical complexes. *Chem. Commun.* 587–589.

Neuzil J., Gebicki J.M., and Stocker R., 1993. Radical-induced chain oxidation of proteins and its inhibition by chain-breaking antioxidants. *Biochem. J.* 293:601–606.

Okada J., Seo T., Kasahara F., Takeda K., and Kondo S., 1991. New degradation product of des-Gly10-NH$_2$-LH-RH-ethylamide (fertirelin) in aqueous solution. *J. Pharm. Sci.* 80:167–170.

Oliyai C., and Borchardt R.T., 1993. Chemical pathways of peptide degradation. IV. Pathways, kinetics, and mechanism of degradation of an aspartyl residue in a model hexapeptide. *Pharm. Res.* 10:95–102.

Oliyai C., and Borchardt R.T., 1994. Chemical pathways of peptide degradation. VI. Effect of the primary sequence on the pathways of degradation of aspartyl residues in model hexapeptides. *Pharm. Res.* 11:751–758.

Oliyai C., Patel J.P., Carr L., and Borchardt R.T., 1994. Chemical pathways of peptide degradation. VII. Solid state chemical instability of an aspartyl residue in a model hexapeptide. *Pharm. Res.* 11:901–908.

Patel K., and Borchardt R.T., 1990a. Chemical pathways of peptide degradation. II. Kinetics of deamidation of an asparaginyl residue in a model hexapeptide. *Pharm. Res.* 7:703–711.

Patel K., and Borchardt R.T., 1990b. Chemical pathways of peptide degradation. III. Effect of primary sequence on the pathways of deamidation of asparaginyl residues in hexapeptides. *Pharm. Res.* 7:787–793.

Patel K., and Borchardt R.T., 1990c. Deamidation of asparaginyl residues in proteins: A potential pathway for chemical degradation of proteins in lyophilized dosage forms. *J. Parent. Sci. Technol.* 44:300–301.

Pogocki D., and Schöneich C., 2002. Redox properties of Met[35] in neurotoxic β-amyloid peptide. A molecular modeling study. *Chem. Res. Toxicol.* 15:408–418.

Robinson N.E., and Robinson A.B., 2001. Molecular clocks. *Proc. Natl. Acad. Sci. USA* 98:944–949.

Ronsein G.E., Oliveira M.C.B., Miyamoto S., Medeiros M.H.G., and Di Mascio P., 2008. Tryptophan oxidation by singlet molecular oxygen [O$_2$ ($^1\Delta_g$)]: Mechanistic studies using ^{18}O-labeled hydroperoxides, mass spectrometry, and light emission measurements. *Chem. Res. Toxicol.* 21:1271–1283.

Roos G., and Messens J., 2011. Protein sulfenic acid formation: From cellular damage to redox regulation. *Free Rad. Biol. Med.* 51:314–326.

Schöneich C., Aced A., and Asmus K-D., 1993. Mechanism of oxidation of aliphatic thioethers to sulfoxides by hydroxyl radicals. The importance of molecular oxygen. *J. Am. Chem. Soc.* 115:11376–11383.

Schöneich C., Hageman M.J., and Borchardt R.T., 1997. Stability of peptides and proteins. In: *Controlled Drug Delivery*, Park, K., Ed., American Chemical Society, Washington, DC, 205–228.

Schöneich C., Pogocki D., Hug G.L., and Bobrowski K., 2003. Free radical reactions of methionine in peptides: Mechanisms relevant to β-amyloid oxidation and Alzheimer's disease. *J. Am. Chem. Soc.* 125:13700–13713.

Schöneich C., and Yang J., 1996. Oxidation of methionine peptides by Fenton systems: The importance of peptide sequence, neighbouring groups, and EDTA. *J. Chem. Soc. Perkin Trans.* 2:915–924.

Sohma Y., Chiyomori Y., Kimura M., Fukao F., Taniguchi A., Hayashi Y., Kimura T., and Kiso Y., 2005. 'O-Acyl isopeptide method' for the efficient preparation of amyloid beta peptide 1-42 mutants. *Bioorg. Med. Chem.* 13:6167–6174.

Stadtman E.R., 1993. Oxidation of free amino acids and amino acid residues in proteins by metal-catalyzed reactions. *Annu. Rev. Biochem.* 62:797–821.

Stevenson C.L., Donlan M.E., Friedman A.R., and Borchardt R.T., 1993a. Solution conformation of Leu27 hGRF(1-32)NH2 and its deamidation products by 2D NMR. *Int. J. Pept. Prot. Res.* 42:24–32.

Stevenson C.L., Friedman A.R., Kubiak T.M., Donlan M.E., and Borchardt R.T., 1993b. Effect of secondary structure on the rate of deamidation of several growth hormone releasing factor analogs. *Int. J. Pept. Prot. Res.* 42:497–503.

Strambini G.B., and Cioni P., 1999. Pressure–temperature effects on oxygen quenching of protein phosphorescence. *J. Am. Chem. Soc.* 121:8337–8344.

Surdhar P.A., and Armstrong D.A., 1987. Reduction potentials and exchange reactions of thiyl radicals and disulfide anion radicals. *J. Phys. Chem.* 91:6532–6537.

Taube H., 1965. Mechanisms of oxidation with oxygen. *J. Gen. Physiol.* 49:29–50.

Trivedi M., Davis R.A., Shabaik Y., Roy A., Verkhivker G., Laurence J.S., Middaugh C.R., and Siahaan T.J., 2009a. The role of covalent dimerization on the physical and chemical stability of the EC1 domain of human E-cadherin. *J. Pharm. Sci.* 98:3562–3574.

Trivedi M.V., Laurence J.S., and Siahaan T.J., 2009b. The role of thiols and disulfides on protein stability. *Curr. Prot. Pept. Sci.* 10:614–625.

Vanhooren A., De Vriendt K., Devreese B., Chedad A., Sterling A., Van Dael H., Van Beeumen J., and Hanssens I., 2006. Selectivity of tryptophan residues in mediating photolysis of disulfide bridges in goat a-lactalbumin. *Biochemistry* 45:2085–2093.

Volkin D.B., and Klibanov A.M., 1987. Thermal destruction processes in proteins involving cystine residues. *J. Biol. Chem.* 262:2945–2950.

Von Sonntag C., 1987. *The Chemical Basis of Radiation Biology*. Taylor & Francis, London.

Wakankar A.A., and Borchardt R.T., 2006. Formulation considerations for proteins susceptible to asparagine deamidation and aspartate isomerization. *J. Pharm. Sci.* 95:2321–2336.

Walrant P., and Santus R., 1974. N-formyl-kynurenine, a tryptophan photooxidation product, as a photodynamic sensitizer. *Photochem. Photobiol.* 19:411–417.

Wang W., Vlasak J., Li Y., Pristatsky P., Fang Y., Pittman T., Roman J., Wang Y., Prueksaritanont T., and Ionescu R., 2011. Impact of methionine oxidation in human IgG1 Fc on serum half-life of monoclonal antibodies. *Mol. Immunol.* 48:860–866.

Wardman P., and Von Sonntag C., 1995. Kinetic factors that control the fate of thioyl radicals in cells. *Meth. Enzymol.* 251:31–45.

Wasylaschuk W.R., Harmon P.A., Wagner G., Harman A.B., Templeton A.C., Xu R., and Reed R.A., 2007. Evaluation of hydroperoxides in common pharmaceutical excipients. *J. Pharm. Sci.* 96:106–116.

Wentworth JR. P., Jones L.H., Wentworth A.D., Zhu X., Larsen N.A., Wilson I.A., Xu X., et al., 2001. Antibody catalysis of the oxidation of water. *Science* 293:1806–1811.

Wentworth JR. P., Mcdunn J.E., Wentworth A.D., Takeuchi C., Nieva J., Jones T., Bautista C., et al., 2002. Evidence for antibody-catalyzed ozone formation in bacterial killing and inflammation. *Science* 298:2195–2199.

Wentworth JR. P., Wentworth A.D., Zhu X., Wilson I A.D., Janda K.D., Eschenmoser A., and Lerner R.A., 2003. Evidence for the production of trioxygen species during antibody-catalyzed chemical modification of antigens. *Proc. Natl. Acad. Sci. USA* 100:1490–1493.

Werber J., Wang Y.J., Milligan M., Li X., and Ji J.A., 2011. Analysis of 2,2'-azobis (2-amidinopropane) dihydrochloride degradation and hydrolysis in aqueous solution. *J. Pharm. Sci.* 100:3307–3315.

Wigginton K.R., Menin L., Montoya J.P., and Kohn T., 2010. Oxidation of virus proteins during UV_{254} and singlet oxygen mediated inactivation. *Environ. Sci. Technol.* 44:5437–5443.

Winterbourn C.C., and Metodiewa D., 1999. Reactivity of biologically important thiol compounds with superoxide and hydrogen peroxide. *Free Rad. Biol. Med.* 27:322–328.

Xie M., Aube J., Borchardt R.T., Morton M., Topp E.M., Vander Velde D., and Schowen R.L., 2000. Reactivity toward deamidation of asparagine residues in beta-turn structures. *J. Pept. Res.* 56:165–171.

Yao J., Dokuru D.K., Noestheden M., Park S.S., Kerwin B.A., Jona J., Ostovic D., and Reid D.L., 2009. A quantitative kinetic study of polysorbate autoxidation: The role of unsaturated fatty acid ester substituents. *Pharm. Res.* 26:2303–2313.

Yin J., Chu J., Speed Ricci M., Brems D.N., Wang D.I.C., and Trout B.L., 2005. Effects of excipients on the hydrogen peroxide-induced oxidation of methionine residues in granulocyte colony-stimulating factor. *Pharm. Res.* 22:141–147.

Zhu X., Wentworth JR P., Wentworth A.D., Eschenmoser A., Lerner R.A., and Wilson I.A., 2004. Probing the antibody-catalyzed water-oxidation pathway at atomic resolution. *Proc. Natl. Acad. Sci. USA* 101:2247–2252.

6 Physical Instability of Peptides and Proteins

Marco van de Weert and Theodore W. Randolph

CONTENTS

6.1 INTRODUCTION

The biological function of peptides and proteins is highly dependent on their three-dimensional structure. Changes in that structure, which may arise due to chemical or physical processes, may alter or abolish that function, or even result in toxicity. Thus, it is of importance that a pharmaceutical formulation of therapeutic peptides and proteins retains the normal (native) structure of those peptides or proteins, or that any changes are fully reversible upon administration to the patient.

A major difference between proteins and low molecular weight drugs is the complexity of the three-dimensional structure and concomitant sensitivity toward external stress factors. The three-dimensional structure of proteins is mostly held together by noncovalent interactions, such as hydrogen bonds, salt bridges, and van der Waals forces. Any stress factor may alter these noncovalent interactions, possibly leading to new intra- or intermolecular interactions which may not be reversible upon removing the stress factor.

In this chapter, we will discuss the noncovalent interactions that result in the formation of the specific three-dimensional fold of most proteins, the most important stress factors that may cause changes in that protein fold, and the resulting physical instability of the protein. It should be noted that we will discuss this physical instability as a separate issue to the chemical instability discussed in Chapter 5. In reality, the two are highly interdependent, as chemical destabilization may lead to physical destabilization, and vice versa. This chapter will end with potential stabilization strategies to prevent physical instability of proteins in pharmaceutical formulations. Several of these strategies will be discussed with more specific examples in other chapters.

Throughout the chapter, a number of semantic issues will be discussed, which are of importance when reading the literature. The commonly used terminology within the field of protein structure, folding, and stability is not always strictly defined, and definitions may differ over time and depending on the context. Unfortunately, the definitions used in a particular scientific paper are often not explicit, which may lead to confusion when the reader is insufficiently aware of the different descriptions that are in use.

6.2   PROTEIN STRUCTURE

6.2.1   Peptides, Polypeptides, and Proteins

All peptides, polypeptides, and proteins are considered condensation polymers of amino acids, resulting in a linear backbone of alternating amide, C–C, and C–N bonds. However, the distinction between peptide, polypeptide, and protein is rather diffuse. One may find at least three different and partly overlapping descriptions, rather than definitions, of the difference between peptide and protein alone. The currently most common description refers to any peptide chain of more than 50 amino acids as a protein. Others refer to peptides as proteins whenever the peptide has a biological function. This could then even include several simple dipeptides (i.e., two amino acids linked together), which can have a biological function. Finally, the absence or presence of a well-defined tertiary structure has been used to distinguish peptides from proteins. Also, this distinction is not without problems; there are proteins that are referred to as "natively unfolded," so called because they do not have a specific tertiary structure. In addition, some "peptides" can form fully reversible multimeric structures, such as glucagon (forming trimers) (Formisano et al., 1977), which involves the formation of a defined three-dimensional structure. The term "polypeptide" generally overlaps with that of "peptide" and "protein." The reader may thus encounter all three terms used in connection with the same biological compound. For practical purposes, we have used the first definition, calling every compound with more than 50 amino acids a protein.

6.2.2   Protein Structure: Primary, Secondary, Tertiary, and Quaternary Structure

The three-dimensional structure of proteins is often subdivided into four types of structure, referred to as the primary, secondary, tertiary, and quaternary structure. The primary structure refers to the amino acid sequence within the polymer chain.

Intra- or interchain cross-links are also prevalent, usually through cysteine residues (forming a cystine or S–S bridges).

The secondary structure refers to the folding of this backbone into specific structures, which are defined by the bond angles and hydrogen bonding pattern of the amide bond. The secondary structure can roughly be subdivided into four classes: helical structures such as the alpha-helix, pleated structures such as the beta-strand and beta-sheet, turn structures such as the beta-turns, and loop structures. The latter are often referred to as "random" structures.

The three-dimensional alignment of the secondary structural elements is known as the tertiary structure of a protein. This alignment often results in an almost ideal close packing of the amino acids, particularly in the core of the protein molecule. Protein tertiary structures can be described by commonly appearing architectures such as barrels or alpha-helix bundles (Orengo and Thornton, 2005).

Some proteins exist under physiological conditions as specific multimeric proteins linked through noncovalent interactions. This multimerization is known as the quaternary structure of a protein. Examples of proteins with a quaternary structure include hemoglobin, alpha-crystallin, and HIV-1 protease. In general, the biological function of such multimeric proteins depends on this multimerization, but some proteins may also form specific (and reversible) multimers that are not biologically active. Perhaps the best known example of the latter is insulin; insulin forms dimers and hexamers at elevated concentration, especially in the presence of certain divalent metal ions, but is only active as a monomer (Uversky et al., 2003).

The ultimate fold of the protein is usually referred to as the "native" structure. In principle, the latter refers to the functional structure of the protein. However, many proteins change structure during their biological function, which would suggest there are multiple "native" structures. Furthermore, artificially created proteins (e.g., fusion proteins created by genetic engineering techniques) may assemble into well-defined folds but have unknown levels of function. It may therefore be easier to describe the protein structure under physiological conditions (in terms of pH, ionic strength, etc.) as the native structure. The mechanism of protein folding is discussed in the following section.

6.3   PROTEIN FOLDING: WHY DO PROTEINS FOLD?

6.3.1   ROLE OF WATER AND STABILIZING INTERACTIONS

The observation that most proteins are folded into a specific structure in simple aqueous solutions suggests that folding is a thermodynamically favorable process. Many decades of research have been aimed at elucidating why, and how, proteins fold (Anfinsen, 1973). Although there are still several limitations, it is now possible to predict with reasonable accuracy how a protein will fold using computational methods (Kryshtafovych and Fidelis, 2009). In this section, we will discuss the driving forces for folding, starting with a protein in the gas phase before moving to the more complex situation of a protein in solution.

For a single protein molecule in the gas phase, there are four fundamental forces to take into account. The first is the entropy of the amino acid chain, which tends to disfavor folding. That is, folding of the amino acid chain into a specific structure reduces the degrees of freedom for that chain, which results in a loss of entropy. In contrast, hydrogen bonding and van der Waals forces favor folding. Electrostatic interactions may either favor or disfavor folding, depending on the sign of the charges. Experiments with peptides in the gas phase have shown that folding can be spontaneous (Chin et al., 2006); hence, at least in some circumstances the entropy loss upon folding can be overcome by the enthalpy gain from electrostatic interactions, hydrogen bonding, and/or van der Waals forces. Often electrostatic interactions are an important driving force for the folding process in the gas phase (Chin et al., 2006).

Because proteins are typically found in an aqueous environment, these gas-phase experiments offer only limited insights to the understanding of protein folding under solution conditions. The high dielectric constant of water means that the strength of electrostatic interactions is significantly reduced, and therefore is a much less important driving force for folding, if at all. Moreover, the peptide chain now has the ability to form hydrogen bonds with water, as well as to interact with water molecules through van der Waals forces. Thus, intramolecular interactions like van der Waals forces and hydrogen bonds also are not immediately apparent driving forces for folding.

And yet, proteins do fold in water. An important driving force of this folding is the negative effect of solute–water interactions on the interaction between the water molecules themselves. Pure water may be viewed as a collection of oxygen atoms suspended in a sea of hydrogen atoms. On average, four hydrogen atoms surround one oxygen atom in a (imperfect) tetrahedral shape, with two of those hydrogen atoms close enough to describe the bond as covalent and two slightly further away, forming a hydrogen bond. This is, however, a highly dynamic system, and there will be a constant exchange between covalently bound and hydrogen bond-linked hydrogen atoms. In essence, any solute will negatively affect this dynamic system; this is known as the hydrophobic effect (Dill et al., 2005). Whether a solute dissolves in water, and how much, is a matter of accounting: as long as there is a negative change in Gibbs free energy for the system as a whole upon dissolution of the solute, the compound will dissolve. Thus, the negative energetic contribution by distorting the dynamic water network needs to be counterbalanced by the positive contribution of the solute dissolving, which includes increased entropy of the solute upon dissolution as well as hydrogen bonding and van der Waals interactions with the water molecules. Due to the ability to form hydrogen bonds and significant van der Waals interactions, polar (hydrophilic) compounds dissolve to a much larger extent in water than nonpolar (hydrophobic) compounds.

Most proteins contain a significant amount of nonpolar amino acid residues and their dissolution in an aqueous environment would be energetically unfavorable. In contrast, the dissolution of the polar amino acids would be a favorable process. By folding of the amino acid chain such that the hydrophobic amino acids are hidden from the aqueous surroundings, a protein significantly reduces the hydrophobic effect by the nonpolar residues, while maintaining the positive interaction between the polar residues and the water molecules. The hydrophobic effect and the resulting "hiding" of nonpolar amino acids in the core of the protein is believed to be the main driving force for folding (Dill, 1990; Kauzmann, 1959). Further folding and specificity of the fold are then governed by other interactions like hydrogen bonding, salt bridge formation, and van der Waals interactions between tightly packed residues (Rose and Wolfenden, 1993). Finally, the ultimate fold may be stabilized by the formation of cystines.

The importance of the hydrophobic amino acids for protein folding is also suggested by the relatively conservative changes of the core amino acids for similar proteins across species. That is, even though the overall amino acid sequence may be significantly different for a given protein isolated from various species, the differences in the (usually hydrophobic) amino acids forming the core of the protein are usually the smallest (Mirny and Shakhnovich, 2001), and mutations in these amino acids are more likely to yield an inactive protein (Guo et al., 2004). As a result, even proteins with a mere 30–40% similarity in amino acid sequence can yield very similar protein folds.

Considering the above, it should be no surprise that natively unfolded proteins generally do not contain such a core of hydrophobic amino acids. In fact, it is likely the absence of a significant amount of hydrophobic amino acids, along with many charged residues, that allows these proteins to have little tertiary fold (Uversky and Dunker, 2010). However, they often do have a specific secondary structure, which suggests that for amino acid chains, the intrachain hydrogen bonding is more favorable than hydrogen bonding to water.

6.3.2 The Energy Landscape of a Protein Molecule

As discussed above, proteins may spontaneously fold in aqueous solution. That means that the change in Gibbs free energy upon folding is negative, that is, $\Delta G_f < 0$, and thus the change in Gibbs free energy of unfolding is positive ($\Delta G_u > 0$). However, due to the complex interaction between protein and solvent, the Gibbs free energy is not a simple linear function of temperature (Privalov, 1990; Robertson and Murphy, 1997). Let us first examine a simple two-state reversible folding process between a protein in its unfolded state (U) and in a folded state (N) (Scheme 6.1):

$$U \rightleftharpoons N \qquad \text{(Scheme 6.1)}$$

The change in Gibbs free energy for this folding process can be approximated using a modified form of the Gibbs–Helmholtz equation (Equation 6.1), in which the temperature dependence of ΔH and ΔS are approximated by a constant difference in heat capacity between the native and unfolded stated of the protein, ΔC_p. In this equation, T_m is a temperature where ΔG is zero; ΔH_f is the enthalpy change upon folding at this temperature, and $\Delta C_{p,f}$ is the change in heat capacity upon folding.

$$\Delta G_f = \Delta H_f \left(1 - \frac{T}{T_m} \right) + \Delta C_{p,f} \left((T - T_m) - T \ln \frac{T}{T_m} \right) \qquad (6.1)$$

Data on T_m, ΔC_p, and ΔH_f can be obtained, for example, using differential scanning calorimetry (DSC)*. Plotting this data using Equation 6.1 will yield a parabola

---

* Note that in a typical DSC experiment the protein is folded at the start of the experiment. Thus, the ΔH and ΔC_p obtained are those for the unfolding process.

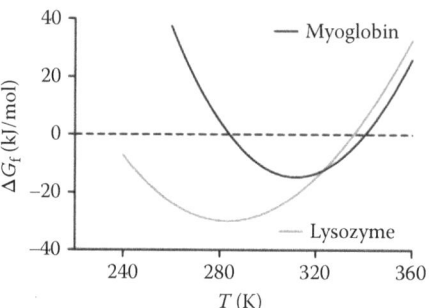

FIGURE 6.1 Graphical representation of the thermodynamic stability of two model proteins as a function of temperature as derived from the modified Gibbs–Helmholtz equation (Equation 6.1). Figure created using numerical data from Anjum et al. (2000), with $T_m = 340.4$ K, $\Delta H_f = -343$ kJ/mol, and $\Delta C_{p,f} = -11.45$ kJ/mol for myoglobin in a pH 6.1 buffer; $T_m = 335.7$ K, $\Delta H_f = -372$ kJ/mol, and $\Delta C_{p,u} = -6.52$ kJ/mol for lysozyme in a pH 4.8 buffer. (Data from Anjum et al., *Biochim. Biophys. Acta* 1476, 2000.)

(Figure 6.1), which is known as the protein stability curve. It has been observed that many proteins have their highest thermodynamic stability around 283 K, independent of their melting temperatures (Rees and Robertson, 2001). This is significantly below physiological temperatures for many organisms, probably because some structural flexibility is required for activity.

Figure 6.1 shows there are two crossings where $\Delta G = 0$, suggesting that proteins can unfold due to both increased as well as decreased temperatures. The latter, cold denaturation (Privalov, 1990), usually occurs at temperatures below 270 K and thus is less likely to be observed in standard analytical techniques due to ice formation. Furthermore, the kinetics of unfolding slows down with decreased temperature, which may result in kinetic trapping of the protein in its folded structure. Finally, it is important to note that under physiological conditions, the magnitude of $\Delta G_{f,max}$ is relatively small, typically ca. 10–50 kJ/mol. This is a rather weak stabilizing interaction, considering that a typical hydrogen bond contributes about 5–30 kJ/mol.

The pathway from unfolded to folded state is, for many proteins, likely not as simple as suggested by Scheme 6.1. Unfolded proteins may assume an enormous number of conformational states; indeed, a simple calculation shows that in a typical sample of unfolded protein molecules, each molecule is likely to be found in a different conformational state.* Yet, proteins can spontaneously fold to their native conformation within a second. This suggests that each protein molecule must necessarily follow a slightly different pathway to the folded state. This complex folding process can be conceptualized in terms of a biased random walk, wherein proteins fold via a large number of small conformational changes, with the likelihood of any conformational change occurring being biased toward those that lower the overall free energy of the protein (Bryngelson et al., 1995). The collection of all possible conformational trajectories and associated free energies forms an "energy landscape." To better visualize

---

* Take, for example, a protein of 100 amino acids, and allow each amino acid only two different conformations. This already yields $2^{100} = 10^{30}$ different potential conformations. This exceeds the number of molecules of a specific protein on Earth.

the high-dimensional space represented by this enormous collection of conformational states and energies, the energy landscape is often conceptualized as a "folding funnel," wherein the vertical position on the funnel is representative of the free energy of a given conformation, and the circumference of the funnel is representative of the number of states having a given free energy (Bryngelson et al., 1995). Thus, the large number of unfolded states would be found at the top of the funnel, and the singular native state conformation would be found at the funnel bottom (Figure 6.2a).

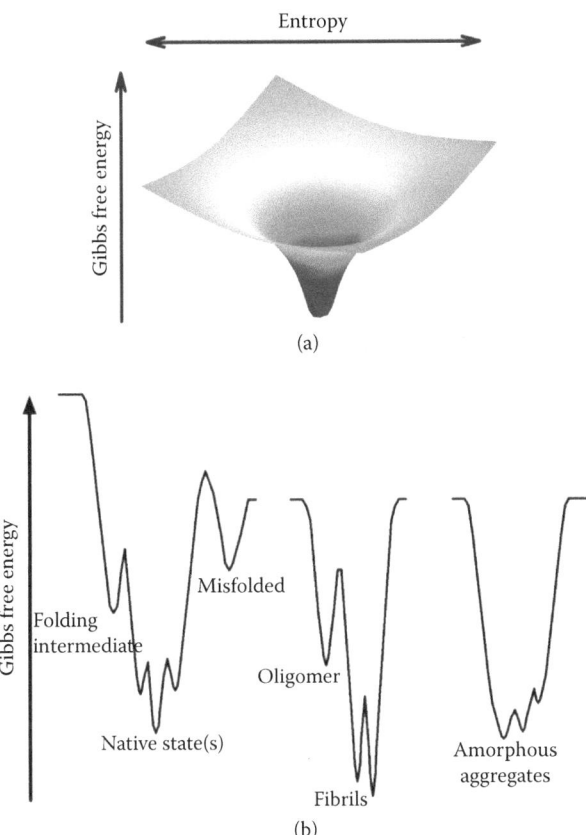

FIGURE 6.2 Energy landscape of a protein. (a) (**See color insert.**) An idealized folding funnel for a single protein molecule. At the high Gibbs free energy end (top of picture), the protein molecule can adopt many different conformations; the width of the funnel can be viewed as a measure of the conformational entropy. At the bottom of the funnel a singular folded state with very limited conformational entropy is present. (b) A more realistic two-dimensional representation of the energy landscape of a protein. On the left-hand side, the folding of a single protein is shown; as depicted, there may be several folds with almost the same low Gibbs free energy. Also, there may be a folding intermediate(s) and misfolded species with higher energy which can be significantly populated due to kinetic barriers. In the middle and on the right-hand side possible energy states are shown for ensembles of protein molecules, resulting in various aggregated species (oligomers, fibrils, and amorphous aggregates). These may have lower Gibbs free energy than the native protein, but may also be populated due to a large kinetic barrier toward refolding.

The energy landscape of proteins can probably be better conceptualized as a jagged funnel (Figure 6.2b), where the protein fold, or a subpopulation of protein molecules, may be kinetically trapped in local minima, rather than in the thermodynamically lowest energy state. Note that the aggregated states, in particular the fibrillar state, are shown in Figure 6.2b as being equally or even more thermodynamically stable than the native state. That is, the protein fold observed for native proteins may well be considered a metastable state, with kinetic barriers preventing rapid population of the aggregated and/or fibrillar state (Baldwin et al., 2011). In the human body, various regulatory processes have developed that are designed to degrade and eliminate improperly folded proteins, and thus prevent and/or reduce the rate of fibril formation. In a pharmaceutical formulation, there are no processes that remove misfolded protein, meaning that aggregation and fibrillation can occur for proteins that are not known to aggregate or fibrillate in vivo, or can occur much faster than observed in vivo. Aggregation and fibrillation is further discussed in Section 6.4.3.

The jagged funnel depicted in Figure 6.2b should not be seen as static; changing the solution conditions will alter the relative magnitudes of the local minima relative to the global minimum, possibly resulting in a new global minimum. This may also occur upon binding of a protein to a ligand or to its receptor, if this involves significant changes in protein structure. The energy barriers between the various states will likely also change, and may either increase or decrease. This will affect the kinetics of the physical degradation processes taking place. As a result, even small changes in solution conditions can have a major impact on the main degradation route.

6.4   PROTEIN PHYSICAL DEGRADATION

The physical degradation of proteins refers to any loss in bioactive protein that does not involve formation or breakage of chemical bonds and is sometimes also referred to as denaturation. It can be subdivided in four, often interrelated processes: unfolding, adsorption, aggregation, and precipitation.

6.4.1   Protein Unfolding

In the previous section, the spontaneous folding of a protein into a specific three-dimensional structure was discussed. Here, we look at the reverse process: the spontaneous *unfolding* of a protein, sometimes also referred to as denaturation. As discussed above, under physiological conditions, the most thermodynamically stable state for a single protein molecule (usually) is the folded state. Any deviation from physiological conditions, for example, a change in temperature, pH, or ionic strength, will change the intramolecular interactions within the protein, as well as the interactions between protein and water. Thus, one may expect a change in the protein folding stability upon changing the environment of the protein. As long as those changes are fully reversible upon removing the stress factor or upon administration to the patient, this may appear irrelevant for a therapeutic protein in a formulation. However, as will be discussed in more detail in Section 6.4.3, (partial) unfolding is commonly the first step in protein aggregation, which is often irreversible. Moreover, (partial) unfolding followed by subsequent refolding may trap the protein in a non-native and thus inactive conformation. Finally, unfolded proteins are often more susceptible to chemical degradation. Protein unfolding is thus, in general, detrimental to a protein's physicochemical stability. At the same time, it should be realized that a pure thermodynamic treatment of Scheme 6.1, which assumes a fully reversible process, suggests that there will always be a population of unfolded protein molecules. This is exemplified in Equation 6.2, where the equilibrium constant K for the folding process is given by the population of the native state (N) divided by that of the unfolded state (U). Although under conditions in which the native protein is stable, U will be significantly smaller than N, it is never zero, and hence there will always be a number of unfolded protein molecules.

$$\Delta G_f = -RT \ln K = -RT \ln \frac{N}{U} \qquad (6.2)$$

As may be obvious from Figure 6.1, temperature is an important factor in determining the thermodynamic stability of a protein. Increasing the temperature from physiological conditions results in increasing levels of unfolded protein due to an increasingly smaller ΔG_f. Ultimately, one will reach a temperature where $\Delta G = 0$. At this temperature, the melting temperature T_m, half of all protein molecules are unfolded. However, some proteins may contain domains that behave as independent units, and such proteins can thus have multiple melting temperatures. Also, the behavior of many proteins in solution does not comply with the required reversibility for Scheme 6.1 and Equation 6.2. For example, some proteins may be kinetically locked into a conformation, thus requiring more energy (and hence higher temperature) than the equilibrium thermodynamics calculations would suggest (Sanchez-Ruiz, 2010). Rapid aggregation upon unfolding will also affect any measurements of the thermodynamics by depleting the solution of the unfolded species. These pitfalls are important to take into account when evaluating protein unfolding data obtained from various analytical methodologies.

Unfolding as a result of thermal stress (but also other stresses) may sometimes proceed through a distinct intermediate, as exemplified in Scheme 6.2:

$$N \rightleftharpoons I \rightleftharpoons U \qquad \text{(Scheme 6.2)}$$

This intermediate is one of the local minima in the energy funnel (Figure 6.2b), and often shows almost completely native secondary structure, but a much less well-defined tertiary structure. It is commonly referred to as a molten globule, and is likely a main starting point for aggregation (Wang et al., 2010).

As noted earlier, cooling a protein solution may also cause unfolding (Privalov, 1990). In most cases, the cold denaturation temperature is much lower than 0°C, and the whole solution will have turned to ice well before that temperature is reached. In addition to the decrease in conformational stability caused by low temperature, proteins may be destabilized as the formation of ice causes both the protein and any cosolutes to become more concentrated (see also Chapter 10). Finally, as also discussed later, proteins may adsorb at the ice–water interface, with concomitant partial unfolding.

Changes in pH are another common cause of protein unfolding. Lowering or increasing the pH, away from its isoelectric point (pI, the pH at which the net charge on the protein is zero) can alter the charge states of ionizable amino acids, resulting in more electrostatic repulsion, which in turn can lead to unfolding. Also, here the unfolding process may proceed through an intermediate, although likely a different intermediate than that observed for thermal unfolding.

The influence of ions and ionic strength on protein unfolding is more complex. An increase in ionic strength will decrease any intramolecular repulsion that may be present, but will also negatively affect favorable salt bridges on the surface of the protein. Thus, both stabilizing and destabilizing effects may occur. For example, sufficiently high concentrations of certain anions at low pH have been shown to counteract the destabilizing effect of the lowered pH (Goto et al., 1990).

Salts may also exert an indirect effect on protein thermodynamic stability, possibly by affecting water structuring. There is an empirical relationship between the type of salt and its (de)stabilizing behavior (cf. Boström et al., 2005; Fesinmeyer et al., 2009; Sedlák et al., 2008; Tadeo et al., 2007; Yang et al., 2010), known as the Hofmeister series (Figure 6.3) (Hofmeister, 1888) (translated version in Kunz et al. [2004]). Originally based on their effect on salting-in and salting-out of proteins, the Hofmeister series salts also often follow roughly the same sequence in a variety of other phenomena. However, solution conditions can have a major impact on the sequence of the series and the Hofmeister series must therefore be used with caution (cf. Boström et al., 2005, 2011). Most importantly, a significant Hofmeister series effect of these salts does not show up until relatively high concentrations (>200 mM), which usually are not encountered in a typical pharmaceutical formulation.

Adding further complexity to the effect of ions is the possible presence of specific binding sites. For example, ribonuclease A has a binding site for phosphate, while many other proteins, such as insulin, and various factors in the blood coagulation process contain metal-binding sites. Thiocyanate, a destabilizing anion according to the Hofmeister series (Figure 6.3), binds to the zinc- and phenol-containing insulin hexamer, thus increasing its thermodynamic stability (Huus et al., 2006). As a result of this complexity, it is often difficult to predict the effect of adding salts on protein stability; this needs to be studied on a case-by-case basis.

Not only organic solutes like ethanol or acetonitrile but also typical preservation agents like benzyl alcohol, phenol, and metacresol will generally decrease the folding stability of a protein at physiological temperature and may thus lead to unfolding or misfolding. This destabilization is mostly due to the reduction in polarity of the solvent, which in turn reduces the hydrophobic effect. However, here also care needs

Decreased protein stability; increased salting-in of protein

| SO_4^{2-} | HPO_4^{2-} | Cl^- | NO_3^- | I^- | SCN^- | Anions |
| NH_4^+ | K^+ | Na^+ | Mg^{2+} | Ca^{2+} | Gdn^+ | Cations |

FIGURE 6.3   The Hofmeister series of anions and cations. From left to right high concentrations (>0.2–0.3 M) of these ions will result in salting-in of the protein, as well as decreased protein thermodynamic stability. Gdn^+ is the guanidinium cation. (From Boström et al., 2005; Fesinmeyer et al., 2009; Sedlák et al., 2008; Tadeo et al., 2007; Yang et al., 2010; Zhang and Cremer, 2006.)

to be taken in the generalization, as some proteins may have specific binding sites for these compounds also (two examples are given in Alford et al., 2011 and Huus et al., 2006). Even when this binding is rather weak, the tendency of the protein to unfold may actually be reduced, contrary to the expectations.

High pressure can also cause protein unfolding (Meersman et al., 2006), but the required high pressures are unlikely to be present in any phase of the protein formulation development. Interestingly, however, slightly more moderate pressures can be used to disaggregate protein aggregates, upon which the protein can then refold into its native state (St. John et al., 1999).

Finally, it is important to note that shear forces are sometimes mentioned as causing protein unfolding (cf. Ashton et al., 2009). However, this is controversial, with several papers contradicting this claim (Bee et al., 2009; Jaspe and Hagen, 2006; Thomas and Geer, 2011). An important confounding factor is that shear may desorb proteins attached to surfaces/interfaces as well as create new and fresh interfaces (e.g., at the air–water interface) where proteins may adsorb. Adsorption is an important cause of protein unfolding, and this will be discussed in the following section.

6.4.2 ADSORPTION

The presence of surfaces and interfaces is an important cause of protein unfolding and subsequent destabilization (Norde, 2008; Pinholt et al., 2011b). During its lifetime, from production to final use, a therapeutic protein encounters a variety of different surfaces and interfaces to which it may adsorb. This includes steel tanks, glass vials, rubber stoppers, Teflon tubing, but also the air–water interface and the ice–water interface. The adsorption will lead to loss of active material, which by itself is undesirable but sometimes unavoidable. More problematic is the potential formation of aggregation-prone intermediates, which may catalyze the formation of aggregates in the bulk solution (see also the following section). In this section, we will, by necessity, discuss a simplified description of the driving forces and interactions that govern protein adsorption. The reader may take up several reviews for a more in-depth discussion of the topic (Haynes and Norde, 1994; Norde, 2008; Pinholt et al., 2011b; Rabe et al., 2011).

Analogous to the protein folding process, water plays an important role as a driving force for protein adsorption. Any interface, be it solid, liquid, or gas, will affect the ability of water molecules to form the same dynamic hydrogen-bonding network as in water. This is most extreme at hydrophobic interfaces, that is, interfaces that do not allow hydrogen bonding, such as the air–water interface. Also the ice–water interface can be described as hydrophobic, as it does not allow the same dynamic hydrogen-bonding network as bulk water. Adsorption of proteins to such interfaces can thus "hide" the hydrophobic surface from the aqueous bulk, exchanging it with the more hydrophilic surface of the protein. Moreover, the protein may (partially) unfold, exposing its hydrophobic core to the hydrophobic surface, and thus release some of the conformational strain within the protein itself. Proteins that are relatively flexible, so-called "soft proteins," may even adsorb to surfaces with the same charge. That is, the electrostatic repulsion is easily overcome by release of water from the surface and the increase in conformational entropy. Proteins with less conformational flexibility ("hard proteins") generally cannot adsorb to surfaces with the same charge. Importantly, however, charge regulation at charged surfaces should also be taken into account (Hartvig et al., 2011). That is, ionizable amino acids at a surface will encounter a different environment than in the bulk solution, particularly in terms of dielectric constant, which may have profound effects on the pK_a of these amino acids. Thus, the charge on a protein may be significantly different when adsorbed, compared to that in solution.

Hydrophilic surfaces are energetically less detrimental to the water molecules adsorbed to these surfaces, but release of water molecules from the surface of the protein and from the hydrophilic surface is probably still an important driving force for protein adsorption. However, unfolding of the protein is most likely governed by electrostatics (Vermeer and Norde, 2000). Charged hydrophilic surfaces will also favor protein adsorption. In such cases, the release of ions from the surface (including that of the protein) will play an important role in the energetics of the adsorption process.

The rate of protein adsorption is normally governed by diffusion, and is thus a function of protein concentration. In many cases a plateau in adsorbed amount is reached, typically around 1 mg/m². This amount sometimes slowly increases further over time, usually due to protein aggregation at the surface. There may also be a slow structural rearrangement of the adsorbed proteins, which alters the amount of protein that fits on the surface. Thus, the adsorbed protein layer often changes its properties over time, which includes a change from a reversibly to an (kinetically) irreversibly adsorbed protein layer (Pinholt et al., 2011b).

An important aspect of protein adsorption is the (in)ability of the adsorbed protein to exchange back into the solution. Most proteins, when given sufficient time, will adsorb so strongly to any interface that they will not or only very slowly exchange back into the solution. However, the presence of other surface-active agents, including an excess of the adsorbing protein, may lead to exchange of the adsorbed protein back into the solution. Especially in the presence of hydrophobic interfaces, this can result in the introduction of misfolded protein into the solution (Vermeer and Norde, 2000). As discussed in the following section, this may well trigger aggregation.

6.4.3 PROTEIN AGGREGATION

The terms "association" and "aggregation" are commonly used to describe various processes in which protein molecules bind together ("associate" or "aggregate") into a larger unit consisting of multiple individual protein molecules. There is significant overlap between the terms in the literature, which may lead to the occasional confusion; for example, some scientists only consider the formation of nonnative (abnormal) protein associates as "aggregation," whereas others also consider native (normal) association to be a form of "aggregation." Attempts are being made to standardize the nomenclature (Joubert et al., 2011). In this section, we use the term protein aggregation for the unwanted noncovalent association of protein molecules into a larger assembly. That is, neither covalent aggregation (see Chapter 5 for a discussion on this subject), nor the association of protein molecules into a functional unit, such as is the case for hemoglobin, is considered.

Protein aggregation is a serious challenge in the development of protein therapeutics. An obvious problem is the loss of active protein due to the aggregation process. Moreover, if the aggregates are of sufficient size, they may clog injection needles and infusion lines. Finally, aggregates are heavily implicated in the development of immunogenicity (see Chapter 12). As a result, there is a general consensus that aggregation should be prevented to the maximum extent possible. Paradoxically, there have also been attempts to use aggregated proteins as a sustained release system, such as for calcitonin (Arvinte et al., 1997). Also the nonnatural peptide degarelix (marketed as Firmagon by Ferring Pharmaceuticals) forms an aggregated peptide gel upon injection, resulting in a sustained release over a month or more (Firmagon product information sheet, Ferring Pharmaceuticals).

6.4.3.1   Aggregation Mechanisms and Kinetics

Various reviews are available on the topic of protein aggregation, which are highly recommended for further background reading (Chi et al., 2003; Cromwell et al., 2006; Mahler et al., 2009; Roberts, 2007; Wang et al., 2010). By necessity, this section will only deal with the general aspects.

It is generally believed that protein aggregation is initiated by the presence of partially unfolded and/or misfolded proteins in the solution, acting as a trigger for assembly. Aggregation can thus be described as a multistep process, typically summarized through the Lumry–Eyring (1954) model in Scheme 6.3. The scheme depicts that a native protein (N) is transformed into an aggregation-prone intermediate (I), a process that may be thermodynamically reversible. Subsequently, the intermediate associates to other intermediates or native protein molecules, and together they form an aggregate (A_n) that may or may not be thermodynamically more stable than the native protein (see also Figure 6.2b and the discussion in Section 6.3.2). Although the main driving factor for aggregation appears to be generic, that is, the exposure of hydrophobic parts of the protein, the interaction between protein molecules forming the aggregate apparently also involves rather specific interactions. For example, coaddition of aggregation-prone populations of protein X and Y generally do not yield mixed aggregates, but rather aggregates of protein X, and aggregates of protein Y (Rajan et al., 2001). However, coaggregates have been observed and can thus not be ruled out (Rajan et al., 2001).

$$N \rightleftharpoons I \rightarrow A_n$$ (Scheme 6.3)

The formation of partially unfolded species can be due to a variety of stress factors, such as the presence of surfaces, a low or high pH, certain salts, or elevated temperature. The type of intermediate that is formed is likely different for each stress factor, and one may thus expect that the types of aggregates, as well as the aggregation kinetics, can differ significantly between these different stresses. That is, the aggregates may differ in size, solubility, and structure of the protein inside the aggregates depending on the stress factor involved. For example, monoclonal antibodies that are heated generally form aggregates that contain a significant amount of intermolecular beta-sheets. In contrast, these same monoclonal antibodies often form aggregates that appear to have native secondary and tertiary structure when subjected to freeze–thaw cycles (Hawe et al., 2009).

Adding to the complexity of protein aggregation in a pharmaceutical formulation is the combination of thermodynamic (conformational) and kinetic (colloidal) stability in determining aggregation kinetics. As shown in Scheme 6.3, the formation of the aggregation-prone intermediate can be described as an equilibrium process, which in most cases is heavily biased toward the native state. In order for the intermediate to trigger aggregation, its lifetime in solution must be large enough to encounter another protein molecule to associate with, forming the first aggregate. This second step in the aggregation process is more complex, as it requires two molecules to collide with sufficient energy to overcome any energy barrier. Strong repulsive interactions between the protein molecules or very high viscosity of the solution may significantly reduce the collisional frequency and energy, thus rendering this second step rate limiting. The latter is generally referred to as the colloidal stability of the protein, whereas the first step is referred to as the conformational stability. Depending on the solution conditions, either the first or second step is rate limiting. When the first step is rate limiting, the nonlinear change in ΔG_u as a function of temperature implies that the aggregation kinetics will likely not follow Arrhenius kinetics. In contrast, if colloidal stability is rate limiting, Arrhenius kinetics are more likely to be obeyed. In practice, these two extreme situations may never be encountered, resulting in some intermediate mechanism.

The conformational stability can be measured using methods such as DSC or unfolding experiments using denaturing agents like urea or guanidine. Characterizing colloidal stability requires methods that can give a measure of protein–protein interactions. This can, for example, be light scattering-based methodology (specifically, static light scattering), which allows the calculation of the so-called second virial coefficient (usually denoted B_{22}). Positive values for B_{22} indicate repulsion and negative values indicate attraction. However, there is as yet no insight into how to predict the relative importance of either colloidal or conformational stability on aggregation rate solely based on the outcome of these analyses, especially in comparisons between different proteins.

6.4.3.2 Fibrillation: A Special Case of Protein Aggregation

A special case of protein aggregation is the formation of protein aggregates in the shape of long fibers. They are best known as assemblies that are associated with a number of serious debilitating diseases, such as Alzheimer's, Parkinson's, and Creutzfeldt–Jakob diseases, in which case they are usually called amyloid fibrils. However, functional fibrils also exist, such as fibrin. Fibril formation of therapeutically relevant peptides and proteins has also been observed, such as glucagon, insulin, and calcitonin. For glucagon, fibril formation is perhaps the single most important challenge in its therapeutic use. For common therapeutic doses of this peptide (1 mg/mL), rapid fibril formation within a day is sometimes observed, putting strict constraints on the ability to process and handle aqueous glucagon formulations.

Amyloid fibrils have a number of common structural characteristics, most importantly the formation of an extensive and highly ordered intermolecular beta-sheet structure, called a cross-beta structure (Chiti and Dobson, 2006). The latter structure can be observed as a number of specific reflections in x-ray fiber diffraction analysis. Moreover, this specific cross-beta structure is also the apparent binding site for dyes like Congo Red (leading to apple-green birefringence) and Thioflavin T (yielding significant enhancement of fluorescence). These dyes generally do not give similar signals when incubated with nonfibrillar protein aggregates. The cross-beta structure may constitute only part of the protein molecule; that is, the whole protein may not have to refold into a cross-beta structure.

Taking a more macroscopic view, the protein fibril structure is built up from two levels of substructure. First, the protein molecules are encompassed in an essentially one-dimensional elongated fibrous shape known as the protofilament. Usually, two of these protofilaments twist around each other in a helical fashion similar to DNA, forming the protofibril. Multiple protofibrils may then twist around each other, forming the fibril itself. It may even be possible for these fibrils to pack together into larger bundles. Significant morphological promiscuity has been observed, depending on the solution conditions during fibrillation, such as temperature, type, and extent of shaking, protein concentration, presence of certain salts, and so on (cf. Pedersen et al., 2006). These morphological differences are likely related to the twisting density of the protofilaments and protofibrils, as well as the number of protofibrils in the fibril itself.

The kinetics of protein fibrillation poses a significant challenge in the development of protein therapeutics. For many cases, these kinetics are best described as a nucleation-dependent process, in which the buildup of a nucleus of specific size (an oligomer) is required before a rapid growth of fibrils is observed (see also Figure 6.2b). Since the nucleus likely is energetically the least favorable species on the pathway to the actual fibril, it may take a significant amount of time to accumulate sufficient amounts of this nucleus for further fibrillation to take place. The ability to slowly build up a sufficient amount of the nucleus is due to the kinetic stability of such aggregates: many different "bonds" need to be broken for the aggregate to fall apart again. During the slow buildup of this nucleus, little to no change may be observed in the solution by most analytical techniques, due to the very low concentration of this nucleus. However, as soon as sufficient nuclei are present, rapid growth is observed, resulting in exponential growth. This ultimately leads to typical sigmoidal kinetics for the fibrillation process (Figure 6.4). The fibrillation process may autocatalyze itself through a process known as secondary nucleation, which further increases the growth rate of the fibrils. In fact, a typical characteristic of protein fibrillation is that already formed fibrils catalyze the fibril formation process when added to a fresh solution of protein, a process called seeding. Cross-seeding between different proteins is generally limited, but has been observed.

It is now generally accepted that the intermediate species in the fibrillation process, the nucleus or species of slightly smaller or bigger size, is likely the most cytotoxic species (Caughey and Lansbury, 2003; Poon et al., 2009; Vetri et al., 2010). Thus, while the formation of fibrils itself is problematic, even the presence of small amounts of nuclei is a concern. As their concentration is generally low, they are difficult to observe using most standard analytical techniques. This means that a protein's ability to fibrillate must be carefully monitored.

6.4.4 Protein Precipitation

A special case of protein physical instability is protein precipitation. Most commonly, protein precipitation is the result of protein aggregation, where the aggregates

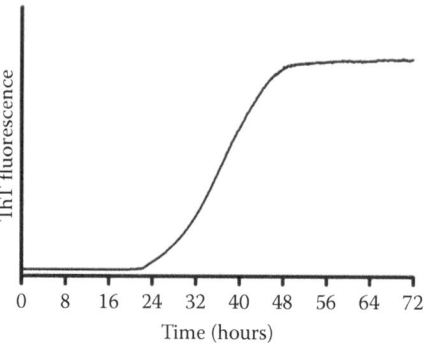

FIGURE 6.4 Typical fibrillation kinetics of a protein, as measured using thioflavin T fluorescence. During the so-called lag phase, most analytical techniques do not show any differences compared to the starting state. The growth phase is characterized by a rapid accumulation of large fibrillar species. At the end of the fibrillation process there may still be a significant amount of native protein present. The data in the example here come from fibrillation of insulin at pH 7.4 when stressed by heat and continuous shaking (Sisse Ellemann-Olesen, unpublished data).

have become too large to be suspended in solution. However, changes in the solution conditions may also result in precipitation of native protein. That is, the solubility limit of the protein can be surpassed, which may easily occur when the solution pH is changed and comes close to the pI (or IEP), or upon addition of certain excipients. Precipitating excipients include salting-out ions (see earlier discussion on Hofmeister series) or high concentrations of preferentially excluded excipients (see Section 6.5). Thus, it is of great importance to determine the nature of any observed particulate species in a protein solution, including the structural characteristics of any precipitated protein. While native protein precipitation can often be easily reversed by, for example, dilution or small adjustments in pH, formation of nonnative protein aggregates is much more problematic.

6.5   STABILIZATION STRATEGIES

A thorough understanding of the underlying mechanism of protein instability is an important benefit in rational protein formulation development. In this section, we will review a number of generic stabilization approaches; the reader is referred to Chapters 7 through 10 for more specific examples or to the literature for the many approaches that have been tested.

Preventing protein unfolding is mostly a matter of removing/preventing the stress factor(s) that causes the unfolding, if this is possible at all. Typically, this involves precautionary measures to prevent exposure to heat or extremes in pH. Note that pH adjustment involving a strong acid or strong base may cause a very low or high pH locally in the solution if not mixed properly, and should thus be avoided.

Another method to prevent protein unfolding is by increasing the stability of the native state versus the unfolded state; that is, increasing the change in Gibbs free energy upon unfolding. There are several approaches to increase the stability of the native state. For example, amino acid mutations may be introduced that increase favorable interactions in the protein fold, or add cystine bridges. However, although protein folding is understood quite well, it is still difficult to accurately predict the outcome of mutations. There will also be parts of the protein chain that should not be modified, as they are important for the activity of the protein. Finally, mutations may lead to immunogenicity due to the formation of nonhuman amino acid sequences (see also Chapter 12). This usually only leaves the core of the protein open to mutations for stability-enhancing mutations. To our knowledge, there are, at the time of writing this chapter, no mutant proteins on the market that were designed to increase thermodynamic stability.

A more practical approach is to increase the thermodynamic stability by adding ligands. In such a situation, protein unfolding will generally require removal of the ligand, an energy-consuming process, before the protein itself can unfold. Well-known examples using this approach include the addition of phenolic ligands to zinc-containing insulin as well as Ca^{2+} in the stabilization of various blood clotting factors. Potential disadvantages of this approach are the need for specific ligands for each individual protein and showing acceptability of these ligands in a pharmaceutical formulation.

Another specific method is to enhance protein self-association, an approach that will only work for those proteins that have a natural tendency to self-associate into specific multimers. Most commonly, this will involve increasing the protein concentration. Analogous to ligand-binding, unfolding in the case of a self-associated protein will first require dissociation of the multimer.

More generic for proteins is increasing the thermodynamic stability through the addition of so-called preferentially excluded excipients. For these excipients, the concentration of the compound is higher in the bulk solution than in the water layer close to the protein surface. This preferential exclusion is energetically unfavorable for the native state, in the sense that the chemical potential of the system goes up and thus solubility goes down. At the same time, an unfolded protein has a larger surface area than the folded protein, and thus the chemical potential of the unfolded protein increases more than that of the folded protein. The net result is an increase in the change in Gibbs free energy for unfolding (Figure 6.5) (Chi et al., 2003; Timasheff, 2002). Excipients known to be preferentially excluded are not only various carbohydrates and polyols, such as sucrose, trehalose, mannitol, and sorbitol, but also certain amino acids and some polymers. Certain salts can be preferentially excluded, but their behavior depends strongly on concentration and surface charge of the protein (see the prior discussion on the Hofmeister series). In addition, the concentrations required to observe any effects of preferential exclusion are usually well above 0.2 M. Such high salt concentrations are generally not desirable in a pharmaceutical formulation, and may also adversely affect the protein's colloidal stability. In contrast, high concentrations of polyols are more acceptable, and these are thus more appropriate.

Note that there are also so-called preferentially bound excipients, which will act opposite to the previously mentioned preferentially excluded excipients. It is important to realize that "bound" in this context does not refer to a direct interaction (i.e., as a ligand), but a preferential accumulation of that compound in the hydration water layer of the protein.

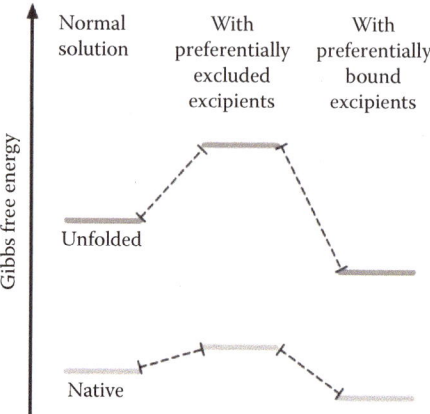

FIGURE 6.5   Effect of preferential exclusion and binding of excipients on protein unfolding thermodynamics. Addition of such excipients increase the Gibbs free energy of the native state (native), but increases the Gibbs free energy of the unfolded state (unfolded) even more. Thus, the change in Gibbs free energy of unfolding is larger in the presence of preferentially excluded excipients (middle). So-called preferentially bound excipients act opposite, reducing the Gibbs free energy of both native and unfolded state (right side), but resulting in a smaller change in Gibbs free energy of unfolding.

In theory, preferentially excluded excipients may increase the aggregation rate of a protein if the aggregation is caused by a partially unfolded species that is formed outside the bulk solution, for example, at an interface. Unfortunately, little work has been done on this area, and the importance of this potential destabilization pathway is thus poorly known.

All the methods described above decrease the propensity of proteins to unfold, and should thus theoretically prevent those protein aggregation mechanisms that depend on the initial formation of an intermediate species within the bulk solution. However, if the aggregation rate is governed mostly by the colloidal stability, increasing the conformational stability will have little stabilizing effect. In such cases, the colloidal stability will need to be manipulated, for example, by a change in ionic strength or a change in pH, both aimed at increasing protein–protein repulsion. An increase in viscosity will also increase a protein's colloidal stability. Reduced protein aggregation may also be obtained by compounds that bind to exposed hydrophobic amino acids, such as cyclodextrins and surfactants, or shield these amino acids in other ways. By doing so, they prevent protein–protein interactions through the exposed hydrophobic patches and thus increase the colloidal stability. Modification with polymers like polyethylene glycol (PEG) or polysaccharides can shield significant parts of the protein from other proteins, and is well known to reduce aggregation tendency. However, usually this is an added benefit from the modification, with the main goal being the improvement of protein pharmacokinetics (see Chapter 7).

The most common way to prevent protein adsorption is by adding excipients that preferentially adsorb to, and thereby block, those surfaces. Most commonly, this means the use of surfactants. Although surfactants are widely used in protein formulations, there are a number of practical issues to take into account. For example, surfactants may also bind to (partially) unfolded proteins; this binding may reduce aggregation tendency (cf. Bam et al., 1996, 1998), but can also decrease the thermodynamic stability. Moreover, surfactants may remove already adsorbed protein, and thus introduce partially unfolded protein molecules into the solution. Finally, many surfactants are toxic at high concentrations. It has been found that nonionic surfactants, such as the polysorbates, are among the best to reduce protein adsorption with minimal toxicity, and that the optimal concentrations are around the critical micelle concentration. A potential problem of polyether containing surfactants, as is the case for the polysorbates, is the formation of peroxides, which can cause oxidation of the protein.

Changing the properties of the surfaces in contact with the protein solution can also be a useful approach to prevent adsorption. This generally means the elimination of contact with hydrophobic surfaces to the maximum extent possible, either by replacing the surface or by modifying the surface to make it more hydrophilic. Reducing or inversing surface charge can be beneficial in reducing protein adsorption, but this may not be an easy task. Selecting proper container material throughout the life cycle of a protein pharmaceutical is thus crucial in assuring protein stability; evaluating container compatibility is thus an important aspect during the preformulation process.

Changing the properties of the protein is yet another possibility to reduce protein adsorption. For example, modifying the protein with polymers like PEG or polysaccharides may reduce the favorable contact with surfaces. However, as discussed earlier, this is an added benefit from the modification, rather than the reason to modify a protein. Moreover, reduced absorption is not necessarily observed for such modified proteins (Pinholt et al., 2011a; Pinholt et al., 2010). Alternatively, mutations may reduce the tendency of proteins to adsorb, for example, by reducing the tendency to unfold upon adsorption. However, also here the rationale for the mutations are usually driven by other goals, due to the concomitant change in various other properties of the mutant compared to the original protein.

6.6   CONCLUDING REMARKS

Protein folding and physical instability are complex phenomena, both in terms of the underlying mechanisms, as well as in their analytical characterization. Added complexity comes from the unique behavior of each individual protein. Even minor differences in amino acid sequence or posttranslational modification may result in significantly different physical instability. A thorough understanding of the behavior of the therapeutic protein through extensive analytical characterization, combined with a good understanding of the fundamental properties of proteins in general, allows the formulation scientist a more rational approach to formulation design and the identification of potential bottlenecks during production, storage, and use. This, in turn, minimizes the need to make ad hoc changes in the formulation during the development process, which minimizes costs.

Nonetheless, the power, and often the necessity, of formulation screening should not be underestimated. For example, a large-scale screening of salmon calcitonin stability in a large number of different buffers showed significant effects of seemingly small differences in buffer composition (Capelle et al., 2009). These observations may be explained by subtle differences in how certain buffer salts interact with the protein, possibly binding to specific patches on the protein, which are at present difficult to predict.

Chapters 7 through 11 of this book discuss formulation approaches in more detail, and also show that it is very much feasible to obtain a protein-containing product that is stable throughout its whole life cycle. This feat should not be underestimated; proteins, and especially those used as therapeutics, have generally not been designed by nature to be stable for such long periods of time. The fact that so many have already been formulated as stable products holds great promise that we can also stabilize those new, as yet unexplored therapeutic proteins.

REFERENCES

Alford, J.R., Fowler, A.C., Wuttke, D.S., Kerwin, B.A., Latypov, R.F., Carpenter, J.F., and Randolph, T.W., 2011. Effect of benzyl alcohol on recombinant human interleukin-1 receptor antagonist structure and hydrogen-deuterium exchange. *J. Pharm. Sci.* 100:4215–4224.

Anfinsen, C.B., 1973. Principles that govern the folding of protein chains. *Science* 181:223–230.

Anjum, F., Rishi, V., and Ahmad, F., 2000. Compatibility of osmolytes with Gibbs energy of stabilization of proteins. *Biochim. Biophys. Acta* 1476:75–84.

Arvinte, T., Cudd, A., and Phillips, J., 1997. Fibrillated calcitonin pharmaceutical compositions. Ciba-Geigy Corporation, 08/102, 117.

Ashton, L., Dusting, J., Imomoh, E., Balabani, S., and Blanch, E.W., 2009. Shear-induced unfolding of lysozyme monitored in situ. *Biophys. J.* 96:4231–4236.

Baldwin, A.J., Knowles, T.P.J., Tartaglia, G.G., et al., 2011. Metastability of native proteins and the phenomenon of amyloid formation. *J. Am. Chem. Soc.* 133:14160–14163.

Bam, N.B., Cleland, J.L., and Randolph, T.W., 1996. Molten globule intermediate of recombinant human growth hormone: Stabilization with surfactants. *Biotechnol. Prog.* 12:801–809.

Bam, N.B., Cleland, J.L., Yang, J., Manning, M.C., Carpenter, J.F., Kelley, R.F., and Randolph, T.W., 1998. Tween protects recombinant human growth hormone against agitation-induced damage via hydrophobic interactions. *J. Pharm. Sci.* 87:1554–1559.

Bee, J.S., Stevenson, J.L., Mehta, B., Svitel, J., Pollastrini, J., Platz, R., Freund, E., Carpenter, J.F., and Randolph, T.W., 2009. Response of a concentrated monoclonal antibody formulation to high shear. *Biotechnol. Bioeng.* 103:936–943.

Boström, M., Parsons, D.F., Salis, A., Niham, B.W., and Monduzzi, M., 2011. Possible origin of the inverse and direct Hofmeister series for lysozyme at low and high salt concentration. *Langmuir* 27:9504–9511.

Boström, M., Tavares, F.W., Finet, S., Skouri-Panet, F., Tardieu, A., and Ninham, B.W., 2005. Why forces between proteins follow different Hofmeister series for pH above and below pI. *Biophys. Chem.* 117:217–224.

Bryngelson, J.D., Onuchic, J.N., Socci, N.D., and Wolynes, P.G., 1995. Funnels, pathways, and the energy landscape of protein folding: A synthesis. *Proteins* 21:167–195.

Capelle, M.A.H., Gurny, R., and Arvinte, T., 2009. A high throughput protein formulation platform: Case study of salmon calcitonin. *Pharm. Res.* 26:118–128.

Caughey, B., and Lansbury, P.T., 2003. Protofibrils, pores, fibrils, and neurodegeneration: Separating the responsible protein aggregates from the innocent bystanders. *Annu. Rev. Neurosci.* 26:267–298.

Chi, E.Y., Krishnan, S., Randolph, T.W., and Carpenter, J.F., 2003. Physical stability of proteins in aqueous solution: Mechanism and driving forces in nonnative protein aggregation. *Pharm. Res.* 20:1325–1336.

Chin, W., Piuzzi, F., Dimicoli, I., and Mons, M., 2006. Probing the competition between secondary structures and local preferences in gas phase isolated peptide backbones. *Phys. Chem. Chem. Phys.* 8:1033–1048.

Chiti, F., and Dobson, C.M., 2006. Protein misfolding, functional amyloid, and human disease. *Annu. Rev. Biochem.* 75:333–366.

Cromwell, M.E.M., Hilario, E., and Jacobson, F., 2006. Protein aggregation and bioprocessing. *AAPS J.* 8: Article 66.

Dill, K.A., 1990. Dominant forces in protein folding. *Biochemistry* 29:7133–7155.

Dill, K.A., Truskett, T.M., Vlachy, V., and Hribar-Lee, B., 2005. Modeling water, the hydrophobic effect, and ion solvation. *Annu. Rev. Biophys. Biomol. Struct.* 34:173–199.

Fesinmeyer, R.M., Hogan, S., Saluja, A., Brych, S.R., Kras, E., Narhi, L.O., Brems, D.N., and Gokarn, Y.R., 2009. Effect of ions on agitation- and temperature-induced aggregation reactions of antibodies. *Pharm. Res.* 26:903–913.

Formisano, S., Johnson, M.L., and Edelhoch, H., 1977. Thermodynamics of the self-association of glucagon. *Proc. Natl. Acad. Sci. USA* 74:3340–3344.

Goto, Y., Takahashi, N., and Fink, A.L., 1990. Mechanism of acid-induced folding of proteins. *Biochemistry* 29:3480–3488.

Guo, H.H., Choe, J., and Loeb, L.A., 2004. Protein tolerance to random amino acid change. *Proc. Natl. Acad. Sci. USA* 101:9205–9210.

Hartvig, R.A., van de Weert, M., Østergaard, J., Jorgensen, L., and Jensen, H., 2011. Protein adsorption at charged surfaces: The role of electrostatic interactions and interfacial charge regulation. *Langmuir* 27:2634–2643.

Hawe, A., Kasper, J.C., Friess, W., and Jiskoot, W., 2009. Structural properties of monoclonal antibody aggregates induced by freeze-thawing and thermal stress. *Eur. J. Pharm. Sci.* 38:79–87.

Haynes, C.A., and Norde, W., 1994. Globular proteins at solid/liquid interfaces. *Colloids Surf. B* 2:517–566.

Hofmeister, F., 1888. Zur Lehre von der Wirkung der Salze. *Arch. Exp. Pathol. Pharmakol.* 24:247–260.

Huus, K., Havelund, S., Olsen, H.B., Sigurskjold, B.W., van de Weert, M., and Frokjaer, S., 2006. Ligand binding and thermostability of different allosteric states of the insulin zinc-hexamer. *Biochemistry* 45:4014–4024.

Jaspe, J., and Hagen, S.J., 2006. Do protein molecules unfold in a simple shear flow? *Biophys. J.* 91:3415–3424.

Joubert, M.K., Luo, Q., Nashed-Samuel, Y., Wypych, J., and Narhi, L.O., 2011. Classification and characterization of therapeutic antibody aggregates. *J. Biol. Chem.* 286:25118–25133.

Kauzmann, W., 1959. Some factors in the interpretation of protein denaturation. *Adv. Protein Chem.* 14:1–63.

Kryshtafovych, A., and Fidelis, K., 2009. Protein structure prediction and model quality assessment. *Drug Discov. Today* 14:386–393.

Kunz, W., Henle, J., and Ninham, B.W., 2004. Zur Lehre von der Wirkung der Salze (about the science of the effect of salts): Franz Hofmeister's historical papers. *Curr. Opin. Colloid Interf. Sci.* 9:19–37.

Lumry, R., and Eyring, H., 1954. Conformation changes of proteins. *J. Phys. Chem.* 58:110–120.

Mahler, H-C., Friess, W., Grauschopf, U., and Kiese, S., 2009. Protein aggregation: Pathways, induction factors and analysis. *J. Pharm. Sci.* 98:2909–2934.

Meersman, F., Dobson, C.M., and Heremans, K., 2006. Protein unfolding, amyloid fibril formation and configurational energy landscapes under high pressure conditions. *Chem. Soc. Rev.* 35:908–917.

Mirny, L., and Shakhnovich, E., 2001. Evolutionary conservation of the folding nucleus. *J. Mol. Biol.* 308:123–129.

Norde, W., 2008. My voyage of discovery to proteins in flatland…and beyond. *Colloids Surf. B* 61:1–9.

Orengo, C.A., and Thornton, J.M., 2005. Protein families and their evolution—a structural perspective. *Annu. Rev. Biochem.* 74:867–900.

Pedersen, J.S., Dikov, D., Flink, J.L., Hjuler, H.A., Christiansen, G., and Otzen, D.E., 2006. The changing face of glucagon fibrillation: Structural polymorphism and conformational imprinting. *J. Mol. Biol.* 355:501–523.

Pinholt, C., Bukrinsky, J.T., Hostrup, S., Frokjaer, S., Norde, W., and Jorgensen, L., 2011a. Influence of PEGylation with linear and branched PEG chains on the adsorption of glucagon to hydrophobic surfaces. *Eur. J. Pharm. Biopharm.* 77:139–147.

Pinholt, C., Fanø, M., Wiberg, C., Hostrup, S., Bukrinsky, J.T., Frokjaer, S., Norde, W., and Jorgensen, L., 2010. Influence of glycosylation on the adsorption of thermomyces lanuginosus lipase to hydrophobic and hydrophilic surfaces. *Eur. J. Hosp. Pharm. Sci.* 40:273–281.

Pinholt, C., Hartvig, R.A., Medlicott, N.J., and Jorgensen, L., 2011b. The importance of interfaces in protein drug delivery—Why is protein adsorption of interest in pharmaceutical formulations? *Expert Opin. Drug Deliv.* 8:949–964.

Poon, S., Birkett, N.R., Fowler, S.B., Luisi, B.F., Dobson, C.M., and Zurdo, J., 2009. Amyloidogenicity and aggregate toxicity of human glucagon-like peptide-1 (hGLP-1). *Protein Peptide Lett.* 16:1548–1556.

Privalov, P.L., 1990. Cold denaturation of proteins. *Crit. Rev. Biochem. Mol. Biol.* 25:281–305.

Rabe, M., Verdes, D., and Seeger, S., 2011. Understanding protein adsorption phenomena at solid surfaces. *Adv. Colloid Interf. Sci.* 162:87–106.

Rajan, R.S., Illing, M.E., Bence, N.F., and Kopito, R.R., 2001. Specificity in intracellular protein aggregation and inclusion body formation. *Proc. Natl. Acad. Sci. USA* 98:13060–13065.

Rees, D.C., and Robertson, A.D., 2001. Some thermodynamic implications for the thermostability of proteins. *Protein Sci.* 10:1187–1194.

Roberts, C.J., 2007. Non-native protein aggregation kinetics. *Biotechnol. Bioeng.* 98:927–938.

Robertson, A.D., and Murphy, K.P., 1997. Protein structure and the energetics of protein stability. *Chem. Rev.* 97:1251–1267.

Rose, G.D., and Wolfenden, R., 1993. Hydrogen bonding, hydrophobicity, packing, and protein folding. *Annu. Rev. Biophys. Biomol. Struct.* 22:381–415.

Sanchez-Ruiz, J.M., 2010. Protein kinetic stability. *Biophys. Chem.* 148:1–15.

Sedlák, E., Stagg, L., and Wittung-Stafshede, P., 2008. Effect of Hofmeister ions on protein thermal stability: Roles of ion hydration and peptide groups? *Arch. Biochem. Biophys.* 479:69–73.

St John, R.J., Carpenter, J.F., and Randolph, T.W., 1999. High pressure fosters protein refolding from aggregates at high concentration. *Proc. Natl. Acad. Sci. USA* 96:13029–13033.

Tadeo, X., Pons, M., and Millet, O., 2007. Influence of the Hofmeister anions on protein stability as studied by thermal denaturation and chemical shift perturbation. *Biochemistry* 46:917–923.

Thomas, C.R., and Geer, D., 2011. Effects of shear on proteins in solution. *Biotechnol. Lett.* 33:443–456.

Timasheff, S.N., 2002. Protein hydration, thermodynamic binding, and preferential hydration. *Biochemistry* 41:13473–13482.

Uversky, V.N., and Dunker, A.K., 2010. Understanding protein non-folding. *Biochim. Biophys. Acta* 1804:1231–1264.

Uversky, V.N., Nielsen Garriques, L., Millett, I.S., Frokjaer, S., Brange, J., Doniach, S., and Fink, A.L., 2003. Prediction of the association state of insulin using spectral parameters. *J. Pharm. Sci.* 92:847–858.

Vermeer, A.W.P. and Norde, W., 2000. CD spectroscopy of proteins adsorbed at flat hydrophilic quartz and hydrophobic teflon surfaces. *J. Colloid Interface Sci.* 225:394–397.

Vetri, V., Carrotta, R., Picone, P., Di Carlo, M., and Militello, V., 2010. Concanavalin A aggregation and toxicity on cell cultures. *Biochim. Biophys. Acta* 1804:173–183.

Wang, W., Nema, S., and Teagarden, D., 2010. Protein aggregation—pathways and influencing factors. *Int. J. Pharm.* 390:89–99.

Yang, Z., Liu, X-J., Chen, C., and Halling, P.J., 2010. Hofmeister effects on activity and stability of alkaline phosphatase. *Biochim. Biophys. Acta* 1804:821–828.

Zhang, Y., and Cremer, P.S., 2006. Interactions between macromolecules and ions: The Hofmeister series. *Curr. Opin. Chem. Biol.* 10:658–663.

7 Peptide and Protein Derivatives

Kristian Strømgaard and Thomas Høeg-Jensen

CONTENTS

7.1 INTRODUCTION

One of the primary barriers of using peptides and proteins as therapeutics is their short half-life in vivo. There are mainly two ways by which peptides and proteins undergo unspecific clearance: either by proteolytic degradation or by kidney-mediated filtration. Generally, the circulation times of peptides and proteins can be increased by modifying (derivatizing) the molecules, which guard them from proteolytic degradation or provide molecular weights above the cutoff for kidney-mediated clearance (approximately 45 kDa). Typically, such derivatizations include addition of large polymers or molecular motifs that impose binding to native proteins such as albumin. Additionally, such derivatizations have demonstrated protective properties regarding immunogenicity issues, and some types of peptides and protein derivatives have also been applied in targeting specific tissues or organs, such as liver or tumor.

Until recently, most peptides and protein derivatives have been generated by chemical modification of the parent peptide or protein, for example, by exploiting the reactivity of the amino terminal or reactive side chains. This is typically achieved by recombinant expression of the peptide and protein, followed by chemical derivatization. Obviously, solid-phase peptide synthesis (SPPS) (see Chapter 1) allows more fundamental changes of a peptide, but the method is generally limited to peptides up to 50 amino acids and is not considered feasible for large-scale production. However, methods that allow more fundamental modifications of larger peptides and proteins are emerging, such as unnatural mutagenesis and peptide/protein ligation strategies, where derivatization can be introduced directly into the protein or a more specific chemical handle can be introduced for further, more selective derivatization.

Methods for residue-specific incorporation of close analogues of natural amino acids have existed for several years, where depletion of one amino acid and addition of another, structurally related unnatural amino acid allows incorporation of this amino acid (Budisa, 2006). This is generally achieved by applying auxotroph strains of bacteria. A typical example is the incorporation of selenomethionine (SeMet in place of methionine), which is used in structural studies of proteins, as the heavy atom selenium may help solve the phase problem in x-ray crystallography (Walden, 2010). Similarly, auxotroph strains have been useful for labeling of recombinant proteins for NMR studies, where ^{15}N- and/or ^{13}C-labeled amino acids are introduced in a residue-specific manner (Whittaker, 2007).

In general, the use of chemical methods, rather than conventional genetic methods, to alter protein structure and function offers exciting possibilities. Genetic methods are generally limited to the use of the 20 proteinogenic amino acids, which contain a finite number of functional groups. Nature has increased the diversity by a large number of posttranslational modifications (PTMs) (Figure 7.1), which normally are not attainable in a site-specific manner by genetic methods. Thus, by combining the principles and tools of chemistry with the synthetic strategies and processes of living organisms, it is possible to generate proteins with a plethora of novel functions. Such proteins can obviously be explored as improved therapeutics.

As previously mentioned, incorporation of PTMs is a key feature for therapeutic properties of proteins, and site-selective introduction of PTMs furthermore allows addressing biological importance of such modification in great detail. PTMs that can be introduced are both group additions, such as phosphorylation, glycosylation, and lipidation, as well as modification of parent amino acids, including methylation, acetylation, and hydroxylation (Figure 7.1) (Walsh, 2006). Another class of modifications is those that incorporate biophysical probes or reactive handles for further derivatization, for example, site-specific labeling with ^{13}C- or ^{15}N-labeled amino acids for biological NMR studies, incorporation of fluorescent amino acids, or amino acids containing photolabile groups such as benzophenone (Figure 7.1). Amino acids with reactive groups for selective derivatization are also of great interest; such groups could be azides or alkyne groups (Figure 7.1) to be used in the Huisgen 1,3-dipolar cycloaddition to furnish 1,2,3-triazoles, also known as "click chemistry" (Meldal and Tornøe, 2008). Another example is the introduction of ketone functionalities that can be selectively modified, for example, with polyethylene glycol (PEG) moieties. Finally, very subtle changes of proteins, such as incorporation of D-amino acids, close analogues of encoded amino acids (Figure 7.1), and modification of the amide backbone can also be introduced. This allows very fine-tuned studies of, for example, ligand–receptor interactions and protein function in general and has been described as "protein medicinal chemistry."

FIGURE 7.1 (See color insert.) Modification and incorporation of amino acids that can be achieved by applying chemically based methods. (a) Posttranslational modifications, such as hydroxylation and phosphorylation. (b) Biophysical probes, such as benzophonene, that is, a photolabile group and an alkyne derivative that can be used in "click chemistry." (c) Close analogues of encoded amino acids, here arginine, where subtle modification of the guanidine group is included.

7.2 TECHNOLOGIES FOR DERIVATIZATION

The primary method for derivatizing peptide and protein therapeutics is to chemically modify the parent peptide or protein. Although a number of developments are currently taking place to significantly expand the protein chemical toolbox (Baslé et al., 2010), a number of challenges remain, most importantly with regard to reactivity and selectivity. In addition, a number of technologies are emerging that address some of these issues and use fundamentally different strategies, that is, rather than modifying the parent peptide or protein, a modified protein is generated directly (Hahn and

Muir, 2005). Applying the latter principles, it should be possible to generate proteins containing, in principle, any functionality. In the following, we will focus on two general methods that allow this: unnatural mutagenesis, which allow site-specific incorporation of unnatural amino acids into protein using the cells' own protein synthesis machinery, and ligation-based strategies, which allow semisynthesis of proteins and thereby incorporation of a wide range of unnatural functionalities into proteins.

7.2.1 UNNATURAL MUTAGENESIS

In 1989, a biosynthetic in vitro method that allowed site-specific incorporation of unnatural amino acids into proteins was introduced based on earlier work on nonsense suppression (Bain et al., 1989; Noren et al., 1989). The term "nonsense suppression" refers to the use of stop (nonsense) codons and suppressor transfer RNA (tRNA) that recognize stop codons. The method is based on the fact that only one of three stop codons in the genetic code is necessary for termination of protein synthesis and the two unused stop codons can then be exploited for introduction of unnatural amino acids.

The primary challenge in this technology is the generation of the modified suppressor tRNA with the unnatural amino acid. Once generated, the aminoacylated-tRNA (aa-tRNA) is recognized by the mRNA carrying the specific stop codon, and the unnatural amino acid is incorporated into the protein at the specific position. Based on this principle, two slightly different methodologies have been developed to incorporate unnatural amino acids site-specifically into proteins: one method applies tRNAs that are chemically aminoacylated with the unnatural amino acid of interest, and the aa-tRNA is subsequently applied in an expression system to generate the protein of interest (Strømgaard et al., 2004). The other method employs development of pairs of orthogonal tRNA and aminoacyl-tRNA synthetases (aaRS), where the latter is developed so that it selectively recognizes and aminoacylate an unnatural amino acid (Liu and Schultz, 2010; Xie and Schultz, 2006).

In the chemical aminoacylation of tRNA, a dinucleotide, 5'-O-phosphoryl-2'-deoxycytidylyl-(3'→5')adenosine (pdCpA), is prepared by chemical synthesis and subsequently aminoacylated with the unnatural amino acid of interest. The aa-tRNA is obtained by ligation of a truncated tRNA, where a dinucleotide at the 3' terminus is missing from the prepared aminoacylated dinucleotide. In an in vitro expression system, the aa-tRNA is added to the media, and when whole cell expression systems are used, the aa-tRNA is injected into the cell. This results in synthesis and surface expression of the target protein containing the unnatural amino acid in the specified position.

The methodology has been used to incorporate a large number of structurally diverse unnatural amino acids, representing a large variety of functionalities, into proteins (Dougherty, 2000; Strømgaard et al., 2004). In most cases, the unnatural amino acids have not only been α-amino acids, but non-α-amino acids and α-hydroxy acids have also been incorporated, the latter introducing an amide-to-ester mutation in the protein backbone. These studies have also demonstrated that translation factors and the ribosome are compatible with a wide range of unnatural amino acids. The method has proven particularly useful in studies of ligand-gated ion channels, where proteins are expressed in *Xenopus* oocytes, and the molecular details of, for example, cation–π interactions (Dougherty, 2008) and cis–trans isomerization of proline residues (Lummis et al., 2005) have been explored.

Although the technology has obvious and wide-ranging potential, it also has substantial limitations. First and foremost, the yields of protein and thereby the amounts generated are very low, and it is therefore hardly feasible to use the method for production of proteins on a therapeutic scale. In addition, highly sensitive detection systems, such as electrophysiology or fluorescence, are generally required for using the methodology in basic studies of proteins, and the generation of aa-tRNAs requires highly skilled persons in both chemistry and molecular biology. Thus, in order to overcome some of these limitations, a modified method for incorporation of unnatural amino acids into proteins in vivo was introduced. In this approach, a custom-made pair of tRNA and aaRS is genetically introduced into a cell, and the aaRS is engineered so that it only recognizes the unnatural amino acid and efficiently acylates the corresponding tRNA (Liu and Schultz, 2010; Xie and Schultz, 2006). Subsequently, the unnatural amino acid, which has to be nontoxic and cell permeable, is added to the growth media, taken up by the host organism, and incorporated into the protein by the specific tRNA/aaRS pair (Wang et al., 2001). This technology has been successfully applied in both yeast and eukaryotic cells (Chin et al., 2003; Liu et al., 2007) and allows generation of proteins with an unnatural amino acid in reasonable to good yields (Ryu and Schultz, 2006). The primary challenge of this technology is that specific aaRS has to be generated for each unnatural amino acid, which is done by extensive mutational studies and rounds of positive and negative selections.

The technology has been applied to a number of model proteins and has been used to site-specifically incorporate a very wide range of different amino acids. In addition, a fully autonomous bacterium, E. coli, has been engineered so it could synthesize p-amino-phenylalanine, and a specific tRNA/aaRS pair was introduced, which allowed incorporation into myoglobin (Mehl et al., 2003). The technology holds substantial prospective for therapeutic proteins, as it allows generation of uniquely modified proteins, such as those containing a ketone functionality (Wang et al., 2003), which subsequently can be selectively derivatized with PEG or other relevant modifying agents. Although site-selective introduction of PTMs by this methodology has proven difficult, acetylated and methylated residues have been introduced, but there are fewer examples of group addition, although introduction of sulfotyrosine has been demonstrated (Liu and Schultz, 2006).

A potential limitation of this technology is that the genetic code only contains three stop codons, which limits the theoretic numbers of different unnatural amino acids that can be incorporated in a single protein to two. Two different approaches have been developed to overcome this: using extended codons and frame shift suppression or by generating orthogonal ribosomes. In the former approach, an mRNA containing an extended codon consisting of four or five bases is read by a modified aa-tRNA containing the corresponding extended anticodon (Sidido et al., 2005). In certain species, some naturally occurring codons are rarely used and the amount of their corresponding tRNA is low. This has been used in the design of four-base codons, which are derived from these rarely used codons, to minimize the competition between the four-base anticodon tRNA and endogenous tRNA. This has been used to incorporate unnatural amino acids into proteins in E. coli as well as to incorporate two different unnatural amino acids into two different sites of a single protein (Taki et al., 2002). Most recently, an ingenious approach has been introduced, whereby orthogonal ribosomes are used to enhance the expression efficiency of proteins with unnatural amino acid (Wang et al., 2007), and in combination with four-base codons, it was also demonstrated that two different amino acids can be introduced into the same protein (Neumann et al., 2010).

7.2.2 PEPTIDE/PROTEIN LIGATION

A conceptually different strategy for modification of proteins is to employ methods based on SPPS (see Chapter 1) for generation of proteins. This would allow incorporation of principally any amino acid and thus circumvent problems of incorporating D-amino acids, which is not feasible by unnatural mutagenesis. SPPS has, in a few cases, been applied for the synthesis of proteins, although yields are generally low. One of the first examples was the synthesis of ribonuclease A (124 residues) (Gutte and Merrifield, 1971) and since then a few other proteins have been prepared by this approach, most notably HIV protease (99 residues), which enabled structural characterization of the protein with inhibitors bound (Schneider and Kent, 1988).

However, SPPS is generally limited to the preparation of up to 50 amino acid peptides, whereas most proteins are considerably larger. Therefore, there has been considerable interest in developing methods that are not confined to these restrictions, and in 1994, a strategy for preparation of proteins from peptide fragments was introduced, called native chemical ligation (NCL) (Figure 7.2) (Dawson et al., 1994). In NCL, two or more unprotected peptide fragments can be ligated together, generating a (native) cysteine residue in the ligation site. The ligation requires a peptide with a C-terminal protein thioester and a peptide with an N-terminal cysteine residue: the thiolate of the N-terminal cysteine attacks the C-terminal thioester to affect transthioesterification, followed by formation of an amide bond after $S{\rightarrow}N$ acyl transfer (Figure 7.2). The reaction, which uses unprotected peptides, takes place in aqueous buffer and generally proceeds in good to excellent yield (Dawson and Kent, 2000).

Thus, NCL is a very useful approach for the total chemical synthesis of proteins and has been used for the preparation of numerous proteins, including glycoproteins. Another example is the synthesis of an analogue of erythropoietin (EPO), which was derivatized with monodisperse polymer moieties in order to improve duration of action in vivo. The 166-residue protein was prepared by ligation of four peptide fragments, two of which were modified with the polymer and the EPO analogue, displaying improved properties in vivo compared to EPO (Kochendoerfer et al., 2003).

In 1998, an extension of the NCL principles was introduced, called expressed protein ligation (EPL) (Muir et al., 1998), which utilizes the same reaction as in NCL, but in contrast to NCL, one of the components is a protein rather than a peptide (Figure 7.2). The protein is expressed as a so-called intein construct, which allows the formation of a protein thioester, which subsequently can be reacted with a peptide with an N-terminal cysteine in an NCL generating a full-length protein (Figure 7.2). Thus, the EPL methodology combines the advantages of molecular biology with chemical peptide synthesis and enables the addition of unnatural functionality to a recombinant protein framework (Muir, 2003; Muralidharan and Muir, 2006).

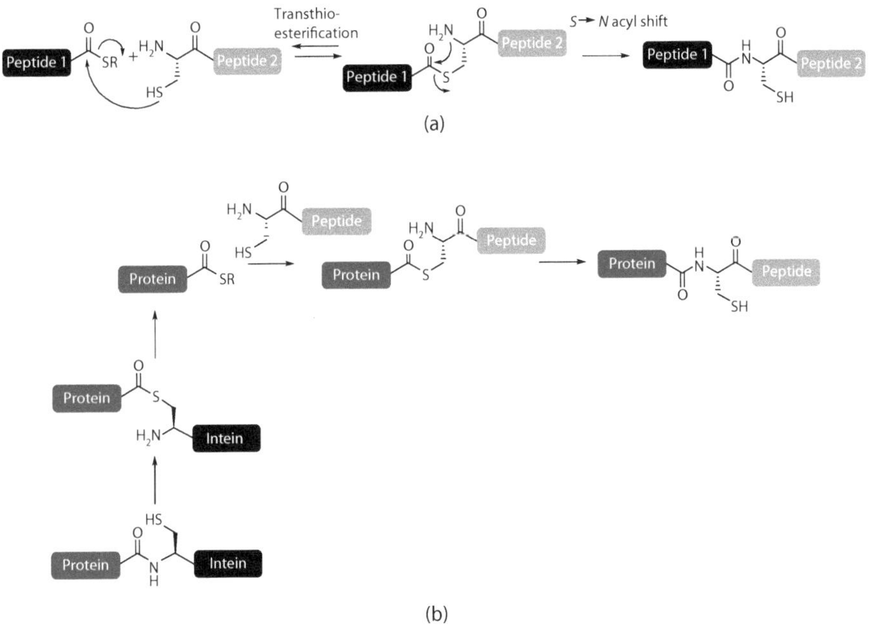

FIGURE 7.2 Principles of native chemical ligation (NCL) and expressed protein ligation (EPL). (a) NCL: A peptide with an *N*-terminal cysteine and another peptide with a *C*-terminal thioester can be ligated together. Initially, a reversible transthioesterification takes place and subsequently *S*→*N* acyl shift, leading to a cysteine in the ligation site. (b) EPL applies the same principles, but one of the reactants is a recombinantly expressed protein, which allows semisynthesis of larger proteins.

EPL has been applied in basic studies of several proteins and shows great potential for the generation of therapeutic proteins. It is particularly useful for the site-specific introduction of PTMs. The latter have been elegantly demonstrated in studies of histones, which are important for storage of DNA and have flexible *N*-terminal tails that are heavily modified by PTMs. A range of site-specific ubiquilated and phosphorylated histones have been prepared by EPL and used to provide seminal information of histones (Chatterjee and Muir, 2010).

7.2.3 CHEMICAL MODIFICATION OF PROTEINS

Besides the two classes of technologies just described, which can be used to alter the very basic structure of proteins, a range of chemistry-based methods exists, which allows modification of the parent protein structure (Baslé et al., 2010). The endogenous protein structure can be exploited for residue- and potentially site-selective derivatization. However, a number of conditions have to be fulfilled for allowing chemical transformation; most importantly, reactions should generally be carried out under aqueous conditions and ambient temperature.

One of the most frequent ways of modifying protein structure is by reacting at cysteine residues. This can often be successfully carried out with either none or minimal changes to the parent protein. The thiol moiety of cysteines is more nucleophilic than a primary amine and can be selectively modified with a range of reactive groups such as maleimides and iodoacetyls (Figure 7.3). Thus, the general advantage is that the thiol of cysteine allows for selective modification relative to the other proteinogenic amino acids, and the frequency by which cysteine occurs in proteins is relatively low, thus often allowing selective modification of specific cysteine residues. Even if

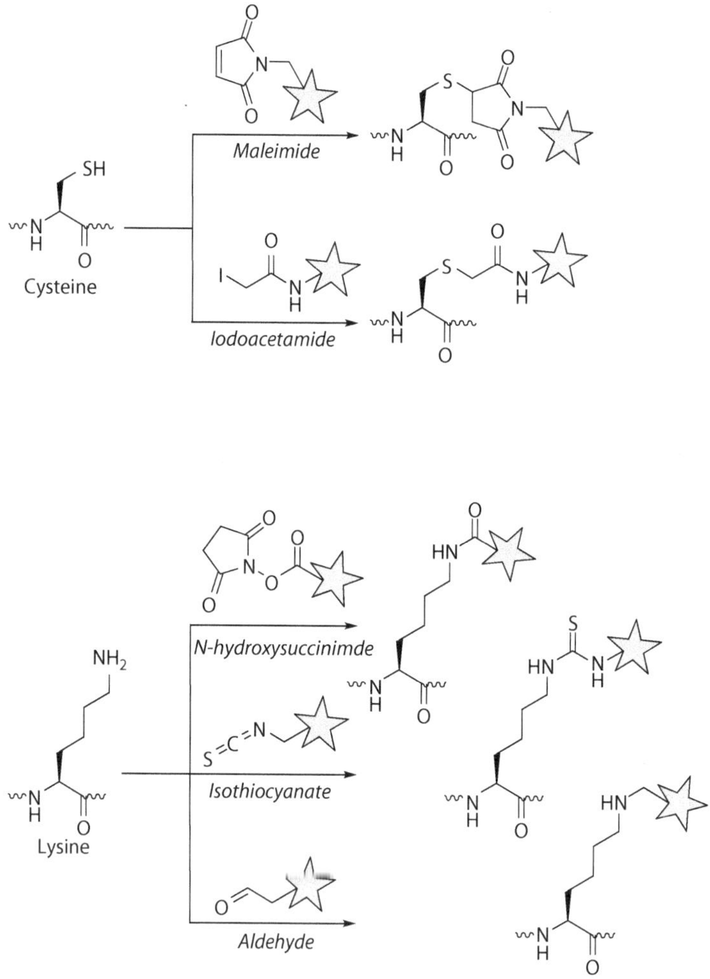

FIGURE 7.3 Examples of common methods for derivatization of cysteine and lysine residues. One of the most common methods for derivatizing proteins, in general, is the reaction between the thiol side chain of cysteine and a substituted maleimide generating a stable thioether linkage. A similar linkage is generated from the alkylation of cysteine by iodoacetamide. The primary amino group of lysine can be derivatized, for example, by reaction with N-hydroxy-succinimide, generating an amide bond, or by reaction with isothiocyanates, generating thioureas. Finally, the primary amine can also react with aldehydes in a reductive amination, generating a secondary amine.

a protein contains more than one cysteine residue, these might have different accessibility, which can allow selective modification of certain residues.

Lysine residues, which bear a side chain with a primary amine, can selectively react with a range of electrophiles such as N-hydroxysuccinimide and isothiocyanates, generating amides and thioureas, respectively (Figure 7.3). In addition, lysine readily undergoes reductive amination by reaction with aldehydes, intermediary generating an imine (Sciffs base) and subsequently a secondary amine. The primary challenge for selective reaction at lysine residues is avoiding reaction at the N-terminal, which can often be managed by carefully controlling pH of the reaction media.

A particularly promising strategy for derivatizing proteins is introduction of PEG moieties, known as PEGylation, which can help in reducing immunogenicity, increasing the circulatory time by reducing renal clearance, and also provide water solubility to hydrophobic proteins. PEGylation is generally performed by reaction of a reactive derivative of PEG with the target protein, typically with side chains of lysine or cysteine, or by reaction at the C- or N-terminal of the protein or peptide. An alternative way of improving protein and peptide therapeutics is by adding lipids to the protein, which likewise can substantially improve the properties of a therapeutic peptide or protein. Addition of lipids to a protein framework can readily be achieved by ligation strategies, but in some cases the differential reactivity of specific residues can be exploited, and this has been used for a range of therapeutic peptides and proteins.

In some cases, site-selective modification of proteins can be achieved by using simple chemical reactions comparable to those used in conventional organic synthesis. Similar to residue-specific reactions, a general requirement is that such reaction should be compatible with the aqueous (buffer) conditions, in which the protein is present, and recently a number of robust and water-compatible reactions have evolved. However, such methods often require introduction of selective handles, as previously described, in order to be sufficiently selective, but once a reactive handle has been incorporated, a wealth of chemical reactions can be performed. The example of "click chemistry," that is, a 1,3-dipolar cycloaddition between an azide and an alkyne providing a 1,2,3-triazole, has gained increasing interest due to the chemoselectivity of the reaction, the low background reaction, and the quantitative yields (Prescher and Bertozzi, 2005). Another prominent example is the Staudinger reaction, which is a phosphine-mediated reduction of an azide to an amine, also known as an aza-Wittig reaction, that has been used particularly in protein glycosylation studies (Prescher and Bertozzi, 2005). However, in both cases, significant challenges remain to apply these in the generation of therapeutic proteins.

A principally different way to modify parent peptides or proteins is to apply enzymes to selectively modify proteins. Enzymes have an inherent advantage in that they efficiently add or remove groups to proteins and they are often highly specific for certain sequences (consensus motifs) of amino acids, so modifications are often site-specific. Enzymes are often also highly substrate specific, that is, kinases add only phosphate groups to serine, threonine, and tyrosine; thus modification of the enzyme is required if other groups have to be introduced. However, some enzymes, such as glycosyltransferases, which transfer carbohydrates to serines or asparagines, have broader substrate specificity, but in this case it can be desirable to modify the enzyme to achieve increased reactivity for specific carbohydrates.

Enzymes are particularly useful for furnishing proteins with tailor-made PTMs, which are often essential for regulation and dynamics of biological activity. For example, many therapeutic proteins are glycosylated, and controlling glycosylation patterns of proteins is a key challenge (Rich and Withers, 2009; Solá and Griebenow, 2010). Glycosyltransferases are enzymes that can catalyze the transfer of a monosaccharide to a protein, and using directed evolution it was possible to modify the transferases so that monosaccharides of interest could be selectively added to a protein framework (Rich and Withers, 2009). Enzymes can also be exploited to introduce polymeric moieties; an example is transglutaminase (TGase), which can be used to obtain selective PEGylation. TGase catalyzes transfer reactions between the γ-carboxamide group of glutamine residues and primary amines, resulting in the formation of γ-amides of glutamic acid and ammonia. Thus, by using an amino-derivative of PEG as substrate for the enzymatic reaction, it is possible to covalently bind the PEG polymer to a therapeutic protein (Fontana et al., 2008).

7.3 CASE STORIES

In recent years, the number of peptide and protein derivatives that have entered the market has experienced a dramatic increase, and the ratio of biopharmaceuticals relative to traditional small molecule drugs is increasing. Thus, in the following representative and recent case stories, general trends in developing new biopharmaceuticals will be given rather than an exhaustive review of all therapeutic peptide and protein derivatives.

7.3.1 PEG-MODIFIED PROTEINS

PEG is the most commonly used size modulator for proteins and has been extensively employed for improvement of therapeutic proteins (Knop et al., 2010; Veronese and Pasut, 2005). PEG moieties with a molecular weight larger than approximately 2 kDa are mixtures of heterogeneous chain lengths; however, these moieties are available in average sizes such as 5, 10, 20, and 40 kDa. The apparent biophysical size of PEGylated proteins is larger than their actual molecular weight (hydrodynamic radius) due to hydration of the PEG moiety.

In order to covalently couple the PEG-chain to the protein, the terminal position of the PEG-chain must be activated, typically as active ester, which can be reacted with protein amino groups, such as lysine side chains or the N-terminal. However, proteins typically contain numerous lysines, and site-specific attachment can therefore be a major challenge. Generally, PEGylated protein products are heterogeneous mixtures both in terms of site(s) of modification as well as PEG chain length. Obviously, PEGylations of different protein sites/residues may affect the biological activities differently. Alternatively, a PEG-maleimide can be reacted with a cysteine thiol, as previously described, as cysteine residues carrying a free exposed thiol are less common in proteins or can be engineered in a given protein by mutation, potentially allowing site-specific chemistry (Weerapana et al., 2010). However, interaction of free cysteine with native disulfide bonds can occur and lead to misfolded protein.

The attractive properties of the polyether PEG are inertness toward metabolism and "hydrophilic" character, which serve in keeping proteins in solution. PEG is highly soluble in water, which supports the notion of PEG as hydrophilic. However PEG is also soluble in organic solvents like dichloromethane, and PEGylated products elute later than their corresponding non-PEGylated products from reverse-phase chromatography columns indicating increased hydrophobicity of PEGylated proteins; hence "amphiphilic" is a more appropriate term for PEG moieties.

The metabolic inertness of PEG can in some cases lead to intracellular accumulation, which may lead to long-term toxicity (Bendele et al., 1998; Pipe, 2010). This issue is most relevant for drugs that are used in chronic treatments, like growth hormone replacement. PEGylated drugs are in some cases circulated mainly via the lymph system, which can be a problem or an advantage depending on the drug target (Gu et al., 2010).

The first marketed PEGylated protein was Adagen®, which entered the market in 1990. This is an adenosine deaminase (40 kDa) used for immunomodulation containing multiple 5 kDa PEG moieties and is administered once a week by intramuscular injection (Ellis et al., 2003). Pegasys® is a PEGylated version of interferon-α-2a (40 kDa), which is used for treatment of hepatitis C, and is heterogeneously derivatized with 40 kDa two-branched PEG moieties (Reddy et al., 2002). Finally, a PEGylated version of the fast-acting insulin lispro (Lys^{B28}, Pro^{B29}-insulin) is in clinical development as basal insulin; that is, the addition of the PEG moiety significantly increases the half-life of the modified insulin (Beals et al., 2009).

An interesting novel, nonheterogeneous PEG-modified protein is ARX-201, which is a long-acting growth-hormone derivative developed by the company Ambrx. Here unnatural mutagenesis has been used to express a recombinant form of human growth hormone that contains a ketone functionality, which allows precise spatial positioning of the site of PEG attachment to human growth hormone (Cho et al., 2011).

7.3.2 ALBUMIN BINDING AND OLIGOMER ENGINEERING

Albumin is a 67 kDa protein, which binds and transports fatty acids and small molecules, including a number of drugs, through the circulation. The plasma concentration of albumin is very high (500–800 µM or 3–5 g/L), and the half-life of albumin is approximately 20 days in humans. Acylation of peptides and proteins with fatty acids and related structures have been shown to impose binding of the acylated derivatives to human serum albumin, resulting in markedly increased circulation times. The binding affinities (K_D) of fatty acid acylated molecules to albumin are in the micromolar range, leaving the free fractions of acylated peptide or protein drugs in the range of only a few percent. The clearance of acylated peptide or protein via this free fraction allows administration of these derivatives to be in the range of once a day to once a week, which is generally a substantial improvement relative to their parent peptide or protein.

Acetylation with fatty acids has, in particular, been explored in the management of diabetes, where both insulin and glucagon-like peptide-1 (GLP1) have been derivatized in various ways. Insulin is a two-chain (A and B chain of 21 and 30 amino acids, respectively) miniprotein, and the predominant physiological role is stimulation of cellular glucose uptake. Native insulin is cleared via insulin receptor binding and cellular internalization, and insulin has a half-life of ca. 1 hour from subcutaneous administration. Most diabetic patients need an external supply of both fast-acting insulin for removing the carbohydrates ingested with a meal as well as long-acting insulin (basal insulin) for controlling glucose levels between meals. Insulin detemir (Levemir®) is an insulin analogue where the C-terminal threonine has been removed (desB30) from human insulin and subsequently the side chain of C-terminal lysine (B29) has been acylated with tetradecanoic (myristic) acid (Havelund et al., 2004) (Figure 7.4). The fatty acid derivatization imposes albumin binding, and in addition the oligomer state of insulin is modulated, which contributes to the prolonged action. The extended half-life of insulin detemir means that it is used as once or twice daily basal insulin. Another advantage of insulin detemir is its solubility in both the formulation and upon injection, as opposed to basal formulations of native insulin, which are suspensions. The glucose metabolic effect of insulin detemir is more reproducible and predictable than with the suspension-based basal insulins, due to a buffering effect from albumin binding and the fully soluble character (Vague et al., 2003). Finally, patients using insulin detemir generally have a smaller weight gain compared to other types of insulin, and insulin detemir is therefore described as "weight neutral" (Fritsche and Häring, 2004), which might be related to the albumin binding and relative liver specificity, that is, that insulin detemir is cleared by liver to a higher extent compared to other insulins (Smeeton et al., 2009).

Insulin degludec is a novel human insulin derivative, which, similar to insulin determir, is truncated in the C-terminal (desB30) and also acylated in the side chain of the C-terminal lysine, but with hexadecanedioic acid via an L-γ-glutamyl linker (Jonassen et al., 2010a) (Figure 7.5). The fatty diacid not only ensures strong albumin affinity, but also mediates formation of multihexamers of insulin degludec upon subcutaneous injection ($M_W > 5$ MDa). The formation of soluble insulin degludec multihexamers provides a depot in the subcutaneous tissue from which it is continuously and slowly absorbed to provide a steady and ultralong action profile, making insulin degludec a very promising basal insulin. Insulin degludec can furthermore be coformulated with fast-acting insulin aspart with retainment of the individual absorption profiles for the fast and slow component (Jonassen et al., 2010b). Similar coformulation with other modern insulins (detemir or glargine/lantus) does not work as desired, since these form mixed hexamers and thus blunted profiles, whereas

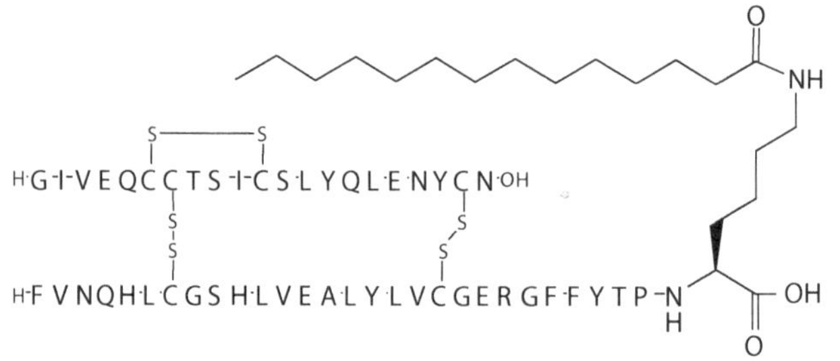

FIGURE 7.4   Insulin detemir, B29-myristoyl desB30 insulin, soluble and long-acting.

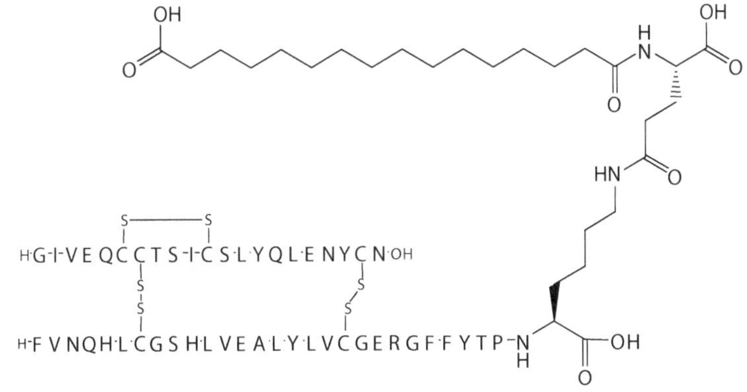

FIGURE 7.5 Insulin degludec, B29-hexadecandioyl-γ-L-glutamyl desB30 insulin, long-acting and suitable for coformulation with fast-acting insulin for meal coverage.

insulin degludec oligomers are discrete from insulin aspart oligomers in coformulations of the appropriate composition. Notably, the fatty *di*acid and the L-γ-glutamyl linker are critical for the described properties of insulin degludec.

Liraglutide (Victoza®) is glucagon-like peptide-1 (GLP-1, 7-37, Arg34) acylated with hexadecanoic (palmitic) acid via an L-γ-glutamyl linker in Lys26 (Knudsen et al., 2000) (Figure 7.6). GLP-1 is an incretin hormone secreted from intestinal L-cells in response to meals, which stimulates insulin secretion from pancreatic beta-cells in a glucose-dependent fashion as well as increases satiety, leading to weight-reducing effects. GLP-1 has in animal models been shown to increase beta-cell mass, which are the insulin-producing cells (Knudsen, 2010). It is therefore a highly promising candidate for the treatment of type 2 diabetes. Native GLP-1 is cleared exceptionally fast with a half-life of only a few minutes via degradation by dipeptidyl peptidase-IV. Liraglutide is protected from enzymatic degradation via binding to albumin and is suitable to be administered once a day. The alternative GLP-1 drug exenatide (a proteolytically stabilized GLP-1 analogue without albumin binding) must be given twice daily, and its glucose and weight-lowering effects are less pronounced (Buse et al., 2009).

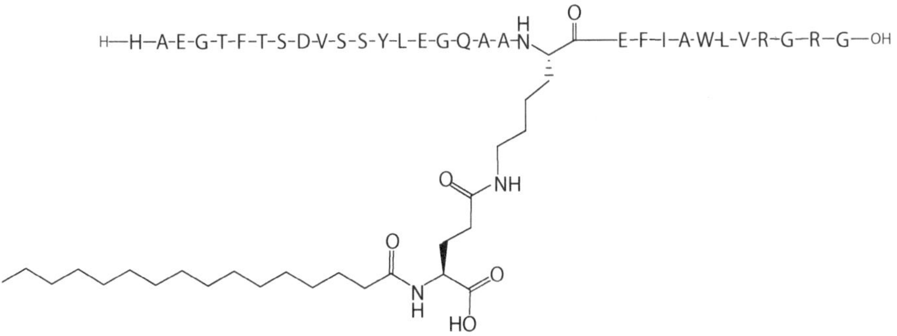

FIGURE 7.6 Liraglutide, Arg34 Lys26-hexadecanoyl-γ-L-glutamyl glucagon-like peptide-1, 7-37, type 2 diabetes drug.

7.3.3 GLYCOSYLATIONS

In general, PTMs are of utmost importance for protein biopharmaceuticals; however, site-specific control of PTMs remains a major challenge (Walsh, 2010). The most common PTM is protein glycosylation, which in several cases has a marked effect on protein properties; one of the best known examples is human blood, where several types exist because of small variations in the glycosylation patterns of hemoglobin. Thus, protein glycosylation has been substantially exploited for engineering of therapeutic proteins, and since glycosylations are rather small or modest (<20 residues/site), the effects result from modified affinities to binding partners and/or increased solubility rather than size modulation.

EPO stimulates red blood cell production and is used against anemia among other cancer chemotherapies, although EPO might be better known for its abuse in sports. EPO with engineered glycosylation patterns, darbepoetin alfa (Aranesp®), which carries increased sialic acid levels, enables prolonged action in vivo, with a half-life of 25 hours in contrast to 8 hours for native EPO when administered intravenously (Maxwell, 2002). The darbepoetin alfa was approved but achieved only limited commercial success and has now been withdrawn from the market in Europe. Antithrombin-III (AT-III) is a circulating protein known to bind heparin and related sulfated polysaccharides. Organon exploited binding of heparylated (Figure 7.7) insulin to AT-III as a means of prolonging insulin action in vivo (de Kort et al., 2008). The affinities are in the nanomolar range, and heparyl-insulin half-lives of 5 hours compared to <1 hour for native insulin were observed (intravenous administration). The concept has not yet matured to clinical applications, but the nanomolar affinities could be advantageous over the micromolar affinities typically observed with fatty acid-albumin systems.

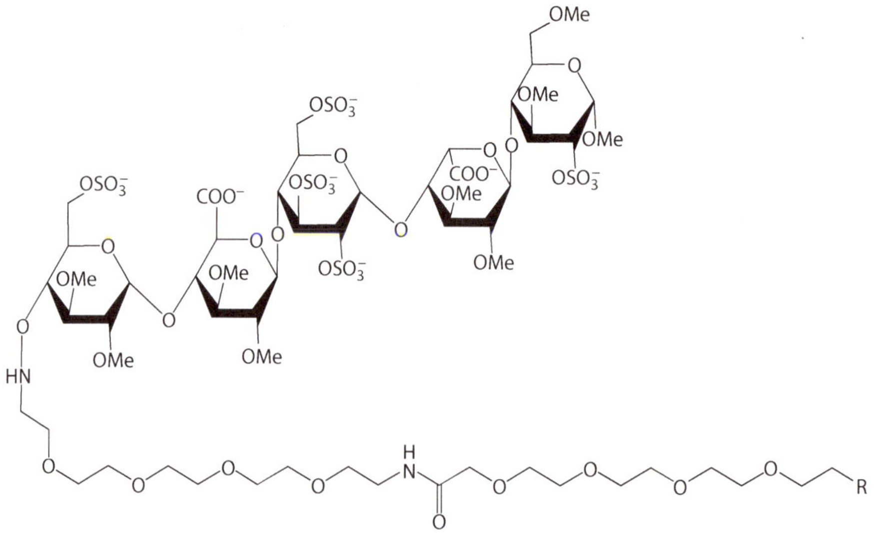

FIGURE 7.7   Heparyl-based antithrombin-III binding motif for prolonged circulation.

7.4 CONCLUSIONS AND OUTLOOK

Peptide synthesis and protein engineering have reached a state where most PTMs can be reproduced in a laboratory scale and where nonnatural amino acids can be introduced by modified recombinant methods. Site-selective derivatization is however still a challenge in many cases, for example, where numerous identical amino acids are present in a peptide or protein. This problem can at times be circumvented by using semisynthetic methods, for example, ligation methods for fusing recombinant sequences with synthetic sequences. Site-selective chemistry can, in many cases, be achieved by using enzymatic modifications with synthetic building blocks, but there is room for further development in this area.

Prolongation of peptide drug action in vivo can be secured by, for example, PEGylation, but retainment of biological activity can be a challenge, and accumulation of PEG moieties can reach unacceptable levels, in particular for chronically administered drugs. Engineering of suitably degradable polymers may solve these problems, but such research is still in the early phase. Some of the methods mentioned above require highly skilled personnel and are not yet suitable for industrial production scale. Recombinant use of nonnatural amino acids and native ligation methods have not yet been used for production of marketed drugs, but the potential is there and such development should not be unrealistic.

REFERENCES

Bain, J. D., Diala, E. S., Glabe, C. G., Dix, T. A. and Chamberlin, A. R. 1989. Biosynthetic site-specific incorporation of a non-natural amino acid into a polypeptide. *J. Am. Chem. Soc.* 111:8013–4.

Baslé, E., Joubert, N. and Pucheault, M. 2010. Protein chemical modification on endogenous amino acids. *Chem. Biol.* 17:213–27.

Beals, J. M., Butler, G. B., Doyle, B., Hansen, R. J., Li, S., Shirani, S. and Zhang, L. 2009. Eli Lilly & Co. Patent application number WO2009152128.

Bendele, A., Seely, J., Richey, C., Sennello, G. and Shopp, G. 1998. Renal tubular vacuolation in animals treated with polyethylene-glycol-conjugated proteins. *Toxicol. Sci.* 42:152–7.

Budisa, N. 2006. *Engineering the Genetic Code.* Weinheim: Wiley-VCH.

Buse, J. B., Rosenstock, J., Sesti, G., Schmidt, W. E., Montanya, E., Brett, J. H., Zychma, M. and Blonde, L. 2009. Liraglutide once a day versus exenatide twice a day for type 2 diabetes: A 26-week randomised, parallel-group, multinational, open-label trial (LEAD-6). *Lancet* 374:39–47.

Chatterjee, C. and Muir, T. W. 2010. Chemical approaches for studying histone modifications. *J. Biol. Chem.* 285:11045–50.

Chin, J. W., Cropp, T. A., Anderson, J. C., Mukherji, M., Zhang, Z. and Schultz, P. G. 2003. An expanded eukaryotic genetic code. *Science* 301:964–7.

Cho, H., Daniel, T., Buechler, Y. J., Litzinger, D. C., Maio, Z., Putnam, A.-M. H., Kraynova, V. S., et al. 2011. Optimized clinical performance of growth hormone with an expanded genetic code. *Proc. Natl. Acad. Sci. USA* 108:9060–9065.

Dawson, P. E. and Kent, S. B. 2000. Synthesis of native proteins by chemical ligation. *Annu. Rev. Biochem.* 69:923–60.

Dawson, P. E., Muir, T. W. Clark-Lewis, I. and Kent, S. B. 1994. Synthesis of proteins by native chemical ligation. *Science* 266:776–9.

de Kort, M., Gianotten, B., Wisse, J. A. J., Bos, E. S., Eppink, M. H. M., Mattaar, E., Vogel, G. M. T., et al. 2008. Conjugation of ATIII-binding pentasaccharides to extend the half-life of proteins: Long-acting insulin. *Chemmedchem* 3:1189–93.

Dougherty, D. A. 2000. Unnatural amino acids as probes of protein structure and function. *Curr. Opin. Chem. Biol.* 4:645–52.

Dougherty, D. A. 2008. Cys-loop neuroreceptors: Structure to the rescue? *Chem. Rev.* 108:1642–53.

Ellis, K. M., Mazzoni, L. and Fozard, J. R. 2003. Role of endogenous adenosine in the acute and late response to allergen challenge in actively sensitized Brown Norway rats. *Br. J. Pharm.* 139:1212–8.

Fontana, A., Spolaore, B., Mero, A. and Veronese, F. M. 2008. Site-specific modification and PEGylation of pharmaceutical proteins mediated by transglutaminase. *Adv. Drug Deliv. Rev.* 60:13–28.

Fritsche, A. and Häring, H. 2004. At last, a weight neutral insulin? *Int. J. Obes.* 28(S2):S41–6.

Gu, B., Xie, C., Zhu, J., He, W. and Lu, W. 2010. Folate-PEG-CKK2-DTPA, a potential carrier for lymph-metastasized tumor targeting. *Pharm. Res.* 27:933–42.

Gutte, B. and Merrifield, R. B. 1971. The synthesis of ribonuclease A. *J. Biol. Chem.* 246:1922–41.

Hahn, M. E. and Muir, T. W. 2005. Manipulating proteins with chemistry: A cross-section of chemical biology. *Trends Biochem. Sci.* 30:26–34.

Havelund, S., Plum, A., Ribel, U., Jonassen, I., Volund, A., Markussen, J. and Kurtzhals, P. 2004. The mechanism of protraction of insulin detemir, a long-acting, acylated analog of human insulin. *Pharm. Res.* 21:1498–504.

Jonassen, I., Havelund, S., Ribel, U., Hoeg-Jensen, T., Steensgaard, D. B., Johansen, T., Haahr, H., Nishimura, E. and Kurtzhals, P. 2010a. Insulin degludec is a new generation ultra-long acting basal insulin with a unique mechanism of protraction based on multi-hexamer formation. *Diabetes* 59(S1):A11.

Jonassen, I., Hoeg-Jensen, T., Havelund, S. and Ribel, U. 2010b. Ultra-long acting insulin degludec can be combined with rapid-acting insulin aspart in a soluble co-formulation. *J. Peptide Sci.* 16(S1):32.

Knop, K., Hoogenboom, R., Fischer, D. and Schubert, U.S. 2010. Poly(ethylene glycol) in drug delivery: Pros and cons as well as potential alternatives. *Angew. Chem. Int. Ed.* 49:6288–303.

Knudsen, L. B. 2010. Liraglutide: The therapeutic promise from animal models. *Int. J. Clin. Pract.* S167:4–11.

Knudsen, L. B., Nielsen, P. F., Huusfeldt, P. O., Johansen, N. L., Madsen, K., Pedersen, F. Z., Thøgersen, H., Wilken, M. and Agersø, H. 2000. Potent derivatives of glucagon-like peptide-1 with pharmacokinetic properties suitable for once daily administration. *J. Med. Chem.* 43:1664–9.

Kochendoerfer, G. G., Chen, S. Y., Mao, F., Cressman, S., Traviglia, S., Shao, H., Hunter, C. L., et al. 2003. Design and chemical synthesis of a homogeneous polymer-modified erythropoiesis protein. *Science* 299:884–7.

Liu, W., Brock, A., Chen, S., Chen, S. and Schultz, P. G. 2007. Genetic incorporation of unnatural amino acids into proteins in mammalian cells. *Nat. Methods* 4:239–44.

Liu, C. C. and Schultz, P. G. 2006. Recombinant expression of selectively sulfated proteins in Escherichia coli. *Nat. Biotechnol.* 24:1436–40.

Liu, C. C. and Schultz, P. G. 2010. Adding new chemistries to the genetic code. *Annu. Rev. Biochem.* 79:413–44.

Lummis, S. C., Beene, D. L., Lee, L. W., Lester, H. A., Broadhurst, R. W. and Dougherty, D. A. 2005. Cis-trans isomerization at a proline opens the pore of a neurotransmitter-gated ion channel. *Nature* 438:248–52.

Maxwell, A. P. 2002. Novel erythropoiesis-stimulating protein in the management of the anemia of chronic renal failure. *Kidney Int.* 62:720–9.

Mehl, R. A., Anderson, J. C., Santoro, S. W., Wang, L., Martin, A. B., King, D. S., Horn, D. M. and Schultz, P. G. 2003. Generation of a bacterium with a 21 amino acid genetic code. *J. Am. Chem. Soc.* 125:935–9.

Meldal, M. and Tornøe, C. W. 2008. Cu-catalyzed azide-alkyne cycloaddition. *Chem. Rev.* 108:2952–3015.

Muir, T. W. 2003. Semisynthesis of proteins by expressed protein ligation. *Annu. Rev. Biochem.* 72:249–89.

Muir, T. W., Sondhi, D., and Cole, P. A. 1998. Expressed protein ligation: A general method for protein engineering. *Proc. Natl. Acad. Sci. USA* 95:6705–10.

Muralidharan, V. and Muir, T. W. 2006. Protein ligation: An enabling technology for the biophysical analysis of proteins. *Nat. Methods* 3:429–38.

Neumann, H., Wang, K., Davis, L, Garcia-Alai, M. and Chin, J. W. 2010. Encoding multiple unnatural amino acids via evolution of a quadruplet-decoding ribosome. *Nature* 464:441–4.

Noren, C. J., Anthony-Cahill, S. J., Griffith, M. C. and Schultz, P. G. 1989. A general method for site-specific incorporation of unnatural amino acids into proteins. *Science* 244:182–8.

Pipe, S.W. 2010. Go long! A touchdown for factor VIII? *Blood* 116:153–4.

Prescher, J. A. and Bertozzi, C. R. 2005. Chemistry in living systems. *Nat. Chem. Biol.* 1:13–21.

Reddy, K. R., Modi, M. W., and Pedder, P. 2002. Use of peginterferon alfa-2a (40 KD) (Pegasys®) for the treatment of hepatitis C. *Adv. Drug Deliv. Rev.* 54:571–86.

Rich, J. R. and Withers, S. G. 2009. Emerging methods for the production of homogeneous human glycoproteins. *Nat. Chem. Biol.* 5:206–15.

Ryu, Y and Schultz, P. G. 2006. Efficient incorporation of unnatural amino acids into proteins in Escherichia coli. *Nat. Methods* 3:263–5.

Schneider, J. and Kent, S. B. H. 1988. Enzymatic activity of a synthetic 99 residue protein corresponding to the putative HIV-1 protease. *Cell* 54:363–8.

Sisido, M., Ninomiya, K., Ohtsuki, T. and Hohsaka, T. 2005. Four-base codon/anticodon strategy and non-enzymatic aminoacylation for protein engineering with non-natural amino acids. *Methods* 36:270–78.

Smeeton, F., Moradie, F. S., Jones, R. H., Westergaard, L., Haahr, H., Umpleby, A. M. and Russell-Jones, D. L. 2009. Differential effects of insulin detemir and neutral protamine Hagedorn (NPH) insulin on hepatic glucose production and peripheral glucose uptake during hypoglycaemia in type 1 diabetes. *Diabetologia* 52:2317–23.

Solá, R. J. and Griebenow, K. 2010. Glycosylation of therapeutic proteins: An effective strategy to optimize efficacy. *BioDrugs* 24:9–21.

Strømgaard, A., Jensen, A. A. and Strømgaard, K. 2004. Site-specific incorporation of unnatural amino acids into proteins. *Chembiochem* 5:909–16.

Taki, M., Hohsaka, T., Murakami, H., Taira, K. and Sisido, M. 2002. Position-specific incorporation of a fluorophore-quencher pair into a single streptavidin through orthogonal four-base codon/anticodon pairs. *J. Am. Chem. Soc.* 124:14586–90.

Vague, P., Selam, J. S., Skeie, S., Leeuw, I. D., Elte, J. W. F., Haahr, H., Kristensen, A. and Draeger, E. 2003. Insulin detemir is associated with more predictable glycemic control and reduced risk of hypoglycemia than NPH insulin in patients with type 1 diabetes on a basal-bolus regimen with premeal insulin aspart. *Diabetes Care* 26:590–6.

Veronese, F. M. and Pasut, G. 2005. PEGylation, successful approach to drug delivery. *Drug Discov. Today* 10:1451–8.

Walden, H. 2010. Selenium incorporation using recombinant techniques, *Acta Crystallogr. D Biol. Crystallogr.* 66:352–7.

Walsh, C. T. 2006. Posttranslational modification of proteins: Expanding nature's inventory. Greenwood Village, CO: Roberts and Company Publishers, pp. 490.

Walsh, G. 2010. Post-translational modifications of protein biopharmaceuticals. *Drug Discov. Today* 15:773–80.

Wang, L., Brock, A., Herberich, B., and Schultz, P. G. 2001. Expanding the genetic code of Escherichia coli. *Science* 292:498–500.

Wang, K., Neumann, H., Peak-Chew, S. Y. and Chin, J. W. 2007. Evolved orthogonal ribosomes enhance the efficiency of synthetic genetic code expansion. *Nat. Biotechnol.* 25:770–7.

Wang, L., Zhang, Z., Brock, A. and Schultz, P. G. 2003. Addition of the keto functional group to the genetic code of Escherichia coli. *Proc. Natl. Acad. Sci. USA* 100:56–61.

Weerapana, E., Wang, E., Simon, G. M., Richter, F., Khare, S., Dillon, M. B. D., Bachovchin, D. A., Mowen, K., Baker, D. and Cravatt, B. F. 2010. Quantitative reactivity profiling predicts functional cysteines in proteomes. *Nature* 468:790–5.

Whittaker, J. W. 2007. Selective isotopic labeling of recombinant proteins using amino acid auxotroph strains. *Methods Mol. Biol.* 389:175–88.

Xie, J. and Schultz, P. G. 2006. A chemical toolkit for proteins—An expanded genetic code. *Nat. Rev. Mol. Cell. Biol.* 7:775–82.

Entropy
Gibbs free energy

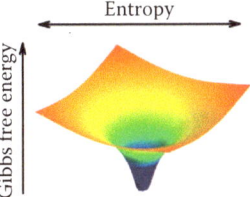

FIGURE 6.2 (a) Energy landscape of a protein. An idealized folding funnel for a single protein molecule. At the high Gibbs free energy end (top of picture), the protein molecule can adopt many different conformations; the width of the funnel can be viewed as a measure of the conformational entropy. At the bottom of the funnel a singular folded state with very limited conformational entropy is present.

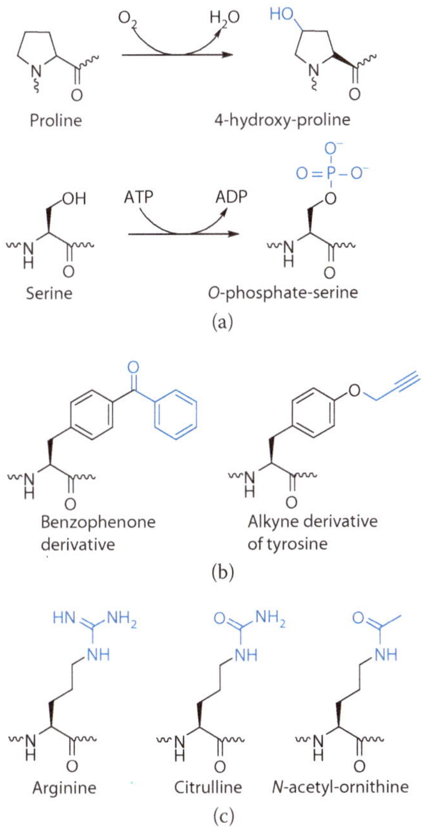

Proline
4-hydroxy-proline
ATP
ADP
Serine
O-phosphate-serine
(a)
Benzophenone derivative
Alkyne derivative of tyrosine
(b)
Arginine
Citrulline
N-acetyl-ornithine
(c)

FIGURE 7.1 Modification and incorporation of amino acids that can be achieved by applying chemically based methods. (a) Posttranslational modifications, such as hydroxylation and phosphorylation. (b) Biophysical probes, such as benzophonene, that is, a photolabile group and an alkyne derivative that can be used in "click chemistry." (c) Close analogues of encoded amino acids, here arginine, where subtle modification of the guanidine group is included.

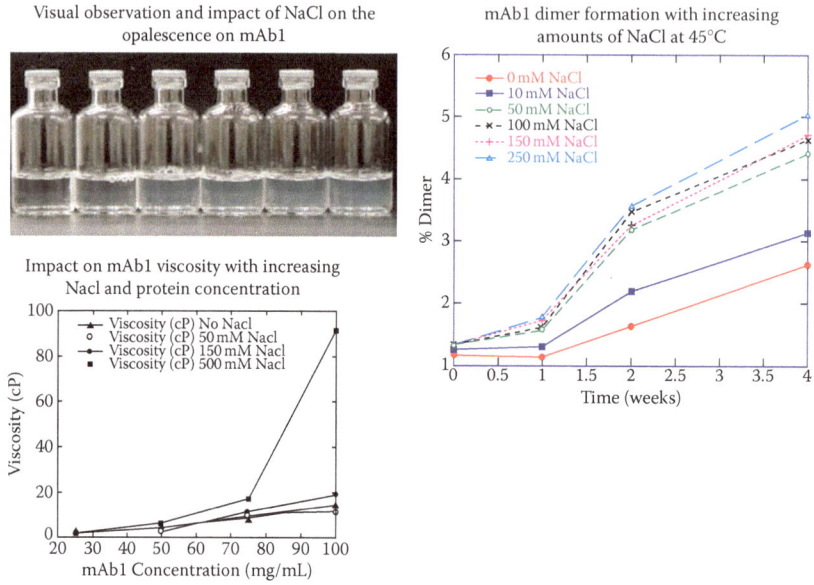

FIGURE 8.2 MAb1 dimer formation with increasing amounts of NaCl at 45°C. (Reprinted from Wang, N., et al., *BioPharm* April, 36–47, 2009. With permission.)

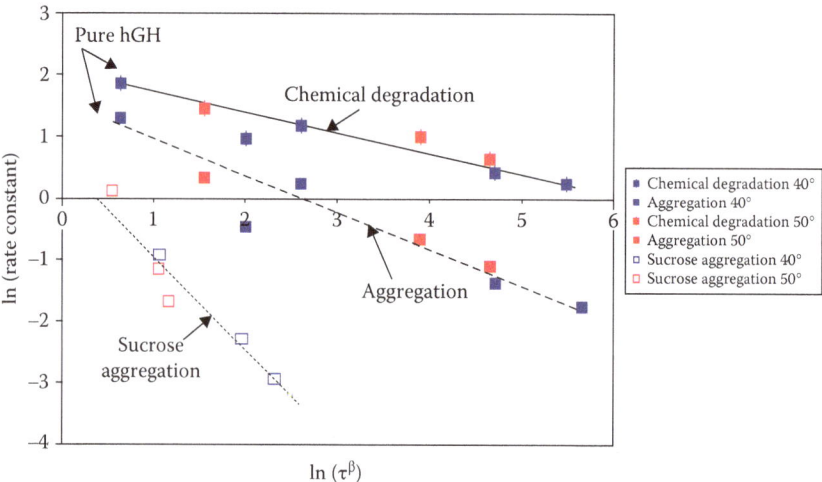

FIGURE 10.6 Correlation between stability and structural relaxation time constant (τ^β) of hGH formulations. Formulations studied include pure hGH, stachyose (1:1), and trehalose (1:1, 1:3, and 1:6). These ratios are weight ratios of protein–saccharide. The triangles represent aggregation data in sucrose formulations. The rate constant is from square root of time (months) kinetics, and τ is in hours. Note that sucrose formulations are more stable than expected based upon the structural relaxation time constant. (Reproduced from Pikal, M.J., et al., *J. Pharm. Sci.* 97, 5106, 2008.)

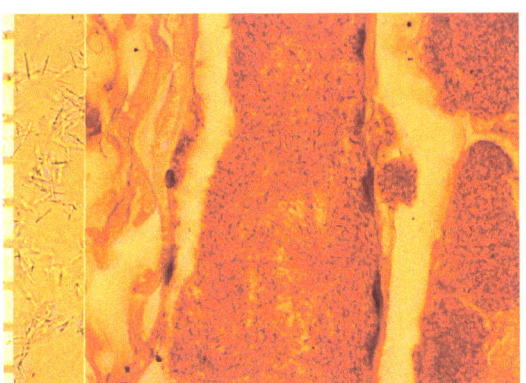

FIGURE 13.4 A formulation of U200 NPH insulin is shown in the left panel. The same formulation 1 hour following administration into the subcutaneous tissue in a pig is shown in the right panel (same scale).

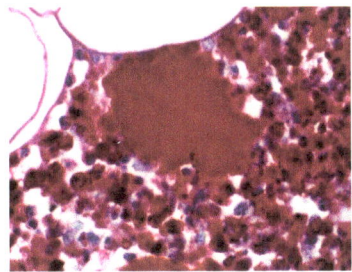

FIGURE 13.5 The NPH insulin heap (solid red) is surrounded by macrophages (purple) and adipose tissue (white).

FIGURE 13.6 The *moose* and the *pine* represent two identical "doses" (number of pixels) of NPH insulin in the subcutaneous depot. The two heaps have different "surface areas" (boundary toward white pixels) resulting in different kinetics. The figure shows the dissolution of the two NPH insulin heaps over time. The corresponding plasma insulin curves are given in Figure 13.7.

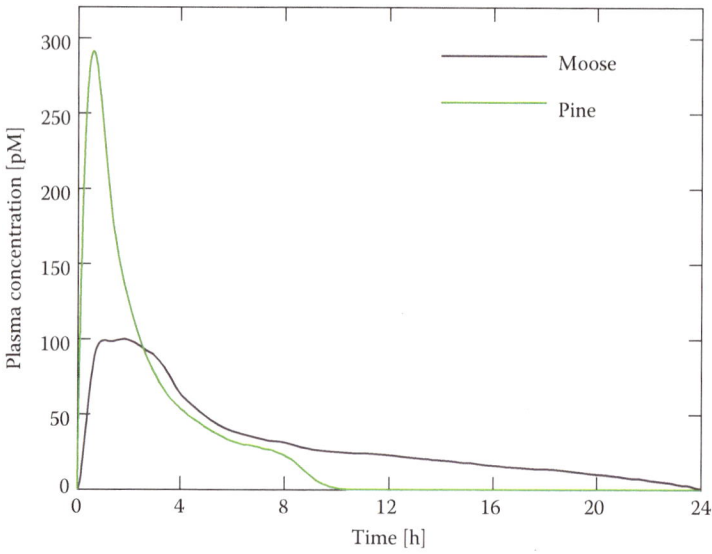

FIGURE 13.7 Plasma insulin profiles corresponding to the *moose* and the *pine* shown in Figure 13.6.

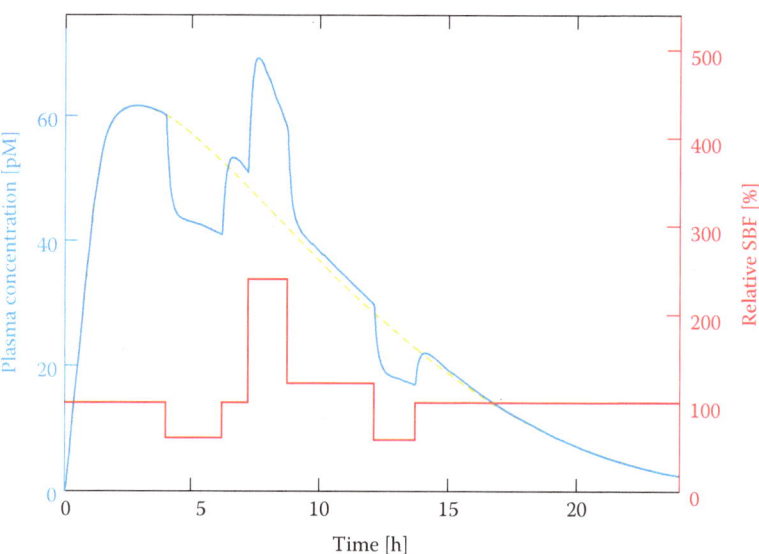

FIGURE 13.8 Simulated insulin plasma profile and the corresponding relative subcutaneous blood flow (SBF) during five phases of sleep following a subcutaneous injection of NPH insulin 4 hours prior to the first phase of sleep. (Modified from Claessen, S., and Mortensen, T., Local subcutaneous blood flow and the effect on insulin absorption. Technical University of Denmark, Copenhagen, 2009.)

8 Peptides and Proteins as Parenteral Solutions

Michael J. Akers and Michael R. DeFelippis

CONTENTS

8.1   OVERVIEW AND INTRODUCTION

The purpose of this chapter is to provide practical guidance to formulation scientists charged with the development of stable, manufacturable, and elegant sterile solution dosage forms of peptides and proteins (henceforth referred to as biopharmaceuticals). Ready-to-use solution dosage forms comprise the largest percentage of sterile dosage forms in the marketplace. The solution formulation must be resistant to both physical and chemical degradation and maintain all microbiological quality attributes (sterility and freedom from pyrogenic and particulate matter contamination) throughout the shelf-life of the product.

This chapter will cover the basics of chemical stabilization, physical stabilization, and microbiological quality of biopharmaceuticals in solutions. We will place more emphasis on the approaches used to solve the problems of protein/peptide solution formulation rather than discussing the nature of degradation mechanisms that is covered elsewhere in this text and many other references. We also will have some coverage of packaging and manufacturing of biopharmaceutical solution dosage forms in the spirit of emphasizing that scientists developing these dosage forms must be equally concerned with the formulation, the package, and the manufacturing process.

The global market for therapeutic biopharmaceuticals (therapeutic proteins, monoclonal antibodies, and vaccines) has skyrocketed since 1982, when the first protein of recombinant DNA technology was approved by FDA. There are over 100 unique protein products in the market; approximately half are stored as ready-to-use solutions and half as freeze-dried powders. Approximately 300 peptides and proteins are being studied in the clinic, most of which are monoclonal antibodies. Biopharmaceuticals now represent about 50% of all newly approved drug products. In 2006 biopharmaceuticals accounted for about 10% of total pharmaceutical product revenue; in 2015, they will account for about 15%. In 2009 the biologics market was valued at $124 billion; in 2015 the sales of biopharmaceutical products is projected to be above $182 billion. Growth of monoclonal antibodies will be nearly fourfold that of therapeutic proteins and double that of vaccines between 2010 and 2014 (Bain and Shortmoor, 2010).

It is reasonable to assume that each of these peptide and protein products, commercial or in clinical study, has had to overcome and control stability issues in solutions. The type of stability issue and the degree of complexity of the degradation mechanism differ for each type of biopharmaceutical molecule. Yet, approaches to resolve instability issues in solutions are relatively universal.

Basic guidelines to consider in the development of injectable solutions of biopharmaceuticals are summarized as follows:

1. A thorough understanding of the physical and chemical properties of the protein or peptide bulk drug substance should be acquired. Well-documented and valid analytical techniques are available for studying these properties in solutions (see Chapter 4). Effects of temperature, pH, shear, oxygen, buffer type and concentration, ionic strength, and protein/peptide concentration must be understood. From preformulation studies,

protein/peptide chemical and physical degradation pathways can be understood so that the final formulation, manufacturing process, and packaging system will be rationally developed.

2. The route of administration must be known in order to select the final dosage form, vehicle, volume, and tonicity requirements for the product. For example, if the primary route of administration is intravenous, the vehicle has to be water although some water-miscible cosolvents can be used. The volume can be limitless (unless an antimicrobial preservative [AP] is part of the formulation, in which case the volume is limited to 15 mL [European Pharmacopoeia, 2012]), and the tonicity does not necessarily have to be isotonic because the injected solution will be rapidly diluted. However, if the route of administration will be subcutaneous or intramuscular, then the vehicle can be aqueous or nonaqueous, the volumes are limited (usually no more than 2 mL for subcutaneous and 3 mL for intramuscular), and the tonicity of the product needs to be more tightly controlled since the product is neither quickly nor readily diluted. The rate of injection also is a factor to be considered while selecting final formulation ingredients in that some ingredients, including the protein/peptide itself, can be irritating and even cause local inflammatory reactions if injected too quickly and/or at too high a concentration.

3. Careful screening and choice of solutes for solubilization, stabilization, preservation, and tonicity adjustment should take place. These aspects will be the thrust of this chapter.

4. Potential effects of the manufacturing process on the stability of the protein/peptide in the final formulation must be understood. Proteins/peptides cannot withstand terminal sterilization techniques (heat, gas, and radiation) and, thus, should be prepared by aseptic filtration. The filter used must be qualified so that it does not bind the protein/peptide or produce leachates. The effect of flow rate during filtration and filling on solution stability must be studied. Also, the effect of shear (mechanical stress) that is encountered during manufacturing should be known. Time limitations must be established from the time the protein solution is compounded until it is sterile filtered in order to avoid any increase in endotoxin levels from whatever the bioburden, however small, maybe in the nonsterile solution. Further, since so many biopharmaceuticals have major stability limitations in solutions at ambient temperatures, there is a common term now used in the industry called "TOR" (time outside refrigeration) that manufacturers must abide by with respect to how much time a biopharmaceutical can be exposed to temperatures above refrigeration temperatures during the entire manufacturing process. Studies by Harwood et al. (1993), Nail and Akers (2002), Akers (2010), and Nema and Ludwig (2010) are excellent references that thoroughly deal with all aspects of the manufacturing of sterile protein and peptide dosage forms.

5. Selection of the most compatible container–closure system is tremendously important. Formulation scientists must appreciate that the container and closure system is just as important as the final solution formulation in

assuring long-term stability and maintenance of sterility and other quality parameters of the product. Proteins and peptides are well known to adsorb to glass, so experiments must be designed to study this possibility and, if adsorption occurs significantly, additives such as recombinant human serum albumin should be considered to reduce the adsorption. Glass leachates and particulates are possible and the formulator must be aware of this. Experiments must be conducted to assure elimination of this potential problem. The choice of rubber closure is particularly important because of known potential for the closure to leach some of its own ingredients into a solution, to adsorb components of the protein/peptide formulation, to core (rubber particulates) when penetrated by a needle, to generate particulates, and to leak because of problems with the fitment on the glass vial or resealability of the elastomer after needle penetration. Studies on adsorption of the protein to plastic surfaces will be necessary if the final product will be a plastic container. Even if plastic is not part of the primary container, protein–plastic compatibility studies should be conducted since plastic tubing such as silicone or polyvinyl chloride will be used in pharmaceutical process equipment (e.g., filling machines) and the final dosage form might be added to large-volume parenteral solutions contained in plastic bags.

6. Studies must be conducted to understand the effects of distribution and storage on the stability of the final product. Temperature excursions during shipping, mechanical stress, exposure to light, and other simulated shipping and storage conditions must be studied. From these studies, appropriate procedures for distribution and long-term storage of these relatively unstable dosage forms can be developed.

Table 8.1 provides a summary of the key steps in the development of solution dosage forms of peptides and proteins.

8.2   OPTIMIZING HYDROLYTIC STABILITY

The effect of solution pH on stability is a very important factor to study in early protein solution development. Figure 8.1 schematically depicts expected stability problems of proteins as a function of pH. Preformulation stability studies are conducted very early in the product development cycle to understand relative protein solubility and stability over an appropriate pH range (normally, pH 3 to pH 10). The relationship of stability versus solubility at various pH values usually follows a pattern of (1) higher solubility, lower chemical stability or (2) lower solubility, lower physical stability. Protein solubility is generally minimum at its isoelectric point. Insulin, for example, has an isoelectric point of 5.4, and at this pH it is quite insoluble in water (<0.1 mg/mL). Adjusting the solution pH to less than 4 or greater than 7 greatly increases insulin solubility (>30 mg/mL, depending on zinc concentration and species source of insulin), but it also increases the rate of deamidation at these pH ranges (Brange and Langkjaer, 1993; Brange et al., 1992a, 1992b). In dosage form development, the scientist must first determine which pH range provides acceptable solubility of the protein for proper dosage and then determine whether this pH range also provides acceptable stability.

TABLE 8.1
Development Strategy for Protein and Peptide Parenteral Solution Dosage Form

Formulation and Package Development Studies

Final strategy/objectives
Development of final formulation
- Justification of choice of excipients, pH, specification

Selection of container–closure
- Extractables
- Container–closure integrity
- Glass leachates, particulates

Stability and compatibility studies
- Effects of light, oxygen, high temperature, freezing
- Interaction of excipients with active components
- Long-term stability studies of final container formulation in final container–closure system
- Temperature/shipping excursions

Microbiological characteristics
- Antimicrobial properties
- Preservative efficacy
- Endotoxin control

Process Studies

Optimization studies of excipients, pH, other possible variations
Process development
- Process control (e.g., time, temperature during each processing step)
- Filter selection/validation
  - Microbial retention
  - Adsorption
  - Extractables
- Effect of terminal sterilization
- Justification of excess
- Process validation
  - Sterilization of components
  - Aseptic process
  - Cleaning
  - Filling

Establishment of critical process parameters
Establishment of control strategy

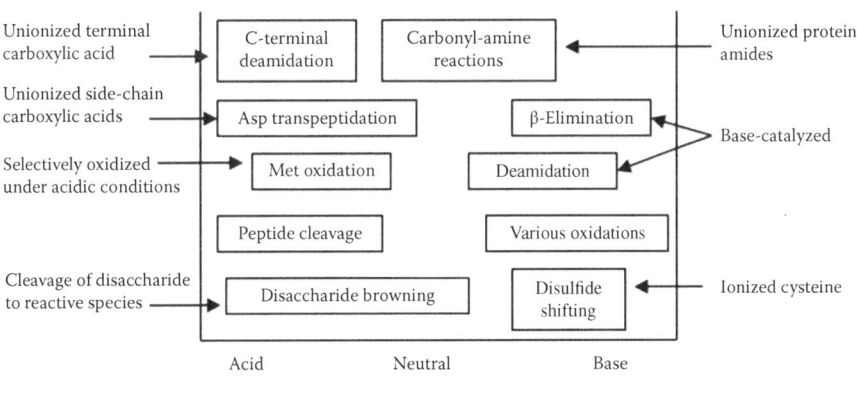

FIGURE 8.1 Protein reactions as a function of pH. (Courtesy of Dr. Lee Kirsch.)

There is usually a give-and-take relationship between solubility and stability, and it is up to the scientist to identify which pH is optimal for both.

Hydrolysis or deamidation occurs with peptides and proteins containing susceptible Asn and Gln amino acids, the only two amino acids that are primary amines. The side chain amide linkage in Gln and Asn residues may undergo deamidation to form free carboxylic acid. Deamidation can be promoted by a variety of factors, including high pH, temperature, and ionic strength (Manning et al., 2010). The rate of deamidation, at neutral to alkaline pH, is affected by amino acid sequence, particularly the amino acid immediately following Asn or Gln amino acids. Deamidation rates for peptides will be the highest when Asn is immediately followed by a Gly amino acid since Gly has no side chain, thus no opportunity to sterically hinder the hydrolysis reaction. For larger protein structures, it is more difficult to determine the effects of adjacent amino acid sequences on the deamidation rates of Asn and Gln owing to the three-dimensional complexities of these structures. However, it certainly is intuitive that adjacent amino acids and their size will have some effect on Asn and Gln deamidation regardless of the size of the total protein.

Oliyai and Borchardt (1994) studied the influence of primary amino acid sequence on the degradation of Asp residues under both acidic and alkaline conditions. As expected, the rate of intramolecular formation of the cyclic imide, the first step in the hydrolytic degradation pathway (Patel and Borchardt, 1990), was most affected by the size of the amino acid on the C-terminal side of the Asp residue. C-terminal substitution of Gly with increasingly more bulky residues (Ser and Val) inhibits the amount of cyclic imide produced. However, with respect to Asp amide bond hydrolysis with adjacent amino acids either before or after Asp, such structural changes had little or no effect.

Hydrolytic instability of peptides and proteins can be minimized through one or more of the following approaches:

1. Optimization of amino acid sequence, that is, engineering protein structures to remove unstable amino acids or insert amino acids that sterically hinder Asn or Gln deamidation, as long as this does not affect protein activity, potency, toxicity, or any other quality attribute.
2. Formulate at optimal solution pH. For example, human epidermal growth factor 1–48 demonstrates interesting pH-dependent stability in that at pH < 6, succinimide formation at Asp[11] is favored, while at pH > 6, deamidation of Asn[1] is favored (Senderoff et al., 1994). The optimal pH therefore is right at pH = 6.
3. Store at low temperatures, although this will always create difficulties in complying with the requirement during distribution and long-term storage of the product. However, it seems that the market prefers ready-to-use solutions that should be refrigerated over lyophilized dosage forms that can be stored at room temperature. Storage condition therefore does not seem to be a major obstacle as once believed, at least in the United States.
4. Optimize the effects of ionic strength.

As Manning et al. (2010) highlight in their review article, many review articles have been published on deamidation and related reactions; the most recent among them are Wakankar and Borchardt (2006), Dehart and Anderson (2007), and Topp et al. (2010).

8.2.1 BUFFERS

Buffers are used to prevent small changes in solution pH that can affect protein solubility and stability. Buffers are composed of salts of ionic compounds, the most common of which are acetate, citrate, and phosphate. Buffer systems acceptable for use in parenteral solutions are listed in Tables 8.2 and 8.3.

TABLE 8.2
Most Common Buffers Used in Sterile Drug Solutions

Buffer System	pKa	Typical Buffer pH Range
Lactic acid/lactate	3.1	2.0–4.0
Tartaric acid/tartrate	3.0, 4.2	2.0–5.3
Glutamic acid/glutamate	2.1, 4.3, 9.7	2.0–5.3
Malic acid/malate	3.4, 5.1	2.5–5.0
Citric acid/citrate	3.1, 4.8, 5.2	2.5–6.0
Gluconic acid/gluconate	3.6	2.6–4.6
Benzoic acid/benzoate	4.2	3.2–5.2
Succinic acid/succinate	4.2, 5.6	3.2–6.6
Acetic acid/acetate	4.8	3.5–5.7
Histidine	1.8, 6.1, 9.2	5.5–7.4
Phosphoric acid/phosphate	2.1, 7.2, 12.7	6.0–8.2
Glycine/glycinate	2.4, 9.8	6.5–7.5
Trometamine (TRIS, THAM)	8.1	7.1–9.1
Diethanolamine	8.0	8.0–10.0
Carbonic acid/carbonate	6.4, 10.3	5.0–11.0

TABLE 8.3
Dissociation Constants of Amino Acids Used as Buffers in Sterile Drug Solutions, Especially Monoclonal Antibody Products

Amino Acid	α-Carboxylic Acid	α-Amino Group	Side Chain
Alanine	2.35	9.87	—
Arginine	2.01	9.04	12.48
Aspartic acid	2.10	9.82	3.86
Cysteine	2.05	10.25	8.00
Glycine	2.35	9.78	—
Histidine	1.77	9.18	6.10
Lysine	2.18	8.95	10.53

The proper selection of buffer type and concentration is determined by performing solubility and stability studies as a function of pH and buffer species. Normally, the pH of maximum solubility is not the pH of maximum stability. However, a pH range that is a good compromise between solubility and stability can be selected, and that pH range can be maintained with the proper selection of the appropriate buffer component.

General acid and/or general base buffer catalysis can accelerate the hydrolytic degradation of the protein. Yoshioka et al. (1993) reported that the inactivation rate of β-galactosidase increased with increasing concentrations of phosphate buffer up to 0.5 M, and then decreased, presumably because of higher buffer components causing a reduction in water mobility. Cleland et al. (1993) cite several examples where the rate of protein deamidation was markedly dependent on the buffer anion. Capasso et al. (1991) compared the deamidation rate of a small peptide using different buffers and found that the peptide was most unstable in a phosphate buffer while most stable in Tris buffer.

Solution pH strongly influences the potential for protein aggregation since pH determines the net charge of the protein structure and subsequent electrostatic interactions (Chi et al., 2003). Since pH is controlled by buffer components, there are many published examples of the effect of buffer systems and concentrations on protein stability, especially aggregation. Here are a few examples:

- Buffer type and concentration affected aggregation of basic fibroblast growth factor depending on pH (Wang et al. 1996). At pH 5, aggregation increased as citrate buffer concentration increased. Citrate buffer at pH 3.7 caused aggregation, whereas acetate buffer at pH 3.8 did not. At pH 5.5–5.7, phosphate, acetate, and citrate buffers all showed similar aggregation rates.
- Aggregation of interferon-tau was found to be much greater in phosphate buffer formulations than in formulations at the same pH buffered with Tris buffer or histidine buffer (Katayama et al., 2006).
- Aggregation of recombinant human interferon α2b also was increased in the presence of phosphate buffer while decreased in the presence of citrate buffer (Ruiz et al., 2006).
- Studies on a new monoclonal antibody developed by Eli Lilly found that different buffers and different pH values had no effect on protein content or aggregation, but asparagine deamidation was dependent on pH and buffer type, with phosphate buffer formulations having higher deamidation rates than citrate buffer formulations (Zheng and Janis, 2006).
- Six different buffer species were compared for their protection against aggregation of a humanized IgG antibody (Kameoka et al., 2007). The antibody in phosphate and citrate buffer formulations showed high aggregation propensity, whereas lower aggregation propensity was shown in 2-(N-morpholino) ethane sulfonate (MES), 3-(morpholino) propane sulfonate (MOPS), acetate, and imidazole buffer formulations. The authors suggested that the buffer molecule and the Fc domain of IgG were involved in the aggregation propensity.

Histidine has been found to be an excellent buffer component for monoclonal antibodies (e.g., Synagis®, Herceptin®, Xolai®, and Raptiva®) maximally stable around pH 6. The pKa of histidine is 6.0, which makes it an ideal buffer at pH of 6.0. Histidine is the only amino acid with pH 7.4 within its buffering range; therefore, it has found importance in parenteral formulations requiring buffering in the physiological pH range (Daugherty and Mrsny, 2006).

High concentrations of monoclonal antibodies ($\geq$50 mg/mL) have the ability to self-buffer (Gokam et al., 2008). IgG_2 was found to be more stable at pH 5 after accelerated stability studies as a self-buffered formulation than in formulations containing conventional buffers such as acetate, glutamate, and succinate.

Several potential problems are associated with using buffers in parenteral solutions. For example, in large-scale manufacturing it should not be expected that the compounded solution containing the buffer will always result in the exact pH specified. Dilute solutions of strong acids (hydrochloric acid) or bases (sodium hydroxide) usually are required to fine-tune the final solution pH. Excessive use of the pH adjustment solutions may alter the buffer capacity and ionic strength of the buffered solution. Increasing buffer capacity to better control pH could significantly increase ionic strength, which, in turn, may cause increased potential of pain upon injection due to the increase in solution osmolality.

8.2.2 IONIC STRENGTH

Ionic strength is a measure of the intensity of the electrical field in a solution. Ionic strength depends on the total concentration of ions in a solution and the valence of each ion. The ionic strength of a 0.1-M solution of sodium chloride is 0.1. The ionic strength of a 0.1-M solution of sodium sulfate is 0.3, because sulfate ions have a valence of 2 added to the valence of 1 for the sodium ions. Ionic strength is well known to affect solubility of drugs and may have an effect on protein stability in a solution. Typically, but not always, increasing ionic strength will decrease drug solubility via a salting-out effect. The Debye–Huckel theory predicts that increasing ionic strength would be expected to decrease the rate of degradation of oppositely charged reactants, but increase the rate of degradation of similarly charged reactants.

Ionic strength will affect the stability of a protein, but in which direction (increase or decrease stability) differs with different proteins. For example, increasing ionic strength will increase the stability of recombinant alpha$_1$ antitrypsin (Vemuri et al., 1993). Conversely, increasing ionic strength will increase the rate of deamidation of human growth hormone (hGH) (Pearlman and Bewley, 1993) and bovine somatotropin (bSt) (Davio and Hageman, 1993), and lead to opalescence and higher viscosity of a monoclonal antibody (Wang et al., 2009) (Figure 8.2). Opalescence was a result of irreversible dimer formation caused by using sodium chloride to increase ionic strength. Higher viscosity could correlate to molecular association and could adversely affect the rate and extent of sterile filtration of the antibody.

Adjusting ionic strength of protein solutions typically is necessary to render the solution isotonic although preformulation studies may reveal either positive or negative effects of adjusting ionic strength. Such information will dictate how much or how little salt (usually sodium chloride, although other salts could be used) to add to the formulation to assure appropriate solubility and stability.

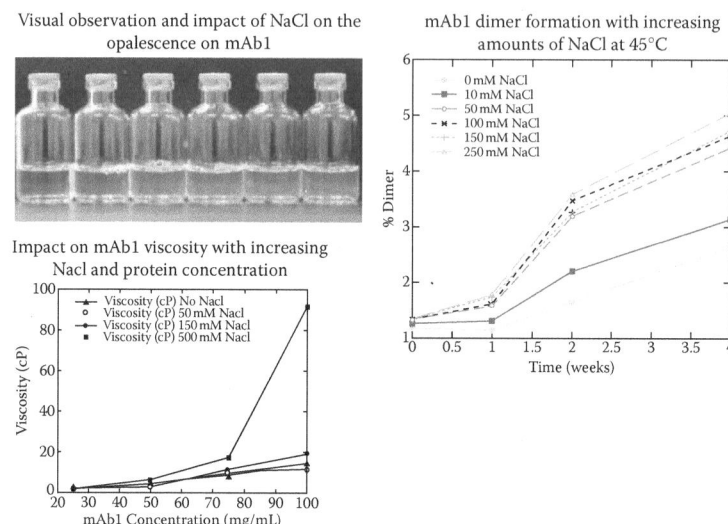

FIGURE 8.2 **(See color insert.)** MAb1 dimer formation with increasing amounts of NaCl at 45°C. (Reprinted from Wang, N., et al., *BioPharm* April, 36–47, 2009. With permission.)

8.3 OPTIMIZING OXIDATIVE STABILITY

Proteins containing methionine, cysteine, cystine, histidine, tryptophan, and tyrosine may be sensitive to oxidative and/or photolytic degradation depending on the conformation of the protein and resultant exposure of these amino acids to the solvent and environmental conditions such as presence of oxygen, light, high temperature, metal ions, and various free-radical initiators. Oxidation of amino acids (e.g., methionine and cysteine) will lead to disulfide bond formation and loss of biological activity. The free thiol group that is present in a Cys residue of any native biologically active protein not only may oxidize to produce an incorrect disulfide bridge, but also can result in other degradation reactions, such as alkylation, addition to double bonds, and complexation with heavy metals.

hGH, chymostrypsin, lysozyme, parathyroid hormone, human granulocyte-colony stimulating factor, insulin-like growth factor I, acidic and basic fibroblast growth factors, relaxin, the monoclonal antibody OKT3, interleukin 1β, and glucagon are examples of proteins that will degrade by this mechanism. Hemoglobin, with its oxygen-carrying properties dependent on the reduced state of ferrous iron, is very sensitive to oxidation and must contain antioxidants to maintain stability of the heme groups (Kerwin et al., 1999).

For protection against oxidation, choice of an effective antioxidant is one of several precautions that must be practiced in formulation development and final product manufacture. Other factors that potentially may protect oxygen-sensitive proteins include

- Preparation and storage at low temperatures
- Use of chelating agents to eliminate metal catalysis
- Increasing ionic strength

- Elimination of peroxide and metallic contaminants in formulation additives
- Protection from light
- Awareness of possible interaction of light exposure and phosphate buffer in forming free radicals (Fransson and Hagman, 1996)
- Replacing oxygen with nitrogen or argon during manufacturing
- Removing oxygen from the headspace of the final container
- Formulation at the lowest pH possible while still maintaining desired protein solubility and hydrolytic stability, since there is an inverse correlation between oxidative stability and pH (Akers, 1982)
- Use of a container–closure system that allows no oxygen transmission through the package during distribution and storage
- Assurance that phenolic or other oxidizing cleaning agent residues are minimal in the production environment, including the freeze-dry chamber (Kirsch et al., 1993)

Fransson (1997) studied methionine oxidation and covalent aggregation in aqueous solutions containing human insulin-like growth factor-I (hIGF). Oxidation of methionine 59 was catalyzed by light and by ferric ions in combination with disodium ethylenediaminetetraacetic acid (DSEDTA). Fransson suggested that DSEDTA actually activates ferric ions by stabilizing the transfer of electrons from ferric ions to ferrous ions. Methionine in this protein was radicalized by light and then oxidized to methionine sulfoxide. Light may also trigger generation of hydroxyl radicals by decomposition of water that may oxidize the methionine.

Several oxidation problems with monoclonal antibody candidates prompted scientists at Genentech to investigate the mechanism of oxidation of methionine, tryptophan, and histidine residues and how to stabilize such degradations (Ji et al., 2009). Methionine is oxidized by the presence of hydrogen peroxide through a nucleophilic reaction, whereas tryptophan is oxidized by a free-radical mechanism. Oxidation of histidine requires the presence of copper. The sources of oxidation include peroxide from degraded polysorbate and sanitization agents and heavy metals from stainless steel processing surfaces. Minimization/elimination of oxidation of these sensitive residues required the addition of free methionine that serves as a protectant for methionine residues and addition of free-radical scavengers such as free tryptophan, pyrodixone, mannitol, DSEDTA, and/or Trolox (6-hydroxy-2,5,7,8-tetramethylchroman-2-carboxylic acid).

Formulators should be aware of the potential of polysorbate surfactants to adversely affect the oxidative stability of proteins. Polysorbates 20 and 80 are commonly used surface active agents in protein formulations to minimize surface aggregation problems. However, they can undergo autocatalyic degradation to produce peroxides that can oxidize methionine and cysteine residues. This phenomenon has been reported in studies involving formulation development of Neupogen® (Herman et al., 1996), recombinant human ciliary neurotrophic factor (Knepp et al., 1996), and a Fab protein (Lam et al., 2011).

8.3.1 Antioxidants

There are several choices of antioxidants that can be used in protein formulations. Those that have been used most frequently are ascorbic acid, salts of sulfurous acid (sodium bisulfite, sodium metabisulfite, or sodium thiosulfate), and thiols such as thioglycerol and thioglycolic acid. Dithiothreitol, reduced glutathione, acetylcysteine, mercaptoethanol, and thioethanolamine are thiols that usually oxidize too readily to be of practical use in pharmaceutical formulations requiring long-term storage.

Sodium thiosulfate, methionine, catalase, or platinum, as antioxidants, were effective in reducing the oxidation of recombinant monoclonal antibody HER2 (Lam et al., 1997b). Replacing air with nitrogen in sample vials also was effective in reducing antibody oxidation. The authors proposed that thiosulfate and methionine either inhibit free-radical induced oxidation by terminating the chain reaction or compete with the methionine residues in the antibody for reaction with the free hydroxyl radicals. Catalase and platinum serve as free-radical scavengers.

Precautions must be applied when considering ascorbic acid as an antioxidant in protein formulations. Li et al. (1993) found that ascorbate in the presence of Fe^{3+} and oxygen actually induces the oxidation of methionine in small model peptides. Ascorbate is a powerful electron donor that is readily oxidized to dehydroascorbate. It also generates highly reactive oxygen species such as hydrogen peroxide and peroxyl radicals. These, in turn, will accelerate the oxidation of methionine. Phosphate buffer compared to other buffer systems (e.g., TRIS, HEPES, and MOPS) accelerated the degradation of methionine in the presence of ascorbic acid. The addition of DSEDTA did not enhance stability even though ferric ion and other transition metals were components in the formulation, either purposely added or as trace components of the buffer and peptide. This pro-oxidant effect of ascorbate on methionine oxidation was concentration dependent and occurred most readily at pH 6–7.

It is also known that sodium bisulfite can cause stability problems with certain drugs and proteins. For example, bisulfite will rapidly destroy insulin (Asahara et al., 1991). These authors studied the compatibility of human insulin in solutions containing sodium bisulfite since human insulin and sodium bisulfite were being used in some intravenous admixtures in their hospital practice. Bisulfite was found to cleave the interchain disulfide bonds of insulin. The addition of glucose to these solutions stabilized human insulin in the presence of sodium bisulfite with the stabilization postulated to be the formation of a bisulfite-glucose adduct in solutions.

8.3.2 CHELATING AGENTS

Chelating agents are used in protein formulations to aid in inhibiting free-radical formation and resultant oxidation of proteins caused by trace metal ions such as copper, iron, calcium, manganese, and zinc. While organic buffer salts, such as sodium citrate, have some capability of binding trace heavy metal contaminants in protein solutions, the major chelating agent used is DSEDTA. The concentration of DSEDTA usually is very low, for example, ≤0.04%. DSEDTA tends to dissolve slowly and is usually among the first of formulation ingredients to be dissolved during compounding before adding other ingredients, including the protein.

DSEDTA should not be used in formulations of metalloproteins such as insulin, hemoglobin, and fibrolase (Pretzer et al., 1993), as the chelating agent will attack the metal that is part of the stable conformation of the protein. DSEDTA will accelerate the oxidative degradation of methionine in hIGF-I solutions (Fransson, 1997). It was suggested that DSEDTA actually enables ferric ions to be active by stabilizing the transfer of electrons from ferric ions to ferrous ions. Methionine in this protein is radicalized by light and then oxidized to methionine sulfoxide. Light may also trigger the generation of hydroxyl radicals by decomposition of water that may oxidize the methionine. Thus, the formulator must not indiscriminately include DSEDTA in protein formulations without carefully determining that its presence aids in oxidative stabilization of the protein.

DSEDTA was shown to complex heavy metals originating from stainless steel surfaces to help stabilize the methionine amino acids of parathyroid hormone from oxidative degradation due to heavy metal catalysis (Junyan et al., 2009).

8.3.3 INERT GASES

Inert gases are frequently used in production of protein dosage forms. The most commonly used inert gas is nitrogen. Other inert gases that can be used, although not very often primarily because of expense, include argon and helium. Argon, however, has been shown to be more efficient in displacing oxygen because it is heavier than air and will more readily stay in the vial than nitrogen, which is lighter than air (Harwood et al., 1993). The normal use of inert gases in protein formulation and production involves the following two ways:

1. Added to water and compounding solutions before aseptic filtration to saturate the solution and minimize the level of dissolved oxygen. However, oxygen is never completely displaced with an inert gas when the solution is sparged (saturating the solution with the gas). Many manufacturers use a dual needle that permits simultaneous filling of a liquid and purging of gas at the same time.
2. Introduced into the headspace (usually 10–15% of the internal capacity of a container) of a filled vial right before the vial is stoppered with a rubber closure, thereby, theoretically, displacing oxygen in the headspace. Again, a dual needle can be used to fill solutions and purge gas into the final container at the same time.

The inert gas must be high-quality grade and must be sterilized, usually with a 0.45-μm hydrophobic membrane filter. The integrity of the gas filter is tested before and after use by diffusion flow methods.

8.3.4 PACKAGING AND OXIDATION

All the appropriate formulation and processing procedures can be in place for stabilizing protein solutions against oxidation, but if the packaging system is inadequate from an integrity standpoint, the product will readily degrade (not to mention the potential for microbial contamination). Most protein products are packaged in glass vials with rubber closures although prefilled syringes are becoming very popular for protein solutions. The rubber–glass interface and the oxygen transmission coefficient of the rubber closure will dictate the quality of the container–closure system. The integrity of the rubber–glass interface can be tested by a variety of container–closure integrity techniques (Guazzo, 2010).

Oxygen transmission coefficients are determined for a particular rubber closure formulation by the rubber closure manufacturer. Rubber formulations having the lowest oxygen transmission coefficients are the synthetic butyl and halobutyl types.

The formulator should determine from the rubber manufacturer how the halobutyl rubber is cured (shaped and molded) since common curing agents are zinc oxide, aluminum, and peroxide, which potentially can leach out of the rubber formulation with time and catalyze oxidative degradation (Boyett and Avis, 1976; Danielson et al., 1984; Liebe, 1995; Milano et al., 1982; Sacha et al., 2010). Special coatings on rubber closures help to reduce, even eliminate, extractables and leachables (see discussion under Section 8.9).

8.3.5　OTHER CHEMICAL STABILIZERS

Sugars and polyols, such as ethylene glycol, glycerol, glucose, and dextran, at high concentrations can inhibit the metal-catalyzed oxidation of human relaxin (Li et al., 1996). All but dextran act as chelating agents in complexing transition metal ions, whereas dextran, which has a higher binding affinity to metal ions and undergoes depolymerization in a metal-catalyzed oxidation, protects relaxin by a radical scavenging mechanism.

Fibroblast growth factors, both acidic and basic, possess nearly identical three-dimensional structures of 12 antiparallel β-strands arranged with approximately threefold internal symmetry (Tsai et al., 1993). Acidic fibroblast growth factor's tendency to aggregate in solutions was inhibited by a variety of polyanionic additives, such as inositol hexasulfate and sulfated β-cyclodextrin, and by a number of commonly used excipients, such as sucrose, dextrose, trehalose, glycerol, and glycine. The polyanionic additives interacted with the polyanion binding site of the protein, whereas the nonspecific agents were thought to be preferentially hydrated (authors did not state exclusion). In all cases, these interactions between acidic fibroblast growth factor and various excipients resulted in an increase in the protein's Tm, the midpoint of the temperature of the transition from the folded to unfolded protein.

A variety of cosolvents can stabilize proteins in solutions because the cosolvent is preferentially excluded from surface interaction with the protein (Arakawa et al., 1991). Cosolvents behaving this way include glycerol and sorbitol. Bhat and Timasheff (1992) found that proteins studied were preferentially hydrated in a range of polyethylene glycol (PEG) sizes (molecular weight 200–6000) and that PEGs were preferentially excluded from protein domains that enabled PEGs to serve as stabilizers except perhaps at higher temperatures.

8.4　OPTIMIZING PHYSICAL STABILITY

Unlike small molecules where physical instability is rarely encountered except for poorly water-soluble compounds, proteins, because of their unique ability to adopt three-dimensional forms, tend to undergo a number of structural changes, independent of chemical modifications. Physical instability of proteins is sometimes a greater cause for concern and is more difficult to control than chemical instability. All protein structures are hydrophobic to some extent. Many proteins, particularly when exposed to stressful conditions, for example, extremes in temperature, will unfold such that the hydrophobic portions become exposed to the aqueous environment. Such exposure will promote aggregation or self-association, possibly leading to physical instability and loss of biological activity, since the interaction with the receptor site requires folded structures with correct conformation.

8.4.1 Denaturation

Denaturation is unique to proteins and occurs when their native quarternary, tertiary, and, frequently, secondary structure is disrupted. Denaturation can lead to unfolding and the unfolded polypeptide chain may undergo further reactions. Such inactivation could be associated with surfaces and/or interaction with other protein molecules, leading to aggregation and precipitation. Denaturation is of two types: (1) reversible denaturation, caused by temperature or exposure to chaotropic agents (urea and guanidine hydrochloride), where, if the denaturing condition is removed, the protein may regain its native state and maintain its activity although sometimes irreversible aggregation will occur, and (2) irreversible denaturation, where the protein, once unfolded, will not regain its native form and activity. However, there are several instances where a protein that is irreversibly denatured is returned to its native state by the use of denaturant followed by dialysis. One example is T4 lysozyme, where its lost activity can be restored by renaturation with guanidine hydrochloride (Wetzel et al., 1988). When a protein recovers its activity by the addition of denaturants such as guanidine hydrochloride or urea and by subsequent dialysis, such a process may indicate the influence of events like aggregation, precipitation, and adsorption. Regardless of the observed phenomenon, the approach is of little use in the solution formulation design.

For stabilizing proteins against denaturation in solutions, Arakawa and Timasheff (1982, 1984) and Arakawa et al. (1993) have shown that reversible denaturation can be decreased by the use of additives such as salts that bind to nonspecific binding sites on the proteins. Dahlquist et al. (1976) demonstrated increased thermal stability of thermolysin by the binding of ions to specific sites on the protein. Roe et al. (1988) showed the ability of $Zn(NO_3)_2$ to significantly increase the thermal stability of superoxide dismutase. Gekko and Timasheff (1981) concluded that the preferential hydration of proteins observed at all conditions in the presence of a glycerol–water mixed solvent system is a prerequisite for stabilizing the native structure of several globular proteins. Pace and Grimsley (1988) found the stability of ribonuclease T_1 to increase in the presence of 0.1 M NaCl, $MgCl_2$, and Na_2HPO_4, respectively. Through genetic engineering they were able to introduce appropriate amino acid substitutions in ribonuclease T_1, creating specific cation/anion binding sites on the protein. The stability profile of ribonuclease T_1 was enhanced considerably with this approach. Pantoliano et al. (1988) were able to introduce negatively charged side chains such as Asp in the vicinity of the weak Ca^{2+} binding site of subtilisin. Such modifications caused an increase in binding affinity for Ca^{2+}, thereby increasing the thermal stability of subtilisin. These engineering approaches are acceptable as long as protein activity and other quality aspects are not adversely affected.

8.4.2 Protein Aggregation

Peptides and proteins are aggregated mainly by hydrophobic interactions that eventually lead to denaturation. When the hydrophobic region of a partially or fully unfolded protein is exposed to water, this creates a thermodynamically unfavorable situation because the normally buried hydrophobic interior is now exposed to a hydrophilic aqueous environment. Consequently, the decrease in entropy from structuring water molecules around the hydrophobic region forces the denatured protein to aggregate, mainly through the exposed hydrophobic regions. Thus, solubility of the protein may also be compromised. In some cases self-association of protein subunits, either native or misfolded, may occur under certain conditions, and this may lead to precipitation and loss of activity (Brange and Langkjaer, 1993; Brange et al., 1992a; Mitraki and King, 1989; Shahrokh et al., 1994; Silvestri et al., 1993). The protein antithrombin aggregates when denatured with guanidine hydrochloride, which proceeds through an intermediate partially unfolded state. It is possible to return antithrombin to its native state by dialyzing the partially unfolded form that aggregates slowly. However, once aggregation occurs, the native state cannot be reformed by this approach (Fish et al., 1985). Irreversible aggregation caused by denaturation can be prevented by the use of surfactants, polyols, or sugars. The use of surfactants is elaborated in Section 8.4.4.

Mahler et al. (2009) suggested the following classification of protein aggregation:

- Type of bond—The aggregates are either noncovalent (weak electrostatic forces) or covalent (e.g., disulfide bridges).
- Reversibility—The aggregates are either reversible or irreversible.
- Size—Several classifications:
  - The aggregates are small and soluble oligomers (dimers, trimers, tetramers, etc.).
  - The aggregates are larger soluble oligomers ($\geq$10-mer).
  - The aggregates are much larger soluble oligomers in the diameter range of 20 nm to approximately 1 μm.
  - The aggregates are insoluble particles in the 1–25-μm range.
  - The aggregates are much larger insoluble particles visible to the eye under defined inspection conditions.
- Protein conformation—The aggregates are predominately native structure or they are predominantly nonnative structure (i.e., partially unfolded multimeric species or fibrillar aggregates).

Factors that affect protein aggregation in solutions generally include protein concentration, pH, temperature, freeze-thawing (e.g., Kueltzo et al., 2007) formulation excipients, and mechanical stress. Some factors (e.g., temperature) can be easily controlled during compounding, manufacturing, storage, and use compared to others (e.g., mechanical stress). Formulation studies will dictate appropriate choices of pH and excipients that will not induce aggregation and/or, in fact, will aid in the prevention of aggregation. An example of the effect of pH and composition on the aggregation of recombinant human granulocyte colony-stimulating factor (rhGCSF) as an exemplary protein is shown in Figure 8.3.

Protein concentration is dictated by the required therapeutic dose and this required concentration will determine whether the potential for higher associated states (dimers, tetramers, etc.) exist, which can then lead to aggregation in solutions. Higher required concentrations of monoclonal antibody formulations will result in increased potential of aggregation (Shire et al., 2004). Treuheit et al. (2002) also showed that higher protein concentrations resulted in acceleration of protein

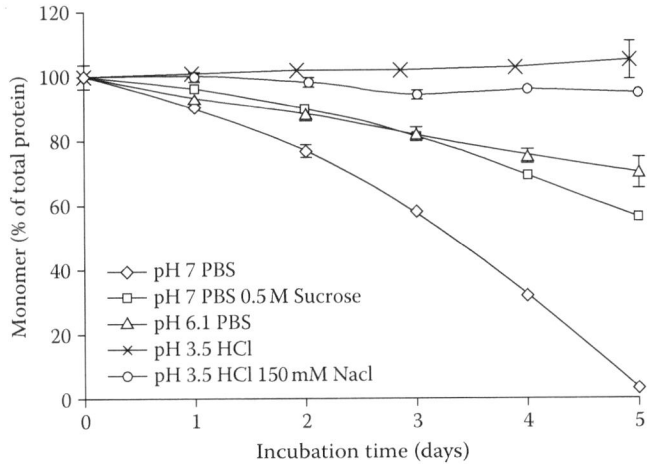

FIGURE 8.3   Aggregation profiles of rhGCSF as a function of pH and type of solution. (Reproduced from Chi, E. Y., et al. *Pharm. Res.* 20, 1325–1336, 2003. With permission.)

aggregation except during agitation conditions where lower protein concentrations showed greater aggregation. Careful studies should be done during formulation development to determine which factors influence protein aggregation and then how these factors can be eliminated or controlled.

8.4.3 ADSORPTION

Proteins exhibit a certain degree of surface activity; that is, they adsorb to surfaces because of their innate nature of being amphiphilic polyelectrolytes. Glass, rubber, or plastic adsorption of proteins is well documented (e.g., Anik and Hwang, 1983; Dong et al., 1987; Hirsch et al., 1977, 1981; Johnston, 1996; Schwarzenbach, 2002; Wang and Chien, 1984). Consequently, biological activity may be either reduced or totally lost if such adsorption occurs during manufacturing, storage, or use of the final product. The process of adsorption in the case of proteins depends on protein–protein interactions, time, temperature, pH and ionic strength of the medium, and the nature of the surface (Absolom et al., 1987). Norde (1995) and Pinholt et al. (2011) reviewed the general principles underlying protein adsorption from aqueous solutions onto a solid surface. Interactions that determine the overall adsorption process between a protein and a surface include redistribution of charged groups in the interfacial layer, changes in the hydration of the sorbent and the protein surface, and structural rearrangements in the protein molecule. Surface denaturation that commonly takes place at the liquid/solid and liquid/air interface has been shown by Lenk et al. (1989) to involve α-helix to β-sheet conformational changes. These structural changes, which are determined by the nature of the interfaces, are similar to those observed with the aggregation phenomenon caused by heat, high pressure, or chemical denaturants. In the case of proteins, sources such as the polymer of the membrane filter, the administration set, agitation that occurs during the purification process, and the method of manufacture are known or at least suspected to cause surface denaturation. Strategies used to overcome protein denaturation due to adsorption are as follows:

- Modify protein structure and conformation (Karlsson et al., 2005).
- Increase protein concentration during filtration and/or using extra volume to saturate the filter with protein solutions.
- Working with the glass manufacturer, modify the surface of the glass containers, providing a resistant barrier to protein–surface interaction (Henderson, 2010).
- Decrease the rate of mixing when it is known that shear will affect protein adsorption.
- Add excipients such as surfactants that have higher surface activity (Johnston, 1996). Surfactants must be of recombinant or synthetic origin and relatively free from peroxide contamination.
- Add macromolecules such as recombinant human serum albumin (Oshima, 1989), the most commonly used competitive binder for protein solutions to minimize protein adsorption (Table 8.4).

Insulin is a classic example of a protein that readily adsorbs onto the surfaces of delivery pumps, glass containers, and to the inside of the intravenous bags (Brennan et al., 1985; Lougheed et al., 1983; Mitrano and Newton, 1982; Sato et al., 1984; Twardowski et al., 1983a, 1983b, 1983c). Insulin adsorption usually is finite once binding sites are covered, and such adsorption is usually not clinically significant.

TABLE 8.4
Examples of Commercial Protein Dosage Forms Containing Human Serum Albumin

Brand (®)	Generic	HSA in Product
Aranesp	Darbepeotin alfa	2.5 mg/mL
Cerdase	Alglucerase	1.0%
Avonex	Interferon-beta-1a	15.8 mg/mL (after reconstitution)
Betaseron	Interferon-beta-1b	15 mg (lyo)
Bioclate	Factor VIII	12.5 mg/mL
Epogen/Eprex	Erythropoietin	0.25%
Roferon-A	Interferon alpha-2a	0.5%
Intron-A	Interferon alpha-2b	0.1% (after reconstitution)
Abbokinase	Urokinase	5.0% (after reconstitution)
Aralast	Alpha-1-proteinase	0.5% (after reconstitution)
Recombinate	Antihemophilic factor	1.0% (after reconstitution)
Myobloc	Botulinum toxin	0.05%
Rebif	Interferon-beta-1a	2 or 4 mg/0.5 mL
Streptase	Streptokinase	2.0% (after reconstitution)
Halozyme	Hyaluronidase	0.1%
Zevalin	Ibritumomab tiuxetan	750 mg

Also, albumin can preferentially bind to surfaces in place of insulin as was studied using sophisticated analytical procedures by Henry et al. (2008).

8.4.4 SURFACTANTS

Surfactants, by their name, are surface active agents that can exert their effect at surfaces of solid–solid, solid–liquid, liquid–liquid, and liquid–air because of their chemical composition containing both hydrophilic and hydrophobic groups. These materials reduce the concentration of proteins in dilute solutions at the air–water and/or water–solid interfaces where proteins can be adsorbed and potentially aggregated. Surfactants can bind to hydrophobic interfaces in protein formulations and packaging. Proteins on the surface of water will aggregate, particularly when shaken, because of unfolding and subsequent aggregation of the protein monolayer (Arakawa et al., 1991).

Surfactants can not only denature proteins, but also stabilize them against surface denaturation. In general, ionic surfactants can denature proteins. However, nonionic surfactants usually do not denature proteins even at relatively high concentrations (1% w/v) (Cleland et al., 1993). Most parenterally acceptable nonionic surfactants come from either the polysorbate (sorbitol-polyethylene oxide polymers) or polyether (polyethylene oxide-polyproplyene oxide block copolymers) groups. Polysorbate 20 and 80 are effective and acceptable surfactant stabilizers in marketed protein formulations (see Table 8.5) (see also Deechongkit et al., 2009).

TABLE 8.5
Examples of Surfactant-Containing Protein Formulations

Product	API Concentration	Dosage Form	Polysorbate Concentration	Polysorbate (%)
Aranesp	25–500 mcg/mL	Liquid	0.05 mg/mL	0.005
ReoPro	2 mg/mL	Liquid	0.001%	0.001
Humira	40 mg/mL	Liquid	0.8 mg/mL	0.08
Avastin	25 mg/mL	Liquid	1.6 mg/mL	0.16
Remicade	10 mg/mL	Lyophilized	0.05 mg/mL	0.05
Aralast	600 mg/mL	Lyophilized	0.05 mg/mL	0.05
Activase	1 mg/mL	Lyophilized	0.09 mg/mL	0.09
Koate	1.5 mg/mL	Lyophilized	0.025 mg/mL	0.0025
Advate	250–1500 IU	Lyophilized	0.17 mg/mL	0.017
Kogenate	1000 IU	Lyophilized	600 mcg	—
Novoseven	0.6 mg/mL	Lyophilized	0.1 mg/mL	0.01
BeneFix	250–1000 IU	Lyophilized	0.01%	0.01
Tissel VH	45 mg	Lyophilized	60 mg	—
WinRho SDF	600–5000 IU	Lyophilized	0.01%	0.01
PEG-Intron	0.106–0.317 mg/mL	Lyophilized	0.106 mg/mL	0.016
Retavase	1.81 mg/mL	Lyophilized	0.52 mg/mL	0.052
TNKase	5.25 mg/mL	Lyophilized	0.43 mg/mL	0.043
Herceptin	22 mg/mL	Lyophilized	0.09 mg/mL	0.009

Without the presence of polysorbate 80, recombinant human hemoglobin at the large volumes required for therapy would have no chance of being physically stable in solutions even at refrigerated conditions (Figure 8.4). Other surfactants used in protein formulations for clinical studies and/or found in the patent literature include polysorbate 20, Pluronic F68, and other polyoxyethylene ethers (e.g., the "Brij" class) (Kerwin, 2009).

Surfactants are well known to prevent the denaturation and aggregation of insulin (Kerwin, 2009; Lougheed et al., 1983; Sato et al., 1984). However, the choice of surfactant and the final concentration optimal for stabilization is quite dependent on a variety of factors, including other formulation ingredients (e.g., sugars), protein concentration, headspace in the container, the type of container, and test methodology.

Recombinant human growth hormone (rhGH) will aggregate readily under mechanical and thermal stress. Aggregation from mechanical stress can be substantially reduced in the presence of surfactants (Katakam et al., 1995). Mechanical stress may cause proteins to be more exposed to air–water interfaces, where denaturation

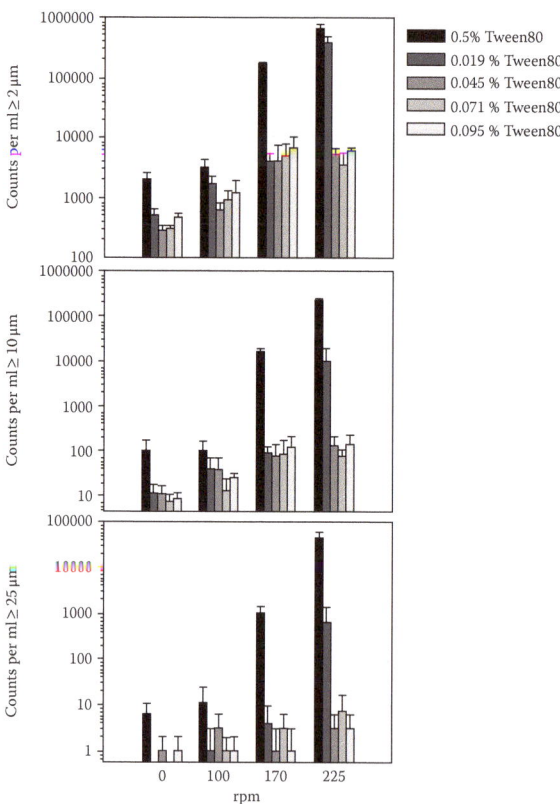

FIGURE 8.4   Effect of Tween 80 concentration on particle formation in solutions of recombinant human hemoglobin as a function of shear stress. (Reproduced from Kerwin, B. A., et al., *J. Pharm. Sci.* 88, 79–99, 1999. With permission.)

is more likely to occur than in the bulk phase of water. Surfactants will preferentially compete with proteins for accumulation at the air–water interface and keep the protein from undergoing interfacial denaturation resulting from mechanical stress. Pluronic F-68 and Brij 35 will stabilize hGH at their critical micelle concentrations (cmc) (0.1% and 0.013%, respectively) while stabilization with polysorbate 80 requires a concentration of 0.1%, higher than the cmc value for polysorbate 80 of 0.0013%. The reasons for these differences in stabilizing concentrations are not clear, but simply reflect differences in interactions between different surfactants and proteins. These surfactants do not stabilize hGH from aggregation due to high temperature stress.

Further substantiation of the important role of surfactants, particularly polysorbate 80, in protecting proteins against surface-induced denaturation during freezing was reported by Chang et al. (1996). They found a strong correlation between freeze denaturation (quick freezing of the protein) and surface denaturation (shaking the protein in solutions). Proteins that tend to denature under these conditions are protected by the addition of polysorbate 80 (0.1%). Other surfactants—Brij 35, Lubrol-px, Triton X-10, and even the ionic surfactant, sodium dodecyl sulfate—also protected the protein from denaturation although these surfactants have not been approved for use in injectable formulations. The authors pointed out that surfactants may be needed to protect proteins from denaturation during the freezing step only and that other stabilizers, for example, sucrose, may be needed to further protect the protein during freeze-drying.

bSt presents an example where surfactants are not effective in preventing protein aggregation and precipitation in solutions at elevated temperatures,* whereas other stabilizers such as sucrose are effective (Davio and Hageman, 1993). This phenomenon also contrasts with the effective use of surfactants to stabilize rhGH (as discussed above). Figure 8.5 shows the effect of polysorbate 80 on bSt precipitation at 54°C, where bSt is more stable without the presence of polysorbate 80, and increasing the amount of polysorbate 80 increases the extent of bSt precipitation. Interestingly, for bSt, increasing the concentrations of sucrose has a positive effect on stabilizing bSt against aggregation and precipitation.

Peroxides are known contaminants of nonionic surfactants. Knepp et al. (1996) reported on the peroxide levels of polysorbate 80 obtained from different manufacturers using a colorimetric titration method. Levels ranged from less than 1 mEq/kg to more than 27 mEq/kg. They found that peroxide levels increased upon storage at ambient temperatures probably due to headspace oxygen and/or the container–closure interface allowing ingress of air. Peroxides in polysorbate can result in oxidative degradation of proteins. Formulators need to screen sources of polysorbate 80 or other polymeric additives used in protein formulations for peroxide contamination and establish peroxide specifications for using the additive. Also, as

---

* Although polysorbate 80 was not effective in stabilizing bSt at elevated temperatures, it was effective when the applied stress was agitation. Also, the authors noted that destabilization of bSt by polysorbate 80 was not observed at ambient or refrigerated temperatures as other decomposition pathways; for example, deamidation became more predominant at lower temperatures.

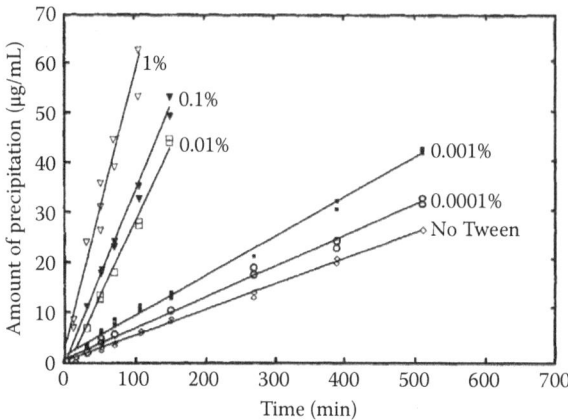

FIGURE 8.5 Effect of Tween 80 concentration on precipitation of recombinant bovine somatotropin (rbSt) as a function of thermal stress at 54°C. (Reprinted from Hageman, M. J., et al., *Pharm. Res.* 10, S-85, 1993. With permission.)

a precaution, incorporation of an antioxidant can help to overcome the potential for nonionic surfactants to serve as oxidative catalysts for oxygen-sensitive proteins.

Studies on protein–surfactant interactions where electron paramagnetic resonance (EPR) spectroscopy (Bam et al., 1995; Jones et al., 1999; Tabak et al., 2006), analytical ultracentrifugation (Jones et al., 1999), and fluorescence spectroscopy and isothermal titration calorimetry (Chou et al., 2005) have been performed to determine the binding stoichiometry of the surfactant to the protein. Knowing the surfactant to protein stoichiometry will help to know the optimal amount of surfactant to use to stabilize the protein against surface denaturation and other physical instability reactions.

8.4.5 Cyclodextrins

Cyclodextrins are cyclic (α-1,4)-linked oligosaccharides of α-ᴅ-glucopyranose containing a relatively hydrophobic central cavity and a hydrophilic outer surface (Loftsson and Brewster, 1996). Cyclodextrins come in a wide variety of structural derivatives, the most common being α-, β-, and γ-cyclodextrins, which consist of six, seven, and eight glucopyranose units, respectively. Two parenteral cyclodextrins are Encapsin™, a hydroxypropyl-β-cyclodextrin, and Captisol™, a sulfobutly ether β-cyclodextrin. They have been used widely for increasing the solubility, stability, and bioavailability of small drug molecules (Thompson, 1997). Peptides and proteins can also be stabilized in cyclodextrin complexes. Irwin et al. (1994) used β-cyclodextrins at a 25-fold excess to stabilize leucine enkephalin against enzymatic degradation in sheep nasal mucosa. Hydroxypropyl-β-cyclodextrin at a 1% concentration was shown to enhance the reconstituted solution stability of keratinocyte growth factor (Zhang et al., 1995). Brewster et al. (1991) presented several examples (ovine growth hormone, interleukin-2, and bovine insulin) where 2-hydroxypropyl-β-cyclodextrin enhanced the solubility and physical stability of proteins in solutions. Johnson et al. (1994) used chemically modified cyclodextrins to solubilize a tripeptide, and Matilainen et al. (2007) reported on the stabilization of glucagon when complexed with gamma-cyclodextrins.

8.4.6 ALBUMIN

Serum albumin is a widely used stabilizer in protein formulations for minimizing protein adsorption to glass and other surfaces (Chuang et al., 2002). Albumin preferentially competes with other proteins for binding sites on surfaces, but why this is so is not clear. It is also used as a protectant in several lyophilized formulations and has stabilizing effects on other proteins. However, which mechanism makes it an effective lyoprotectant stabilizer is not yet understood. Examples of commercial protein formulations containing albumin can be found in Table 8.4.

Because albumin is a natural protein, concerns were raised about potential contamination of albumin with human prion protein, which is thought to be the infectious agent in bovine spongiform encephalopathy (BSE) (Pharmaceutical Research and Manufacturers of America, 1998). Animal source materials included not only albumin but also gelatin, glycerol, and polysorbate 80. However, today the use of synthetic versions of these materials has eliminated concerns over potential disease transmission.

8.4.7 OTHER PHYSICAL COMPLEXING/STABILIZING AGENTS

Polyethylene glycol (PEG) is a common cosolvent for solubilizing small nonproteinaceous molecules, yet it also has been reported to minimize the aggregation of several peptides and proteins (Arakawa and Timasheff, 1985; Bhat and Timasheff, 1992; Lee and Lee, 1987; Powell et al., 1991). The concentration of PEG needs to be fairly low (<1%, w/v) to serve as a stabilizer; at higher concentrations (>10% w/v), it can cause precipitation (Cleland and Randolph, 1992). PEG modification of proteins for sustained-release purposes is beyond the scope of this chapter (see Francis et al., 1992; also see discussion in Chapter 7).

Poly(vinylpyrrolidone) (PVP) can at low concentrations stabilize proteins, whereas at high concentrations, it may lead to protein aggregation and precipitation. Gombotz et al. (1994) reported that PVP at low concentrations (≤2.0%) effectively stabilizes human immunoglobulin M (IgM) monoclonal antibody against heat-induced aggregation, whereas PVP concentrations ≥ 5.0% will cause aggregation.

Fibroblast growth factors, acidic and basic, are prone to acid and thermal inactivation and can be stabilized by a number of heparin and heparin-like molecules (Tsai et al., 1993). Human keratinocyte growth factor, also prone to aggregation at high temperatures, is stabilized by heparin, sulfated polysaccharides, anionic polymers, and citrate ion (Chen et al., 1994).

A new class of alkylsaccharide excipients originally intended to dramatically enhance transmucosal absorption of peptide and protein drugs was found to be highly effective in preventing protein aggregation (Maggio, 2008). These alkylsaccharide excipients at 0.125% concentrations stabilize and reduce aggregation of peptides or proteins in therapeutically useful insulin or growth hormone formulations, and they may provide solutions for aggregation-related manufacturing or formulation problems and/or unwanted immunogenicity.

8.5 PARTICULATE MATTER AND BIOPHARMACEUTICAL SOLUTIONS

Accurate measurement of particles in therapeutic biopharmaceutical solutions is especially important because of the potential safety problems associated with the effects of small particles on protein aggregation. Protein aggregation is believed to be one of the causes of immune responses in patients administered with these products (Section 8.6 and further discussion in Chapter 12).

Protein and other biopharmaceutical molecules form particles with a huge range of sizes (1 mm down to 1 nm, a range of one-million size units) and shapes. A major challenge is to find particle counting and sizing methods that can comprehensively characterize this huge range in actual biopharmaceutical solution dosage forms.

United States Pharmacopoeial (USP) and European Pharmacopoeial (EP) tests for subvisible particulate matter (light obscuration and light microscopy) are not sufficiently adequate to detect submicron particles that may/will form in biopharmaceutical solutions that could result in the eventual formation of large subvisible and eventual visible protein aggregates. Thus, especially during early development of biopharmaceutical solutions, other methods for detection of particulate matter formation might be more useful. Such methods include laser diffraction particle analyzers, polarization intensity differential scattering, dynamic image analysis, and Raman spectroscopy (Das and Nema, 2008). Also, of course, during formulation development of biologic products, methods such as size-exclusion chromatography and dynamic light scattering are very important to measure soluble protein aggregates as a function of formulation, processing effects, final packaging, and storage stability.

While compendial and regulatory standards exist for particulate matter in solutions released from manufacture for use commercially or in clinical studies, there are no specific standards for particulate matter in solutions just before injectable administration. A survey of commercial biopharmaceutical products' package inserts contained in the Physicians Desk Reference revealed a wide variety of statements for the acceptability of visible particulate matter and use of syringe or in-line filters just before product injection (Table 8.6). Some summary remarks from this table are as follows:

- Different product solutions require different filter porosity sizes, ranging from 0.2 to 15 μm.
- Some products are incompatible with in-line filters.
- Some products require rejection if visible particles are seen, but most do not require rejection; rather use the described filter.
- There is no standard of language or grammar used for filter descriptions or particle descriptions.

A 2007 special article in *Hospital Pharmacy* listed 53 commercial drug products that require a filter for their preparation in the hospital pharmacy and/or administration of the product to a patient (Anon., 2007). Updated articles will likely appear over time in this same journal.

TABLE 8.6
Sterile Product Package Insert Statements Regarding Particulate Matter in and/or Filter Usage for Solutions Prior to Being Injected

Excerpts of Statements Regarding Use of Filters

Withdraw solution from vial using an enclosed sterile filter needle.

Withdraw solution into a syringe through a low protein-binding 0.2- or 0.22-μm filter.

Withdraw solution into a syringe and filter the injection using a sterile, nonpyrogenic, low protein-binding 0.2- or 0.22-μm filter (filter vendor and type may be provided).

PVC infusion set equipped with an in-line, low protein-binding 0.2-μm filter.

Colorless solution that may contain translucent particles; use a 0.22-μm low protein-binding filter to be in-line between the syringe and the infusion port.

Reconstituted solution is not transparent; any undissolved particulate matter is difficult to see when inspected visually. Therefore, terminal filtration through a sterile 0.45-μm or smaller filter is recommended.

The reconstituted solution can be filtered through a 0.8-μm or larger-pore-size filter.

A separate IV line equipped with a low protein-binding 1.2-μm terminal filter should be used for administration of the drug.

Occasionally, a very small number of gelatinous fiber-like particles may develop on standing. Filtration through a 5.0-μm filter during administration will remove the particles with no resultant loss in potency.

Withdraw the necessary amount of product from the ampoule into a syringe; filter with a sterile, low-protein binding, nonfiber releasing 5-μm filter before dilution.

Use a 15-μm filter.

Administered through an intravenous line using an administration set that contains an in-line filter (pore size 15 μm). A smaller in-line filter (0.2 μm) is also acceptable.

Reconstituted product does not need to be filtered. If a filter is used, it should be a 15-μm filter or larger.

Do not use in-line filters.

Do not use an in-line filter.

Do not filter the reconstituted solution.

Excerpts from Statements Regarding Appearance of Particles

Solution should be clear to slightly opalescent and colorless to pale yellow. A few translucent particles may be present. Do not use if there is particulate matter in the solution.

Parenteral drug products should be inspected for visible particulate matter and discoloration before administration. If particulate matter is present or the solution is discolored, the vial should not be used.

Parenteral drug products should be inspected visually for particulate matter and discoloration before administration whenever solution and container permit. If visibly opaque particles, discoloration, or other foreign particulates are observed, the solution should not be used.

If visibly opaque particles, discoloration or other foreign particulates are observed, the solution should not be used.

There should be no visible gel-like particles in the solution. Do not use if foreign particles are present.

Solution should be clear immediately after reconstitution. Do not inject if the reconstituted product is cloudy immediately after reconstitution or after refrigeration (2–8°C/36–46°F) for up to 14 days. Occasionally, after refrigeration, small colorless particles may be present in the solution. This is not unusual for proteins like (protein name).

Because the product is a protein, shaking can result in a cloudy solution. The solution should be clear immediately after reconstitution. Do not inject if the reconstituted product is cloudy immediately after reconstitution or refrigeration. Occasionally, after refrigeration, small colorless particles may be present in the solution. This is not unusual for proteins like (protein name).

(Continued)

TABLE 8.6   (*Continued*)

Sterile Product Package Insert Statements Regarding Particulate Matter in and/or Filter Usage for Solutions Prior to Being Injected

The solution should be clear immediately after removal from the refrigerator. Occasionally, after refrigeration, you may notice that small colorless particles of protein are present in the solution. This is not unusual for solutions containing proteins. If the solution is cloudy, the contents must not be injected. There should be no visible gel-like particles in the solution. Do not use if foreign particles are present. It is acceptable to have small bubbles or foam around the edge of the vial. Do not use if the contents of the vial do not dissolve completely in 40 minutes.

Other Examples of Package Insert Statements Related to Particles or Filters

Some loss of potency has been observed with the use of a 0.2-μm filter.

After reconstitution, product should be inspected visually before use. Because this is a protein solution, slight flocculation (described as thin translucent fibers) occurs occasionally after dilution. The diluted solution may be filtered through an in-line low protein-binding 0.2-μm filter during administration. Any vials exhibiting opaque particles or discoloration should not be used.

Thin translucent filaments may occasionally occur in reconstituted product vials, but do not indicate any decrease in potency of this product. To minimize formation of filaments, avoid shaking the vial during reconstitution. Roll and tilt the vial to enhance reconstitution. The solution may be terminally filtered, for example, through a 0.45-μm or smaller cellulose membrane filter.

- Since reconstituted product is not transparent, any undissolved particulate matter is difficult to see when inspected visually. Therefore, terminal filtration through a sterile 0.45-μm or smaller filter is recommended.

Source: Physicians' Desk Reference (2004) Montvale, NJ, Medical Economics.

Note: Use of the word ("product"), "protein name" (or "solution") replaces actual product stated in insert.

8.6 FOREIGN PARTICLES, PROTEIN AGGREGATION, AND IMMUNOGENICITY

The reality of protein aggregation has raised concerns about such aggregates, even at subvisible levels, leading to an immune response resulting from antibody-mediated neutralization of the protein's activity or alterations in bioavailability (Carpenter et al., 2009; Rosenberg, 2006). Among many causes for protein aggregation are protein particles resulting either from bioprocessing (Cromwell et al., 2006) or from heterogeneous nucleation on foreign micro- or nanoparticles originating from the manufacturing process (mixing tanks, process tubing, filter systems, filling machines [Tyagi et al., 2009], or any other stainless steel, rubber, glass, or plastic surface) (Sharma, 2007) and from the container–closure system (Tyagi et al., 2009). Silicone oil, used as a lubricant for rubber closures on vials on rubber plungers in prefilled syringes and to coat the inner surface of glass syringes and cartridges, also can induce protein aggregation (Esfandiary et al., 2010; Jones et al., 2005; Sharma et al., 2009; Thirumangalathu et al., 2009; and further discussion under Section 8.9).

The potential for protein aggregation in silicone-coated glass syringes drives the marketing of plastic syringes for biopharmaceutical products since the plastic surface does not require silicone for facile movement of the rubber plunger and plunger rod through the plastic barrel. Manufacturers of plastic syringes have developed alternatives to silicone to provide lubricity within the plastic composition of the syringe to achieve acceptable functional performance. However, use of plastic syringes has not yet been proven to be a suitable replacement for glass syringes for biopharmaceuticals, and there are other challenges to overcome using plastic containers with biopharmaceuticals (leachate potential, sterilization of components, cost, and reliability). Until plastic syringes without the presence of silicone become more common, continuous improvements in the consistent application and distribution of silicone in glass syringe barrels must be pursued.

Large protein aggregates are subvisible particles (<10 μm) that are not currently monitored and quantified by compendial subvisible particulate matter measurement systems. Carpenter et al. (2009) have questioned this current practice and have proposed that (1) scientists from industry and academia should work together to define the quantitative capabilities of particle counting instruments for particles as small as 0.1 μm, (2) researchers should develop new particle-counting instruments for more reliable measurement of particles at sizes approaching 0.1 μm, and (3) more studies should be conducted and published on the impact of protein aggregation on immunogenicity including the role of protein class, amount of aggregate, size of aggregates, and protein conformation in aggregates.

Carpenter et al.'s commentary, all coauthors being in academia or working for the FDA, was followed up about a year later with a response commentary from 13 industrial scientists (Singh et al., 2010). In the opinion of these industrial pharmaceutical scientists,

1. A link between aggregation and clinical immunogenicity has not been unequivocally established, although research must be continued to determine conclusively that a link either exists or does not exist.
2. Current limitations in particle size measurement technologies (low precision and reproducibility) at sizes less than 10 μm and variability from container–closure, concentration, viscosity, history, and inherent batch heterogeneity all contribute to the reality that establishing particle number limits at particle sizes less than 10 μm for product release and for stability testing or product comparability testing would be unreasonable and indefensible.
3. Further research and new technology is needed to accurately measure particle levels less than 10 μm, especially if such new technologies could accurately measure proteinaceous subvisible particles such that standards could be established.
4. Likely all marketed protein products contain large but varying numbers of subvisible particles in size less than 10 μm, but these small particles represent protein content in fractions of micrograms. The largest portion of subvisible particles ranges from 10 μm or above that are measured and monitored over time, especially during clinical development of these products.
5. Other protein product quality attributes such as oxidation, deamidation, charge-variants, and glycosylation-variants are neither detected or reported at these low levels and are considered well below safety concerns.

6. Characterization of product quality attributes, including subvisible particulates, is performed on all products during clinical development. Safety and efficacy of biopharmaceutical products are established by appropriate designed nonclinical and clinical research studies. Immunogenicity is monitored and evaluated during clinical studies with product quality attributes, including subvisible particulate matter, that will be relevant for safety and/or efficacy identified and qualified with appropriate and reproducible testing methodologies.

8.7   OPTIMIZING MICROBIOLOGICAL ACTIVITY

AP agents are required for parenteral products intended for multiple dose use. Many protein products, because of their expense and, more importantly, because of their ability to support the growth of microorganisms, are packaged as multiple dose products, the APs either formulated with the protein or, more commonly, formulated in a special diluent to be combined with the protein before use. The most common APs used in protein dosage forms are phenol, meta-cresol, benzyl alcohol, and methyl and propylparaben (Meyer et al., 2007). Examples of use of these preservatives are listed in Table 8.7.

The use of antimicrobial agents requires passing a preservative efficacy test (PET). Unfortunately, the USP and the British and/or European Pharmacopoeial (BP/EP) tests for PET are different in their requirements. Table 8.8 summarizes the differences between the tests. The USP basically requires a bacteriostatic preservative system, whereas the BP/EP requires a bacteriocidal system. For example, the USP requires a 3-log reduction in the bacterial challenge by the 14th day after

TABLE 8.7
Antimicrobial Preservative Agents for Protein Products

Type	Concentration	Antimicrobial activity[a]				pH	Other Comment
		Gram+	Gram−	Fungi	Yeast		
Benzyl alcohol	0.1–3.0%	+++	+++	~	−	3–6	Not effective pH>7
m-Cresol	0.1–0.3%	++	++	++	++	4–10	Most effective
Methylparaben	0.08–0.1%	++	++	++	++	3–9	Slowly soluble
Propylparaben	0.001–0.02%	+++	+++	+++	+++	3–9	Very slowly soluble
Phenol	0.2–0.5%	++	++	++	++	4–10	Most effective
Thimerosal	0.1–0.4%	++	++	++	++	4–8	Japan will not allow

[a] +++, most effective; ++, moderately effective; −, poor.

TABLE 8.8

Comparison of USP 24 and EP 2 Requirements for Preservative Efficacy Testing

Time after Inoculation with Microorganisms	Log Reduction in Microorganism Count		
	USP 24	EP "A" Criteria	EP "B" Criteria
6 hours	Not required	3 (bacteria)	Not required
24 hours	Not required	No recovery (bacteria)	1 (bacteria)
2 days	Not required	No recovery (bacteria)	Not required
7 days	1 (bacteria)	No recovery (bacteria) 2 (fungi)	3 (bacteria)
14 days	3 (bacteria) No increase (fungi)	No recovery (bacteria) No increase (fungi)	1 (fungi)
21 days	No increase	No increase	Not required
28 days	No increase	No recovery (bacteria) No increase (fungi)	No increase (bacteria) No increase (fungi)

inoculation, whereas criteria A of the BP/EP test requires that same 3-log reduction within 24 hours. This great difference in compendial requirements for preservative efficacy has caused many problems in the formulation of protein dosage forms for various markets. Passing the BP/EP preservative efficacy test requires the use of relatively high amounts of phenol or cresol or other AP that may have an impact on the stability of the formulation and could result in sorption of the preservative into the rubber closure. The formulator must keep in mind that increasing the concentration of APs may have a negative impact on protein physical stability (precipitation, aggregation, etc.). Increasing AP levels will increase the hydrophobicity of the formulation and could affect the aqueous solubility of the protein. Increasing AP concentrations also increases the potential for toxicological hazards.

It is well known that APs not only protect insulin formulations against inadvertent contamination, but also may have a significant effect on protein stability. For example, phenolic preservatives have a profound effect on the conformation of insulin in solutions (Wollmer et al., 1987) and the assembly of the specific type of LysPro insulin hexamer (Birnbaum et al., 1997). Furthermore, phenol and/or m-cresol in insulin solutions will have a tendency to be adsorbed by and permeate rubber closures (Brange, 1987). Certain halobutyl rubber closure formulations have proved to be resistant to potential adsorption, absorption, and/or permeation of volatile AP molecules.

APs are known to interact with proteins and can cause stability problems such as aggregation. For example, phenolic compounds will cause aggregation of hGH (Kirsch et al., 1993; Maa and Hsu, 1996). Phenol will produce a significant decrease in the α-helix content of insulinotropin resulting in aggregation into β-sheet-rich structures (Kim et al., 1994). Benzyl alcohol, above certain concentrations

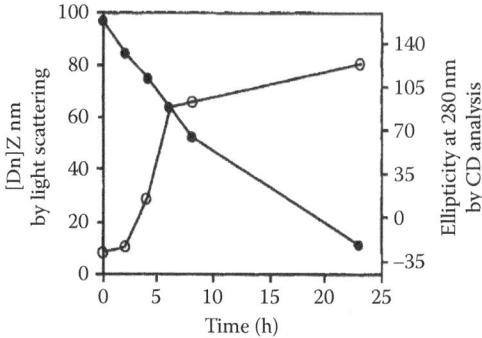

FIGURE 8.6 Time course of aggregate formation of rhIFN-γ 1.0 mg/mL in 5 mM succinate (pH 5.0) in the presence of 0.9% benzyl alcohol as determined by dynamic light scattering (o) and circular dichroism analysis (●). *Note:* Tobler et al. (2004) used hydrogen–deuterium isotope exchange detected by mass spectrometry (MS) to detect tertiary structure changes that involve only a limited part of this protein, still causing irreversible loss of activity. Benzyl alcohol causes protein to unfold, forming very large aggregates. (Reprinted from Lam, X. M., et al., *Pharm. Res.*, 14, 725–729, 1997a. With permission.)

and depending on other formulation factors, will interact with recombinant human interferon-γ (rhIFN-γ), causing aggregation of the protein (Figure 8.6) (Lam et al., 1997a; Tobler et al., 2004). These examples point out the need for the formulation scientist to understand the importance of potential effects of preservative type, concentration, and other formulation additives on the interaction with proteins in solutions while balancing the needs for antimicrobial efficacy.

In determining the appropriate AP agent or agents, the model described by Akers et al. (1984) might be appropriate to follow. The authors used insulin as the protein to be preserved and combined insulin with different types of AP agents either alone or in combination. These formulations were challenged with the five USP preservative efficacy test organisms and *D* values* were determined. The *D*-value determination allows a single quantitative estimate of the AP effectiveness of a certain agent or combination of agents in a specific formulation against a specific microorganism. The AP system with the lowest *D* value in insulin formulations against *Staphylococcus aureus* was 0.2% phenol + 0.3% m-cresol. Also very effective was 0.3% m-cresol alone and 2.0% benzyl alcohol. The least effective AP systems were the parabens and lower concentrations of m-cresol (0.2%) and phenol (0.2%) alone or in combination.

There are instances where a manufacturer, because of concerns regarding aseptic processing and sterility assurance of the product throughout its shelf-life, will add an AP agent in the protein formulation even though it is intended only for a single-dose injection. This is a very controversial practice. Regulatory agencies worldwide object

---

* *D* value = Time required for a one-log reduction in the microbial population due to the effect of the antimicrobial preservative system. The smaller the *D* value, the greater the effect of the preservative on the microorganism in question.

to this approach if, in their opinion, the use of APs in a single-dose injectable product is practiced in order to cover up for inadequate aseptic manufacturing practices and controls.

Many countries require PETs be performed for routine stability protocols and for special stability studies. Also, there may be requests from agencies to do PET on containers that have been used (i.e., penetrated and partial volume withdrawn) to demonstrate that the product can still kill microorganisms. In mid-1995, the Australian Drug Evaluation Committee (ADEC) passed resolutions that in light of safety concerns with contamination and cross-contamination, the use of inject-able products in multidose packages is discouraged. In order to support the use of a multidose product and the shelf-life once a package has been reconstituted or opened for use, AP efficacy data are required for approval.

8.8 OSMOLALITY (TONICITY) AGENTS

Salts or nonelectrolytes (e.g., glycerin) are added to protein formulations in order to achieve an isotonic solution. Nonelectrolytes often are preferred over salts as tonic-ity adjusters because of the potential problems salts cause in precipitating proteins (Pikal, 1990). In general, solutions containing proteins administered intravenously (IV), intramuscularly, or subcutaneously do not have to be precisely isotonic because of immediate effects from dilution by the blood. Intrathecal and epidural injections into the cerebrospinal fluid require very precise specifications for the product to be isotonic and at physiological pH. This is because extremes in osmolality and/ or pH can damage or destroy cells and cerebrospinal cells cannot be reproduced or replaced (Cradock et al., 1977).

8.9 PACKAGING

Solution dosage forms of peptides and proteins most commonly are packaged in glass vials sealed with rubber closures and aluminum seals. Other packaging sys-tems used are cartridges, syringes, and plastic vials and bottles. In all these packag-ing systems, the formulator needs to be very concerned about potential reactivity between the peptide/protein and other ingredients in the formulation (e.g., APs) and the packaging components.

Selection of the packaging system not only depends on compatibility with the product formulation and the convenience to the consumer, but also depends on the integrity of the container–closure interface to assure maintenance of sterility throughout the shelf-life of the product. Container–closure integrity testing has received significant attention recently and usually is an integral part of the regula-tory submission and subsequent regulatory GMP inspections. While it is beyond the scope of this chapter to discuss the various container–closure integrity issues (see Guazzo, 2010), it is emphasized that formulation scientists developing the final product including the final package must appreciate the need to develop appropri-ate testing methods to assure that the selected packaging system indeed has the proper seal integrity to protect the product during its shelf-life from any ingress of microbiological contamination.

8.9.1 SILICONE

Historically, silicone oil or emulsion has been applied to rubber closures for facilitating movement through high speed stoppering machines and to glass syringes and cartridges for facile movement of a rubber plunger through the lumen of the primary container. Siliconization of rubber closures still takes place today for older NDA products, but for the past several years, the need for siliconization of rubber has been replaced by using coated rubber closures (also called rubber laminates) that do not require siliconization. To reduce the problem of leachables, laminates have been applied to the product contact surfaces of closures, with various polymers, the most successful being Teflon® polytetrafluoroethylene (PTFE), FluroTec® (West/Daikyo fluorocarbon film, made from a modified ethylene-tetrafluoroethylene [ETFE]), and Omniflex® (Helvoet, a fluorinated polymer coating).

Coated or laminated stoppers offer the following advantages compared with stoppers that must be siliconized:

- Eliminates the need for adding silicone oil
- Provides lubricity for machinability
- Reduces rubber stopper clumping problems
- Decreases particulate matter levels
- Reduces potential for formulation adsorption and absorption
- Reduces chemical extractable levels

Polymeric coatings have been developed that are claimed to have more integral binding with the rubber matrix, but details of their function are trade secrets.

While siliconization of glass syringes and cartridges is still necessary, alternatives include newly developed coated prefilled syringes (e.g., BD-42, a proprietary coating developed by Becton-Dickinson [Majumdar et al., 2011]), or, for syringes, switching from glass to plastic.

8.9.2 EXTRACTABLES/LEACHABLES

Components of biopharmaceutical formulations potentially will interact with rubber closures (as well as surfaces of manufacturing equipment) to produce leachables (Bee, 2009; Yu, 2010). Using gas chromatography–mass spectrometry and liquid chromatography–mass spectrometry, Yu et al. (2010) showed that formulations containing polysorbate 80 have a significant impact on appearance of leachables in vials stoppered with chlorobutyl rubber. Polysorbate 80 will extract rubber components such as 2-methylpentane, 3-methylpentane, hexane, methylcyclopentane, and cyclohexane. In contrast, trehalose has minimal impact on leachable formation. Formulations containing sucrose and mannitol will lead to leachable levels higher than trehalose formulations, but much less than polysorbate 80 formulations. Extracts from rubber stoppers stored in phosphate buffer solution at pH 6.8 yielded the following leachables: 3-methylpentane, hexane, methylcyclopentane, cyclohexane, and butyl hydroxytoluene. The chelating agent DSEDTA did not significantly affect the amount of leachables, while formulation pH did matter with slightly acidic (pH 5.0) and slightly alkaline (pH 8.2) formulations having similar and higher leachable profiles than neutral (pH 6.8) formulations. The results of this study demonstrated that changes in formulation components can lead to appreciable changes in leachable profiles.

Some proteins are added to intravenous infusion bags containing either 0.9% sodium chloride or 5% dextrose. Chang et al. (2010) reported that an elevated loss of dulanermin monomers was observed in 100-mL polyolefin bags containing 0.9% sodium chloride (but not polyvinyl chloride bags). The plastic bags used were found to leach 2-mercaptobenzothiazole (measured by LC-UV-high resolution MS/MS analysis) and its zinc salt (measured by inductively coupled plasma-MS). These leachables came not from the plastic bag, but from the 13-mm rubber stopper in the Med-Add port of the bag. The manufacturer subsequently corrected this problem by using nonlatex components in the IV bag.

Extractables from rubber closures gained more attention from regulatory authorities and the sterile products industry after the EPREX® incident, where extractables from uncoated rubber plungers in syringes containing EPREX® acted as an adjuvant to stimulate the formation of anti-EPO antibodies causing antibody-positive pure red cell aplasia (PRCA). The aqueous formulation contained polysorbate 80 that facilitated the leaching of small-molecule, aromatic compounds from the uncoated syringe rubber plungers (Pang et al., 2007). Extractables were eliminated by using Fluorotec rubber plungers.

All types of glass have the potential to leach alkali-based substituents into the product as well as glass particles, a phenomenon called delamination (Section 8.9.3). Glass leachables are most problematic when the product solution pH is ≥ 9. Major extractables include silicon, sodium, and boron, while other extractables might include potassium, barium, calcium, and aluminum, depending on the specific glass formulation. Extractables can be minimized using ammonium-sulfate-treated (or other type of treatment) glass.

The presence of phosphate (e.g., phosphate buffer) anions make protein solution formulations especially vulnerable to leachables because of the distinct possibility of phosphate-forming insoluble salts with divalent metal cations (e.g., calcium, iron, zinc, and magnesium) potentially present at the surface of the glass interior. The amount of potential extractable ions at the inner glass surface is a function of how the glass was manufactured, which temperatures were used, and exposure to high temperatures (e.g., glass sterilization).

Incompatibilities between glass and product may include the following:

1. Ion exchange of metal ions if the product contains sodium, magnesium, calcium, aluminum, or lithium
2. Dissolution of glass and resultant particles if the product contains phosphate or citrate
3. Pitting of glass resulting in particles if the product contains a metal-chelating agent such as DSEDTA
4. Adsorption of the active ingredient at the glass surface, a major problem for many biomolecules requiring the use of competitive binding excipients in the formulation

If any of these problems occur during product development, then treated glass must be used or the formulation must be modified to remove or reduce the amount of ingredient reacting with the glass surface.

Glass syringes present an interesting case where an additional extractable did not directly originate from the glass. The inner needle channel in glass syringes is often formed using a tungsten pin (Bee et al., 2009; Jenke, 2008). Residual tungsten can remain on the glass depending on the processing conditions. The residual tungsten can interact with proteins and lead to aggregation.

Scientists at Genentech have developed a thorough and holistic extractable and leachable program that contains elements of risk-assessment, literature review, and consolidation of the best industry practices (Wakankar et al., 2010). Their extractable-leachable program includes six stages:

1. Selection of components
2. Determination of analytical procedures, including extraction conditions and analytical methods, to be used on components identified based on a risk-assessment approach
3. Selection of target leachables
4. Performing the leachable study
5. Health-based risk assessment of leachables
6. Life-cycle management

The bottom line is that any surface to which a biopharmaceutical active component can make contact potentially is a source of either a binding problem or a leachable problem.

8.9.3 DELAMINATION

The phenomenon of small flakes of glass appearing over time in a protein solution is called delamination. Lots from three commercial products—Epogen®, Procrit®, and Hylenex®—were recalled in 2010 because of glass flakes appearing in these solutions. This problem is not new; it has been described in the literature for many years, but it is not understood very well.

Several factors contribute to glass delamination (Ennis et al., 2001; Iacocca et al., 2010):

- Temperatures during glass container formation—Excessive temperatures can create surface imperfections such as pitting and/or deposits that later can produce flaking.
- Glass composition—Tubing vial more likely to delaminate than mold vials.
- Glass treatment—Ammonium-sulfate-treated tubing vials more vulnerable to delamination than nontreated vials.
- Sterilization temperature—Dry heat sterilization of empty containers does not affect delamination potential, but steam sterilization temperatures for products that can be terminally sterilized very well might affect glass delamination.

- Product solution pH and composition—Delamination occurs more readily at alkaline pHs and citrate buffer and sodium chloride can accelerate delamination.
- Mechanical stress during the formation of glass containers.
- Storage time.

Alkali borates can migrate to the inner surface, evaporate during the glass-forming process, then recondense during final glass preparation. Migration can be minimized by use of proper temperatures and times during the glass forming and annealing process. Flaking potential can increase when using sulfur treatment that strips sodium from the surface, especially in the heel. Solutions with alkaline pH dissolve the silica network and can increase flaking potential. Excessive water in vials during dry heat sterilization also can increase flaking potential. Smaller vials are prone to trapping water.

Iacocca et al. (2010) pointed out that delamination is not defined by appearance of visible glass particles in the product solution, but more correctly is defined by the observation of flakes on the interior surface of the glass container, detectable by scanning electron microscopy and/or dynamic secondary ion mass spectroscopy.

8.10 PROCESSING

Unit processes involved in the manufacturing of solution sterile dosage forms include compounding and mixing, filtration, filling, terminal sterilization (although almost always not possible for proteins and peptides), closing and sealing, sorting and inspection, labeling, and final packaging for distribution. Since this chapter deals only with solution dosage forms, one of the most complex processes, lyophilization or freeze-drying, will not be covered (see Chapter 10).

The pharmaceutical scientist must be aware of the various issues involved in the manufacturing arena that can impact the stability and quality of the protein or peptide formulation. Among the more relevant areas of concern are the following:

- Shear rate and stress during compounding, filtration, filling, and distribution (Charm and Wong, 1976; Kiese et al., 2008; Maa and Hsu, 1996; Thurow and Geisen, 1984)
- Adsorption onto process tubing, filter surfaces, and other manufacturing systems (Brophy and Lambert, 1994; Brose and Waibel, 1996; Hawker and Hawker, 1975; Itoh et al., 1995; Nakanishi et al., 2001)
- Particle generation from surfaces of equipment used during manufacturing that can lead to the formation of protein aggregates (Tyagi et al., 2009)
- The effects of time and temperature during each step of the manufacturing process (Hsu et al., 1988)
- The effects of highly viscous solutions on the efficiency of filtration and control of filling operations (Yadav et al., 2010a, 2010b)

Formulation scientists and process engineers must work together to design and implement experiments to determine processing effects on protein stability and establish an appropriate control strategy. In most cases, for example, protein adsorption onto filter surfaces, the potential problems can be avoided or minimized once understood through experimentation by alternative choices of filter material or predicting the amount of solutions to be passed through the filter to saturate the binding sites.

The surge of potential heat labile products from biotechnology and the inability to terminally sterilize these molecules has accelerated the development of barrier/isolator technology. This technology, when perfected, will enable the processing of protein and peptide solutions to occur under a much higher degree of sterility assurance than what is now achievable with conventional aseptic processing. The main features of barrier/isolator technology are the ability to sterilize, not just sanitize, the environment under which sterile solution is exposed during filling and stoppering, and the removal of humans from direct contact with the exposed sterile solution.

8.11 CONCLUSION

Formulation of stable, manufacturable, elegant, and high-quality protein and peptide solution preparations offer significant challenges to the development scientist. However, these challenges can be overcome through sound, rational formulation approaches with manufacturing processes and packaging systems designed to maintain stability and other quality features of the formulation. This chapter has provided current approaches and thinking on formulation science involved in the formulation, package, and process development of commercially viable protein and peptide solution dosage forms.

REFERENCES

Absolom, D.R., Zingg, W., and Neumann, A.W., 1987. Protein adsorption to polymer particles: Role of surface properties. *J. Biomed. Mater. Res.* 21:161–171.

Akers, M.J., 1982. Antioxidants in pharmaceutical products. *J. Parenter. Sci. Tech.* 36:222–228.

Akers, M.J., 2010. *Sterile Drug Products: Formulation, Packaging, Manufacturing, and Quality.* Informa, London, p. 516.

Akers, M.J., Boand, A., and Binkley, D., 1984. Preformulation method for parenteral preservative efficacy evaluation. *J. Pharm. Sci.* 73:903–907.

Anik, S.T., and Hwang, J.Y., 1983. Adsorption of D-Nal(2)6LHRH, a decapeptide, onto glass and other surfaces. *Int. J. Pharm.* 16:181–190.

Anon., 2007. Drugs to be used with a filter for preparation and/or administration. *Hosp. Pharm.* 42:378–382. www.factsandcomparisons.com/assets/./20070401_apr2007_spec.pdf

Arakawa, T., and Timasheff, S.N., 1982. Preferential interaction of protein with salts in concentrated solution. *Biochemistry* 21:6545–6552.

Arakawa, T., and Timasheff, S.N., 1984. Mechanism of protein salting in and salting out by divalent cation salts: Balance between hydration and salt binding. *Biochemistry* 23:5912–5923.

Arakawa, T., and Timasheff, S.N., 1985. Mechanism of poly(ethylene glycol) interaction with proteins. *Biochemistry* 24:6756–6762.

Arakawa, T., Kita, Y., and Carpenter, J.F., 1991. Protein-solvent interactions in pharmaceutical formulations. *Pharm. Res.* 8:285–291.

Arakawa, T., Prestrelski, S.J., Kenney, W.C., and Carpenter, J.F., 1993. Factors affecting short-term and long-term stabilities of proteins. *Adv. Drug Deliv. Rev.* 10:1–29.

Asahara, K., Yamada, H., and Yoshida, S., 1991. Stability of human insulin in solutions containing sodium bisulfite. *Chem. Pharm. Bull.* 39:2662–2666.

Bain, B., and Shortmoor, J., 2010. Pharma market trends 2010. *Pharm. Tech.* 34:S38–S45.

Bam, N.B., Randolph, T.W., and Cleland, J.L., 1995. Stability of protein formulations: Investigation of surfactant effects by a novel EPR spectroscopic technique. *Pharm. Res.* 12:2–11.

Bee, J.S., Chiu, D., Sawicki, S., Stevenson, J.L., Chatterjee, K., Freund, E., Carpenter, J.F., and Randolph, T.W., 2009. Monoclonal antibody interactions with micro- and nanoparticles: Adsorption, aggregation, and accelerated stress studies. *J. Pharm. Sci.* 98:3218–3238.

Bhat, R., and Timasheff, S.N., 1992. Steric exclusion is the principle source of the preferential hydration of proteins in the presence of polyethylene glycols. *Protein Sci.* 1:1133–1143.

Birnbaum, D.T., Kilcomons, M.A., DeFelippis, M.R., and Beals, J.M., 1997. Assembly and dissociation of human insulin and Lys Pro-insulin hexamers: A comparison study. *Pharm. Res.* 14:25–36.

Boyett, J.B., and Avis, K.E., 1976. Extraction rates of marker compounds from rubber closures for parenteral use II. Mechanism of extraction and evaluation of select extraction parameters. *Bull. Parenter. Drug Assoc.* 30:169–177.

Brange, J., 1987. *Galenics of Insulin*. Springer-Verlag, New York, p. 41.

Brange, J., Havelund, S., and Hougaard, P., 1992a. Chemical stability of insulin. 2. Formation of higher molecular weight transformation products during storage of pharmaceutical preparations. *Pharm. Res.* 9:727–734.

Brange, J., and Langkjaer, L., 1993. Insulin structure and stability. In *Stability and Characterization of Protein and Peptide Drugs: Case Histories*, Wang, Y.J., and Pearlman, R., eds., Plenum Press, New York, 315–350.

Brange, J., Langkjaer, L., Havelund, S., and Volund, A., 1992b. Chemical stability of insulin. 1. Hydrolytic degradation during storage of pharmaceutical preparations. *Pharm. Res.* 9:715–726.

Brennan, J.R., Gebhart, S.S., and Balckard, W.G., 1985. Pump-induced insulin aggregation. *Diabetes* 34:353–359.

Brewster, M.E., Hora, M.S., Simpkins, J.W., and Bodor, N., 1991. Use of 2-hydroxypropyl-β-cyclodextrin as a solubilizing and stabilizing excipient for protein drugs. *Pharm. Res.* 8:792–795.

Brophy, R.T., and Lambert, W.J., 1994. The adsorption of insulinotropin to polymeric sterilizing filters. *J. Parenter. Sci. Tech.* 48:92–94.

Brose, D.J., and Waibel, P., 1996. Adsorption of proteins in commercial microfiltration capsule. *Pharm. Tech.* 20:48–52.

Capasso, S., Mazzarella, L., and Zagari, A., 1991. Deamidation via cyclic imide of asparaginyl peptides: Dependence on salts, buffers, and organic solvents. *Pept. Res.* 4:234–238.

Carpenter, J.F., Randolph, T.W., Jiskoot, W., Crommelin, D.J., Middaugh, C.R., Winter, G., Fan, Y.X., et al., 2009. Overlooking subvisible particles in therapeutic protein products: Gaps that may compromise product quality. *J. Pharm. Sci.* 98:1201–1205.

Chang, B.S., Kendrick, B.S., and Carpenter, J.F., 1996. Surface-induced denaturation of proteins during freezing and its inhibition by surfactants. *J. Pharm. Sci.* 85:1325–1330.

Chang, J.Y., Xiao, N.J., Zhang, J., Hoff, E., Russell, S.J., Katt, V., and Shire, S.J., 2010. Leachables from saline-containing IV bags can alter therapeutic protein properties. *Pharm. Res.* 27:2402–2413.

Charm, S.E., and Wong, B.L., 1976. Enzyme inactivation with shearing. *Biotech Bioeng.* 12:1103–1109.

Chen, B., Arakawa, T., Morris, C.F., Kenney, W.C., Wells, C.M., and Pitt, C.G., 1994. Aggregation pathway of recombinant human keratinocyte growth factor and its stabilization. *Pharm. Res.* 11:1581–1587.

Chi, E.Y., Krishnam, S., Randolph, T.W., and Carpenter, J.F., 2003. Physical stability of proteins in aqueous solution: Mechanism and driving forces in nonnative protein aggregation. *Pharm. Res.* 20:1325–1336.

Chou, D.K., Krishamurthy, R., Randolph, T.W., Carpenter, J.F., and Manning, M.C., 2005. Effects of Tween 20 and Tween 80 on the stability of albutropin during agitation. *J. Pharm. Sci.* 94:1368–1381.

Chuang, V.T., Kragh-Hansen, U., and Otagiri, M., 2002. Pharmaceutical strategies utilizing recombinant human serum albumin. *Pharm. Res.* 19:569–577.

Cleland, J.L., 1993. Impact of protein folding on biotechnology. In *Protein Folding In Vivo and In Vitro*, Cleland, J.L., ed., ACS Symposium Series 526, American Chemical Society, Washington, DC, 1–21.

Cleland, J.L., Powell, M.F., and Shire, S.J., 1993. The development of stable protein formulations: A close look at protein aggregation, deamindation and oxidation. *Crit. Rev. Ther. Drug Carrier Sys.* 10:307–377.

Cleland, J.L., and Randolph, T.W., 1992. Mechanism of polyethylene glycol interaction with the molten globule folding intermediate of bovine carbonic anhydrase B. *J. Biol. Chem.* 267:3147–3153.

Cradock, J.C., Kleinman, L.M., and Davignon, J.P., 1977. Intrathecal injections—A review of pharmaceutical factors. *Bull. Parenter. Drug Assoc.* 31:237–247.

Cromwell, M.E.M., Hilario, E., and Jacobson, F., 2006. Protein aggregation and bioprocessing. *AAPS J.* 8:E572–E579.

Dahlquist, F.W., Long, J.W., and Bigbee, W.L., 1976. Role of calcium in thermal stability of thermolysin. *Biochemistry* 15:1103–1111.

Danielson, J.W., Oxborrow, G.S., and Placencia, A.M., 1984. Quantitative determination of chemicals leached from rubber stoppers into parenteral solutions. *J. Parenter. Sci. Tech.* 38:90–93.

Das, T., and Nema, S., 2008. Protein particulate issues in biologics development. *Amer. Pharm. Rev.* 11:52–57.

Daugherty, A.L., and Mrsny, R.J., 2006. Formulation and delivery issues for monoclonal antibody therapeutics. *Adv. Drug Deliv. Rev.* 58:686–706.

Davio, S.R., and Hageman, M.J., 1993. Characterization and formulation considerations for recombinantly derived bovine somatotropin. In *Stability and Characterization of Protein and Peptide Drugs: Case Histories*, Wang, Y.J., and Pearlman, R., eds., Plenum Press, New York, pp. 76–80.

Deechongkit, S., Wen, J., Narhi, L.O., Jiang, Y., Park, S.S., Kim, J., and Kerwin, B.A., 2009. Physical and biophysical effects of polysorbate 20 and 80 on darbepoetin alfa. *J. Pharm. Sci.* 98:3200–3217.

Dehart, M.P., and Anderson, B.D., 2007. The role of cyclic imide in alternate degradation pathways for asparagines-containing peptides and proteins. *J. Pharm. Sci.* 96:2667–2885.

Dong, D.E., Andrade, J.D., and Coleman, D.L., 1987. Adsorption of low density lipoproteins onto selected biomedical polymers. *J. Biomed. Mater. Res.* 21:683–700.

Ennis, R.D., Pritchard, R., Nakamura, C., Coulon, M., Yang, T., Visor, G.C., and Lee, W.A., 2001. Glass vials for small volume parenterals: Influence of drug and manufacturing processes on glass delamination. *Pharm. Dev. Tech.* 6:393–405.

Esfandiary, R., Joshi, S.B., Vilivalam, V., and Middaugh, C.R., 2010. Characterization of protein aggregation and adsorption on prefillable syringe surfaces. www.westpharma.com/na/en/support/Scientific%20Posters/Characterization%20of%20Protein%20Aggreagation%20and%20Adsorption%20on%20Prefillable%20Syringe%20Systems.pdf (accessed August 10, 2010).

European Pharmacopoeia 7.5, 2012. Monograph 0520 *Parenteral Preparations Parenteralia'*. (Refer text under the section on Injections subheading Definition.)

Fish, W.W., Danielsson, A., Nordling, L., Miller, S.H., Lam, C.F., and Bjork, I., 1985. Denaturation behavior of antithrombin in guanidinium chloride: Irreversibility of unfolding caused by aggregation. *Biochemistry* 24:1510–1517.

Francis, G.E., Delgado, C., and Fisher, D., 1992. PEG-modified proteins. In *Stability of Protein Pharmaceuticals, Part B: In Vivo Pathways of Degradation and Strategies for Protein Stabilization*, Ahern, T.J., and Manning, M.C., eds., Plenum Press, New York, pp. 235–263.

Fransson, J., and Hagman, A., 1996. Oxidation of human insulin-like growth factor I in formulation studies 2. Effects of oxygen, visible light, and phosphate on methionine oxidation in aqueous solution and evaluation of possible mechanisms. *Pharm. Res.* 13:1476–1481.

Fransson, J.R., 1997. Methionine oxidation and covalent aggregation in aqueous solution. *J. Pharm. Sci.* 86:1046–1050.

Gekko, K., and Timasheff, S.N., 1981. Mechanism of protein stabilization by glycerol: Preferred hydration in glycerol-water mixtures. *Biochemistry* 20:4667–4676.

Gokam, Y.R., Kras, E., Nodgaard, C., Dharmavaram, V., Fesinmeyer, R.M., Hultgen, H., Brych, S., Remmele, R.L. Jr., Brems, D.N., and Hershenson, S., 2008. Self-buffering antibody formulations. *J. Pharm. Sci.* 97:3051–3066.

Gombotz, W.R., Pankey, S.C., Phan, D., Drager, R., Donaldson, K., Antonsen, K.P., Hoffman, A.S., and Raff, H.V., 1994. The stabilization of a human IgM monoclonal antibody with poly(vinylpyrrolidone). *Pharm. Sci.* 11:624–632.

Guazzo, D., 2010. Container-closure integrity testing. In *Sterile Drug Products—Formulation, Packaging, Manufacturing, and Quality*, Akers, M.J., ed., Informa Healthcare, London, 455–472.

Hageman, M.J., Tinwalla, A.Y., and Bauer, J.M., 1993. Kinetics of temperature-induced irreversible aggregation/precipitation of bovine somatotropin as studied by initial rate methods. *Pharm. Res.* 10:S-85.

Harwood, R.J., Portnoff, J.B., and Sunbery, E.W., 1993. The processing of small volume parenterals and related sterile products. In *Pharmaceutical Dosage Forms: Parenteral Medications*, vol. 2, 2nd ed., Avis, K.E., Lieberman, H.A., and Lachaman, L., eds., Marcel Dekker, New York, pp. 70–73.

Hawker, R.J., and Hawker, L.M., 1975. Protein losses during sterilizing by filtration. *Lab. Pract.* 24:805–807.

Henderson, O., 2010. Primary container and closure selection for biopharmaceuticals. In *Formulation and Process Development Strategies for Manufacturing Biopharmaceuticals*, Jameel, F., and Hershenson, S., eds., Wiley, Hoboken, NJ, 881–896.

Henry, M., Dupont-Gillain, C., and Bertrand, P., 2008. Characterization of insulin adsorption in the presence of albumin by time-of-flight secondary ion mass spectrometry and x-ray photoelectron spectroscopy. *Langmuir* 24:458–464.

Herman, A.C., Boone, T.C., and Lu, H.S., 1996. Characterization, formulation, and stability of Neupogen® (Filgrastim), a recombinant human granulocyte-colony stimulating factor. In *Formulation, Characterization, and Stability of Protein Drugs: Case Histories*, Pearlman, R., and Wang, Y.J., eds., Plenum Press, New York, pp. 324–325.

Hirsch, J.I., Fratkin, M.J., Wood, J.H., and Thomas, R.B., 1977. Clinical significance of insulin adsorption by polyvinylchloride infusion systems. *Am. J. Hosp. Pharm.* 34:583–588.

Hirsch, J.I., Wood, J.H., and Thomas, R.B., 1981. Insulin adsorption to polyolefin infusion bottles and polyvinylchloride administration sets. *Am. J. Hosp. Pharm.* 38:995–997.

Hsu, C.C., Pearlman, R., and Curley, J.C., 1988. Some factors causing protein denaturation and aggregate formation in pharmaceutical processing. *Pharm. Res. Suppl.* 5:S34.

Iacocca, R.G., Toltl, N., Allgeier, M., Bustard, B., Dong, X., Foubert, M., Hofer, J., Peoples, S., Shelbourn, T., 2010. Factors affecting the chemical durability of glass used in the pharmaceutical industry. *AAPS PharmSciTech.* 11:1340–1349.

Irwin, W.J., Dwivedi, A.K., Holbrook, P.A., and Dey, M.J., 1994. The effect of cyclo-dextrins on the stability of peptides in nasal enzymatic systems. *Pharm. Res.* 11:1698–1703.

Itoh, H., Nagata, A., Toyomasu, T., Sakiyuama, T., Nagai, T., Saeki, T., and Nakanishi, K., 1995. Adsorption of beta-lactoglobulin onto the surface of stainless steel particles. *Biosci. Biotechnol. Biochem.* 59:1648–1651.

Jenke, D., 2008. Suitability-for-use considerations for pre-filled syringes. *Pharm. Tech.* 34:30–33.

Ji, J.A., Zhang, B., Cheng, W., and Wang, Y.J., 2009. Methionine, tryptophan, and histi-dine oxidation in a model protein, PTH: Mechanisms and stabilization. *J. Pharm. Sci.* 99:4485–4500.

Johnson, M.D., Hoesterey, B.L., and Anderson, B.D., 1994. Solubilization of a tripeptide HIV protease inhibitor using a combination of ionization and complexation with chemically modified cyclodextrins. *J. Pharm. Sci.* 83:1142–1146.

Johnston, T.P., 1996. Adsorption of recombinant human granulocyte colony stimulating factor (rGH-CSF) to polyvinyl chloride, polypropylene, and glass: Effect of solvent additives. *PDA J. Pharm. Sci. Tech.* 50:238–245.

Jones, L.S., Cipolla, D., Liu, J., Shire, S.J., and Randolph, T.W., 1999. Investigation of pro-tein surfactant-interactions by analytical ultracentrifugation and electron paramagnetic resonance: The use of recombinant human tissue factor as an example. *Pharm. Res.* 16:808–812.

Jones, L.S., Kaufmann, A., and Middaugh, C.R., 2005. Silicone oil induced aggregation of proteins. *J. Pharm. Sci.* 94:918–927.

Junyan, J.A., Zhang, B., Cheng, W., and Wang, Y.J., 2009. Methionine, tryptophan, and histi-dine oxidation in a model protein, PTH. *J. Pharm. Sci.* 98:4485–4500.

Kameoka, D., Masuzaki, E., Ueda, T., and Imoto, T., 2007. Effect of buffer species on the unfolding and the aggregation of humanized IgG. *J. Biochem.* 142:383–391.

Karlsson, M., Ekeroth, J., Elwing, H., and Carlsson, U., 2005. Reduction of irreversible pro-tein adsorption on solid surfaces by protein engineering for increased stability. *J. Biol. Chem.* 280:25558–25564.

Katakam, M., Bell, L.N., and Banga, A.K., 1995. Effect of surfactants on the physical stability of recombinant human growth hormone. *J. Pharm. Sci.* 84:713–716.

Katayama, D.S., Nayar, R., Chou, D.K., Valente, J.J., Cooper, J., Henry, C.S., Velde, D.G.V., Villarete, L., Lui, C.P., and Manning, M.C., 2006. Effect of buffer species on the ther-mally induced aggregation of interferon-tau. *J. Pharm. Sci.* 95:1212–1226.

Kerwin, B.A., 2009. Polysorbates 20 and 80 used in the formulation of protein biotherapeutics: Structure and degradation pathways. *J. Pharm. Sci.* 97:2924–2935.

Kerwin, B.A., Akers, M.J., Apostol, I., Moore-Einsel, C., Etter, J.E., Hess, E., Lippincott, J., et al., 1999. Acute and long-term stability studies of deoxy hemoglobin and characterization of ascorbate-induced modifications. *J. Pharm. Sci.* 88:79–88.

Kiese, S., Pappenberger, A., Friess, W., and Mahler, H.-C., 2008. Shaken, not stirred: Mechanical stress testing of an IgG1 antibody. *J. Pharm. Sci.* 97:4347–4366.

Kim, Y., Rose, C.A., Liu, Y., Ozaki, Y., Datta, G., and Ut, A.T., 1994. RT-IR and near-infrared FT-Raman studies of the secondary structure of insulinotropin in the solid state: α-Helix to β-sheet conversion induced by phenol and/or high shear force. *J. Pharm. Sci.* 83:1175–1180.

Kirsch, L.E., Riggin, R.M., Gearhart, D.A., Lefeber, D.S., and Lytle, D.L., 1993. In-process protein degradation by exposure to trace amounts of sanitizing agents. *J. Parenter. Sci. Technol.* 47:155–160.

Knepp, V.M., Whatley, J.L., Muchnik, A., and Calderwood, T.S., 1996. Identification of antioxidants for prevention of peroxide-mediated oxidation of recombinant human ciliary neurotrophic factor and recombinant human nerve growth factor. *PDA J. Pharm. Sci. Technol.* 50:163–171.

Kueltzo, L.A., Wang, W., Randolph, T.W., and Carpenter, J.F., 2007. Effects of solution conditions, processing parameters, and container materials on aggregation of a monoclonal antibody during freeze-thawing. *J. Pharm. Sci.* 97:1801–1812.

Lam, X.M., Lai, W.G., Chan, E.K., Ling, V., and Hsu, C.C., 2011. Site-specific tryptophan oxidation induced by autocatalytic reaction of polysorbate 20 in protein formulation. *Pharm. Res.* 28:2543–2555.

Lam, X.M., Patapoff, T.W., and Nguyen, T.H., 1997a. The effect of benzyl alcohol on recombinant human interferon-γ. *Pharm. Res.* 14:725–729.

Lam, X.M., Yang, J.Y., and Cleland, J.L., 1997b. Antioxidants for prevention of methionine oxidation in recombinant monoclonal antibody HER2. *J. Pharm. Sci.* 86:1250–1255.

Lee, J.C., and Lee, L.L.Y., 1987. Thermal stability of proteins in the presence of polyethylene glycols. *Biochemistry* 26:7813–7819.

Lenk, J.R., Ratner, B.D., Gendreau, R.M., and Chittur, K.K., 1989. IR spectral changes of bovine serum albumin upon surface adsorption. *J. Biomed. Mater. Res.* 23:549–569.

Li, S., Schoneich, C., Wilson, G.S., and Borchardt, R.T., 1993. Chemical pathways of peptide degradation. V. Ascorbic acid promotes rather than inhibits the oxidation of methionine to methionine sulfoxide in small model peptides. *Pharm. Res.* 10:1572–1579.

Li, W., Patapoff, T.W., Nguyen, T.H., and Borchardt, R.T., 1996. Inhibitory effect of sugars and polyols on the metal-catalyzed oxidation of human relaxin. *J. Pharm. Sci.* 85:868–872.

Liebe, D.C., 1995. Pharmaceutical packaging. In *Encyclopedia of Pharmaceutical Technology*, vol. 12, Swarbrick, J., and Boylan, J.C., eds., Marcel Dekker, New York, pp. 16–28.

Loftsson, T., and Brewster, M.E., 1996. Pharmaceutical applications of cyclodextrins. 1. Drug solubilization and stabilization. *J. Pharm. Sci.* 85:1017–1025.

Lougheed, W.D., Albisser, A.M., Martindale, H.M., Chow, J.C., and Clement, J.R., 1983. Physical stability of insulin formulations. *Diabetes* 32:424–432.

Maa, Y.F., and Hsu, C.C., 1996. Aggregation of recombinant human growth hormone induced by phenolic compounds. *Int. J. Pharm.* 140:155–168.

Maggio, E.T., 2008. Novel excipients prevent aggregation in manufacturing and formulation of protein and peptide therapeutics. *Bioprocess Int.* November:2–5.

Mahler, H.-C., Friess, W., Grauschopf, U., and Kiese, S., 2009. Protein aggregation: Pathways, induction factor and analysis. *J. Pharm. Sci.* 98:2902–2934.

Majumdar, S., Ford, B.M., Mar, K.D., Sullivan, V.J., Ulrich, R.G., and D'Souza, A.J., 2011. Evaluation of the effect of syringe surfaces on protein formulations. *J. Pharm. Sci.* 100:2563–2573.

Manning, M.C., Chou, D.K., Murphy, B.M., Payne, R.W., and Katayama, D.S., 2010. Stability of protein pharmaceuticals: An update. *Pharm. Res.* 27:544–575.

Matilainen, L., Larsen, K.L., Wimmer, R., Keski-Rahkonen, P., Auriola, S., Järvinen, T., and Jarho, P., 2007. The effect of cyclodextrins on chemical and physical stability of glucagon and characterization of glucagon/γ-CD inclusion complexes. *J. Pharm. Sci.* 97:2720–2729.

Meyer, B.K., Ni, A., Hu, B., and Shi, L., 2007. Antimicrobial preservative use in parenteral products: Past and present. *J. Pharm. Sci.* 96:3155–3167.

Milano, E.A., Waraszkiewicz, S.M., and Dirubio, R., 1982. Extraction of soluble aluminum from chlorobutyl rubber closures. *J. Parenter. Sci. Tech.* 36:116–120.

Mitraki, A., and King, J., 1989. Protein folding intermediates and inclusion body formation. *Nat. Biotechnol.* 7:690–697.

Mitrano, F.P., and Newton, D.W., 1982. Factors affecting insulin adherence to Type I glass bottles. *Am. J. Hosp. Pharm.* 39:1491–1495.

Nail, S.L., and Akers, M.J., eds., 2002. *Development and Manufacture of Protein Pharmaceuticals*, Kluwer Academic/Plenum, New York (now Springer-Verlag).

Nakanishi, K., Sakiyama, T., and Imamura, K., 2001. On the adsorption of proteins on solid surfaces, a common but very complicated phenomenon. *J. Biosci. Bioeng.* 91:233–244.

Nema, S., and Ludwig, J.D., 2010. *Pharmaceutical Dosage Forms: Parenteral Medications*, 3rd ed. Informa, London.

Norde, W., 1995. Adsorption of proteins at solid-liquid interfaces. *Cells Mater.* 5:97–112.

Oliyai, C., and Borchardt, R.T., 1994. Chemical pathways of peptide degradation. VI. Effect of the primary sequence on the pathways of degradation of aspartyl residues in model hexapeptides. *Pharm. Res.* 11:751–758.

Oshima, G., 1989. Solid surface-catalyzed inactivation of bovine alpha-chymotrypsin in dilute solution. *Int. J. Biol. Macromol.* 11:43–48.

Pace, C.N., and Grimsley, G.R., 1988. Ribonuclease T1 is stabilized by cation and anion binding. *Biochemistry* 27:3242–3246.

Pang, J., Blanc, T., Brown, J., Labrenz, S., Villalobos, A., Depaolis, A., Gunturi, S., Grossman, S., Lisi, P., and Heavner, G.A., 2007. Recognition and identification of UV-absorbing leachables in EPREX® pre-filled syringes: An unexpected occurrence at a formulation—component interface. *PDA J. Pharm. Sci. Technol.* 61:423–432.

Pantoliano, M.W., Whitlow, M., Wood, J.F., Rollence, M.L., Finzel, B.C., Gilliand, G.L., Poulos, G., and Bryant, P.N., 1988. The engineering of binding affinity at metal ion binding sites for the stabilization of proteins: Subtilisin as a test case. *Biochemistry* 27:8311–8317.

Patel, K., and Borchardt, R.T., 1990. Chemical pathways of peptide degradation II. Kinetics of deamidation of an asparaginyl residue in a model hexapeptide. *Pharm. Res.* 7:787–793.

Pearlman, R., and Bewley, T.A., 1993. Stability and characterization of human growth hormone. In *Stability and Characterization of Protein and Peptide Drugs: Case Histories*, Wang, Y.J., and Pearlman, R., eds., Plenum Press, New York, 1–58.

Pharmaceutical Research and Manufacturers of America BSE Committee, 1998. Assessment of risk of bovine spongiform encephalopathy in pharmaceutical products, part 1. *BioPharm* 11:20–31.

Physicians' Desk Reference, (2004). Montvale, NJ, Medical Economics.

Pikal, M.J., 1990. Freeze drying of proteins II: Formulation selection. *BioPharm* 3:26–30.

Pinholt, C., Hartvig, R.A., Medicott, N.J., and Jorgensen, L., 2011. The importance of interfaces in protein drug delivery—why is protein adsorption of interest in pharmaceutical formulations? *Expert Opin. Drug Deliv.* 8:949–964.

Powell, M.F., Sanders, L.M., Rogerson, A., and Si, V., 1991. Parenteral peptide formulations: Chemical and physical properties of native luteinizing hormone-releasing hormone (LHRH) and hydrophobic analogues in aqueous solution. *Pharm. Res.* 8:1258–1263.

Pretzer, D., Schulteis, B.S., Smith, C.D., VanderVelde, D.G., Mitchell, J.W., and Manning, M.C., 1993. Fibrolase. A fibrinolytic protein from snake venom. In *Stability and Characterization of Protein and Peptide Drugs: Case Histories*, Wang, Y.J., and Pearlman, R., eds., Plenum Press, New York, pp. 287–314.

Roe, J.A., Butler, A., Schaller, D.M., Valentine, J.S., Marky, L., and Breslauer, K.J., 1988. Differential scanning calorimetry of Cu, Zn-superoxide dismutase, the apoprotein, and its Zn-substituted derivatives. *Biochemistry* 27:950–958.

Rosenberg, A.S., 2006. Effects of protein aggregates: An immunologic perspective. *AAPS J.* 8:E501–E507.

Ruiz, L., Aroche, K., and Reyes, N., 2006. Aggregation of recombinant human interferon alpha 2b in solution: Technical note. *AAPS PharmSciTech.* 7:99.

Sacha, G.A., Saffell-Clemmer, W., Abram, K., and Akers, M.J., 2010. Practical fundamentals of glass, rubber, and plastic sterile packaging systems. *Pharm. Dev. Tech.* 15:6–34.

Sato, S., Ebert, C.D., and Kim, S.W., 1984. Prevention of insulin self-association and surface adsorption. *J. Pharm. Sci.* 72:228–232.

Schwarzenbach, M.S., Reimann, P., Thommen, V., Hegner, M., Mumenthaler, M., Schweb, J., and Guntherodt, H.-J., 2002. Interferon α-2a interactions on glass vial surfaces measured by atomic force microscopy. *PDA J. Pharm. Sci. Tech.* 56:78–89.

Senderoff, R.I., Wootton, S.C., Boctor, A.M., Chen, T.M., Giordani, A.B., Julian, T.N., and Radebaugh, G.W., 1994. Aqueous stability of human epidermal growth factor 1–48. *Pharm. Res.* 11:1712–1720.

Shahrokh, Z., Eberlein, G., and Wang, Y.J., 1994. Probing the conformation of protein (bFGF) precipitates by fluorescence spectroscopy. *J. Pharm. Biomed. Anal.* 12:1035–1041.

Sharma, B., 2007. Immunogencity of therapeutic proteins. Part 2, Impact of container closures. *Biotechnol. Adv.* 25:318–324.

Sharma, D.K., Oma, P., and Krishnan, S., 2009. Silicone microdroplets in protein formulations—detection and enumeration. *Pharm. Tech.* 33:74–79.

Shire, S.J., Shahrokh, Z., and Liu, J., 2004. Challenges in the development of high protein concentration proteins. *J. Pharm. Sci.* 93:1390–1402.

Silvestri, S., Lu, M.Y., and Johnson, H., 1993. Kinetics and mechanism of peptide aggregation I: Aggregation of a cholecystokinin analog. *J. Pharm. Sci.* 92:689–693.

Singh, S.K., Afonina, N., Awwad, M., Bechtold-Peters, K., Blue, J.T., Chou, D., Cromwell, M., et al., 2010. An industry perspective on the monitoring of subvisible particles as a quality attribute for protein therapeutics. *J. Pharm. Sci.* 99:3302–3321.

Tabak, M., de Sousa Neto, D., and Salmon, C.E.G., 2006. On the interaction of bovine serum albumin (BSA) with cethyltrimethyl ammonium chloride surfactant: Electron paramagnetic resonance (EPR) study. *Braz. J. Phys.* 36:83–89.

Thirumangalathu, R., Krishnan, S., Ricci, M.S., Brems, D.N., Randolph, T.W., and Carpenter, J.F., 2009. Silicone oil- and agitation-induced aggregation of a monoclonal antibody in aqueous solution. *J. Pharm. Sci.* 98:3167–3181.

Thompson, D.O., 1997. Cyclodextrins-enabling excipients: Their present and future use in pharmaceuticals. *Crit. Rev. Ther. Drug Carrier Sys.* 14:1–104.

Thurow, H., and Geisen, K., 1984. Stabilisation of dissolved proteins against denaturation at hydrophobic interfaces. *Diabetologia* 27:212–218.

Tobler, S.A., Holmes, B.W., Cromwell, M.E.M., and Fernandez, E.J., 2004. Benzyl alcohol-induced destabilization of interferon-γ: A study by hydrogen-deuterium isotope exchange. *J. Pharm. Sci.* 93:1605–1617.

Topp, E.M., Zhang, L., Zhao, H., Payne, R.W., Evans, G.J., and Manning, M.C., 2010. Chemical instability in peptide and protein pharmaceuticals. In *Formulation and Process Development Strategies for Manufacturing of a Biopharmaceutical*, Jameel, F., and Hershenson, S., eds., John Wiley & Sons, New York, pp. 41–68.

Treuheit, M.J., Kosky, A.A., and Brems, D.N., 2002. Inverse relationship of protein concentration and aggregation. *Pharm. Res.* 19:511–516.

Tsai, P.K., Volkin, D.B., Dabora, J.M., Thompson, K.C., Bruner, M.W., Gress, J.O., Matuszewska, B., et al., 1993. Formulation design of acidic fibroblast growth factor. *Pharm. Res.* 10:649–659.

Twardowski, Z.J., Nolph, K.D., McGary, T.J., and Moore, H.L., 1983a. Influence of temperature and time on insulin adsorption to plastic bags. *Am. J. Hosp. Pharm.* 40:583–586.

Twardowski, Z.J., Nolph, K.D., McGary, T.J., and Moore, H.L., 1983b. Nature of insulin binding to plastic bags. *Am. J. Hosp. Pharm.* 40:579–582.

Twardowski, Z.J., Nolph, K.D., McGary, T.J., Moore, H.L., Collin, P., Ausman, R.K., and Slimack, W.S., 1983c. Insulin binding to plastic bags: A methodologic study. *Am. J. Hosp. Pharm.* 40:575–579.

Tyagi, A.K., Randolph, T.W., Dong, A., Maloney, K.M., Hitcherich, C., and Carpenter, J.F., 2009. IgG particle formation during filling pump operation: A case study of heterogeneous nucleation on stainless steel nanoparticles. *J. Pharm. Sci.* 98:94–104.

Vemuri, S., Yu, C.T., and Roosdorp, N., 1993. Formulation and stability of recombinant α1-antitrypsin. In *Stability and Characterization of Protein and Peptide Drugs: Case Histories*, Wang, Y.J., and Pearlman, R., eds., Plenum Press, New York, pp. 269–270.

Wakankar, A.A., and Borchardt, R.T., 2006. Formulation considerations for proteins susceptible to asparagines deamidation and aspartate isomerization. *J. Pharm. Sci.* 95:2321–2336.

Wakankar, A.A., Wang, Y.J., Canova-Davis, E., Ma, S., Schmalzing, D., Grieco, J., Milby, T., et al., 2010. On developing a process for conducting extractable-leachable assessment of components used for storage of biopharmaceuticals. *J. Pharm. Sci.* 99:2209–2218.

Wang, N., Hu, B., Ionescu, R., Mach, H., Sweeney, J., Hamm, C., Kirchmeier, M.J., and Meyer, B.K., 2009. Opalescence of an IgG1 monoclonal antibody formulation is mediated by ionic strength and excipients. *BioPharm* April:36–47.

Wang, Y.J., and Chien, Y.W., 1984. *Sterile Pharmaceutical Packaging: Compatibility and Stability. Technical Report #5*, Parenteral Drug Association, Inc., Philadelphia.

Wang, Y.J., Shahrokh, Z., Vemuri, S., Eberlein, G., Beylin, I., and Busch, M., 1996. Characterization, stability, and formulations of basic fibroblast growth factor. In *Formulation, Characterization, and Stability of Protein Drugs: Case Histories*, Pearlman, R., and Wang, Y.J., eds., Plenum Press, New York, pp. 164–165.

Wetzel, R., Perry, L.J., Baase, W.A., and Becktel, W.J., 1988. Disulfide bonds and thermal stability of T4 lysozyme. *Proc. Natl. Acad. Sci. USA* 85:401–405.

Wollmer, A., Rannefeld, B., Johansen, B.R., Hejnaes, K.R., Balschmidt, P., and Hansen, F.B., 1987. Phenol-promoted structural transformation of insulin in solution. *Biol. Chem. Hoppe-Seyler* 368:903–911.

Yadav, S., Liu, J., Shire, S.J., and Kalonia, D.S., 2010a. Specific interactions in high concentration antibody solutions resulting in high viscosity. *J. Pharm. Sci.* 99:1152–1168.

Yadav, S., Shire, S.J., and Kalonia, D.S., 2010b. Factors affecting the viscosity in high concentration solutions of different monoclonal antibodies. *J. Pharm. Sci.* 99:4812–4829.

Yoshioka, S., Aso, Y., Izutsu, K., and Terao, T., 1993. Stability of β-glactosidase, a model protein drug, is related to water mobility as measured by ^{17}O nuclear magnetic resonance (NMR). *Pharm. Res.* 10:103–108.

Yu, X., DeCou, D., Wood, D., Zdravkovic, S., Schmidt, H., Stockmeier, L., Piccoli, R., Rude, D., and Ding, X., 2010. A study of leachables for biopharmaceutical formulations stored in rubber-stoppered glass vials. *BioPharm Int.* 23:26–36.

Zhang, M.Z., Wen, J., Arakawa, T., and Prestrelski, S.J., 1995. A new strategy for enhancing the stability of lyophilized protein: The effect of the reconstitution medium on keratinocyte growth factor. *Pharm. Res.* 12:1447–1452.

Zheng, J.Y., and Janis, L.J., 2006. Influence of pH, buffer species and storage temperature on physicochemical stability of a humanized monoclonal antibody LA298. *Int. J. Pharm.* 308:46–51.

9 Peptides and Proteins as Parenteral Suspensions

An Overview of Design, Development, and Manufacturing Considerations

Michael R. DeFelippis and Michael J. Akers

CONTENTS

9.1   INTRODUCTION AND SCOPE

Industry trends for molecules under pharmaceutical development reflect the grow-ing importance of peptides and proteins as therapeutic candidates (Walsh, 2006). As research continues to identify and elucidate the function of new peptides and proteins and biotechnology provides the means to produce them in requisite quantities for clinical evaluation, it will ultimately become necessary to devise viable formulations and delivery modalities to allow marketing of these potential therapies. However, it is widely recognized that peptides and proteins have unique physicochemical prop-erties limiting the formulation approaches and delivery options that can be applied to these molecules (Frokjaer and Otzen, 2005). This chapter is concerned with the formulation of peptides and proteins as parenteral suspension preparations for injec-tion. Suspensions, in general, are perhaps the most challenging formulations to develop because of the need to balance numerous variables to achieve an acceptable product. In addition, the complexity of the molecules being considered and the type of desired formulation exacerbate the problem of developing a peptide or protein suspension.

There are two general approaches that a formulation scientist can undertake to prepare a parenteral suspension of a peptide or protein: (1) crystalline or amorphous particles are produced in situ by combining sterile solutions and (2) sterile drug particles are produced separately, then suspended in an appropriate sterile vehicle. This chapter focuses on concepts related to particle formation methods, excipient selection, optimization of product properties, characterization techniques, and manufacturing considerations. The objective is to provide relevant information and summarize the general principles for pharmaceutical scientists who might be involved in designing and developing parenteral peptide or protein suspensions. Strategies to overcome some of the general challenges associated with suspension formulations and special design features that must be considered are also discussed. For practical reasons, the information provided in this chapter is based mostly on insulin suspensions, since they represent the only examples involving a protein (excluding encapsulated systems) that have been successfully commercialized and licensed as pharmaceutical products to date. Furthermore, the various insulin suspension preparations have an established profile of safety and efficacy over many decades of subcutaneous administration to humans demonstrating the viability of the suspension formulation approach.

9.2   RATIONALE FOR SUSPENSION DEVELOPMENT

The basic reasons for preparing small molecule pharmaceutical suspensions (Akers et al., 1987) are also applicable to peptides and proteins. These include

1. The solubility of the peptide or protein prohibits solution formulation.
2. The stability of the peptide or protein is improved in a suspension formulation.
3. There is a desire to control or retard the release profile of the peptide or protein.

Upon considering these reasons, one might question the practicality of preparing a suspension over some other dosage form given the inherent complexity associated with such formulations. For example, if solubility is an issue, peptides or proteins can readily be solubilized by adjustment of solution conditions, such as pH, or by the addition of appropriate excipients. Alternatively, molecular design strategies enabled with recombinant DNA technology or peptide synthesis can be employed to selectively introduce amino acid modifications that increase the biological potency, so that concentrations required to achieve a therapeutic effect may be reduced or the general solubility of the molecule is improved. Assuming that the resulting peptide or protein solution preparation has an appropriate physicochemical stability over its shelf-life and for the intended in-use period, such approaches would clearly be desirable. Many peptides and proteins are sufficiently stable to be prepared as solutions for parenteral administration (see Chapter 8). However, as highlighted in other chapters, the susceptibility of peptides and proteins to various forms of chemical and physical degradation makes the option of solution preparations not always feasible.

One might then consider a suspension formulation as a means to circumvent chemical or physical stability problems. Indeed, microcrystalline suspensions of insulin demonstrate considerably reduced deamidation at position B3 compared to amorphous or soluble preparations (Brange et al., 1992c); however, this is only one of the many possible degradation reactions known for insulin (Brange, 1994 and the references therein). In reality, a formulation scientist would likely opt for freeze-drying the peptide or protein over a suspension product to obtain the needed stability despite the inconvenience of requiring a suitable diluent for reconstitution. This point is supported by the number of freeze-dried peptide or protein preparations on the market (e.g., human growth hormone [hGH], tissue-type plasminogen activator, erythropoietin, and glucagon). Certain types of stability issues could potentially be overcome using molecular design approaches in early research and development to optimize physico-chemical properties of the peptide or protein selected for clinical development. For example, amino acids susceptible to deamidation or oxidation can be substituted with less vulnerable residues provided other properties of the molecule are not impacted.

Thus, the need to retard or control release of the drug provides the best rationale for choosing to develop a suspension preparation of a peptide or protein. There are two basic motivations for modifying pharmacological properties. One reason relates to convenience for patients in administering the medication. Due to their intrinsic fragility and generally short circulating half-life in vivo, peptides and proteins typically require dosing regimes involving multiple daily injections. Extending the pharmacokinetic profile can reduce the frequency of dosing thereby improving patient acceptance and compliance. The second reason involves a desire to more closely mimic the naturally occurring physiological secretion of an endogenous peptide or protein. Basal insulin secretion in nondiabetic subjects, which is characterized by a continuous and relatively flat profile, offers the best example of a pharmacokinetic response that suspension preparations of insulin attempt to simulate for persons with diabetes; albeit for the time intervals between meals and through the evening over the course of a 24-hour period and typically requiring more than a single injection (see the reviews in Beals et al. [2008] and DeFelippis et al. [2001] along with the references therein for more details). Despite these important considerations there are very few examples of marketed peptide or protein suspension formulations with the long-acting insulin preparations (Table 9.1) still being the best characterized of

TABLE 9.1
Commercial Insulin Suspensions

Product Name	Suspension Characteristics
Insulin protamine suspension NPH	Crystalline (protamine complex)
Insulin Ultralente[a]	Crystalline
Lys[B28], Pro[B29] human insulin analog protamine suspension	Crystalline (protamine complex)
Asp[B28] insulin analog protamine suspension	Crystalline (protamine complex)
Insulin Semilente[a]	Amorphous
Insulin Lente[a]	Amorphous/crystalline mixture
Regular/NPH insulin mixtures	Soluble/crystalline mixture
25% soluble porcine insulin[b]	Soluble/crystalline mixture
75% crystalline bovine insulin[b]	
Lys[B28], Pro[B29] human insulin analog mixtures	Soluble/crystalline mixture
Asp[B28] insulin analog mixtures	Soluble/crystalline mixture

[a] Production discontinued by major manufacturers of insulin products.
[b] Production of animal pancreas-derived insulin has been discontinued by major insulin producers.

this type. The general concepts discussed in the later sections of this chapter heavily leverage the body of knowledge related to insulin pharmaceutical suspensions.

9.3 TYPES OF SUSPENSIONS AND PARTICLE FORMATION

9.3.1 IN SITU PARTICLE FORMATION

Based on the experience with insulin preparations that have been produced for the commercial market, there are at least four approaches for producing particles in situ that can be considered for development of peptide or protein suspension formulations:

1. The suspension is composed exclusively of crystalline material in vehicle.
2. The suspension is composed exclusively of amorphous material in vehicle.
3. The suspension contains a mixture of crystalline and amorphous materials in vehicle.
4. The suspension contains the active ingredient in both the suspension and solution phases.

Each of these categories is discussed in more detail in this section along with a description of fundamental aspects and special considerations that may be applied to other peptides and proteins being considered as suspension products.

9.3.1.1 Crystalline Suspensions

The topic of macromolecular crystallization as it pertains primarily to x-ray crystallography has been reviewed in great detail in the literature (Durbin and Feher, 1996; Gilliland and Bickham, 1990; McPherson, 1982 and references therein); however, it is worth briefly summarizing this information as some basic knowledge of protein crystallization is necessary in order to prepare suspensions composed of crystalline material. One advantage of preparing crystals for pharmaceutical suspensions as opposed to x-ray crystallographic studies is that large, perfect single crystals with dimensions of $0.3 \times 0.3 \times 0.3$ mm, 0.027 mm^3 (Drenth, 1994) are not essential. In fact, a suspension composed of microcrystalline material in the size range of about 1–40 μm will likely have desirable pharmaceutical properties such as resuspendability, syringeability, and injectability based on the experiences with marketed insulin preparations. Even with this slight advantage, it should be recognized that the task of preparing peptide or protein crystals for pharmaceutical preparations is still not trivial. This discussion assumes that a highly purified drug substance is available for formulation and does not cover isolation and purification methods. However, the importance of having a drug substance with reproducible and high purity cannot be overemphasized since impurities will likely influence the crystallization outcome.

There are no theoretical rules that can be followed for preparing peptide or protein crystals and most of what is known is based on empirical observations and experience. The complexity of structure and diversity of chemical groups characterizing these molecules necessitates the elucidation of appropriate conditions that are fairly unique to a given peptide or protein. The objective is to devise conditions that result in supersaturation of the solution so that crystal growth is promoted. Conditions such as pH, ionic strength, temperature, peptide or protein concentration, the presence/absence of auxiliary ions or other chemical entities can all influence solubility. Crystallization is governed by a precise balance among many of these parameters such that in most instances there will be a limited set of conditions that will yield the desired crystals as opposed to the formation of amorphous material. Furthermore, slight variations in the conditions that produce crystals can strongly influence the final polymorph (McPherson, 1982).

Prior to initiating studies to determine crystallization conditions, it is essential to have a good understanding of the precipitation behavior of the molecule of interest. This knowledge can be obtained by conducting solubility experiments (Ducruix and Ries-Kautt, 1990) exploring variables such as concentration and pH. Using this information, various conditions can be screened at small scale that might result in crystal growth. Once potential crystallization conditions are identified, optimization studies can be undertaken to improve the yield or achieve a desired polymorph. For example, a factorial design evaluating salt concentration, pH, and temperature in appropriate ranges could serve as a reasonable starting point. Carter (1990) provided additional information concerning the use of statistically designed experiments for screening crystal growth conditions.

In general, the best quality crystals will form slowly from clear solutions (McPherson, 1982). Amorphous precipitates usually occur when aggregation occurs too rapidly, but crystals may grow from this material over time in some cases. As described later in more detail for insulin suspensions, slow crystal growth and prevention of amorphous precipitation are not critical requirements for preparing crystalline pharmaceutical suspensions. The two commercial microcrystalline insulin suspensions are prepared in batch by rapid formation of an amorphous precipitate that slowly converts into microcrystals. Other crystallization aids suitable for preparing pharmaceutical suspensions include seeding or cocrystallization with other chemical agents.

Crystallization protocols used to prepare samples for x-ray crystallographic studies are readily available from established databases* (Tung and Gallagher, 2009) and these can serve as a starting point in the development of crystalline peptide or protein suspensions. However, it is worth mentioning some practical limitations of this information when applying it to the preparation of pharmaceutical crystalline suspensions. Typical crystallizing solutions often contain small organic molecules, various salts, detergents, and/or metal ions as precipitants or simply because they are necessary to achieve crystal growth. Since these additives may be difficult to completely remove, they can be expected to become part of the final preparation. All ingredients must therefore be pharmaceutically acceptable and proven to have no deleterious toxicological effects.

In addition to limitations on ingredients, techniques such as dialysis, vapor diffusion, or evaporation commonly used to prepare crystals for x-ray crystallography will be difficult to apply in a large-scale manufacturing setting where crystallization may need to be accomplished at volumes ranging from 10 to thousands of liters.

---

* Other databases are available on the World Wide Web and can be accessed using, for example, the search term "macromolecular crystallization."

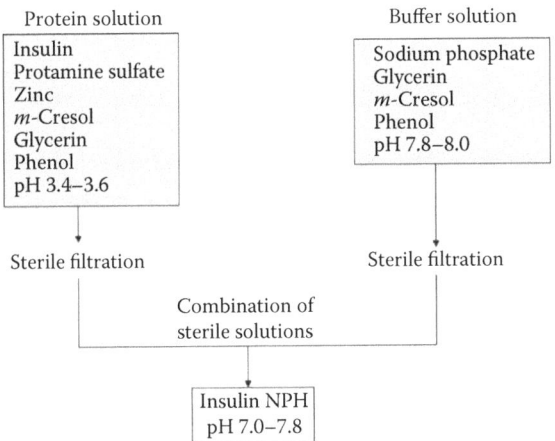

FIGURE 9.1   Insulin NPH pharmaceutical process. The procedure involves cocrystallizing insulin with the basic peptide protamine in the presence of all necessary excipients used in the final suspension preparation.

Complicated sample manipulation procedures, easily performed at the laboratory bench, may be impractical or impossible to accomplish in an aseptic manufacturing facility. Crystallization by temperature changes such as slow cooling from high temperature again may be difficult to accomplish at large scale because efficient sterile processing requires operations to be accomplished in as short a time as possible.*

Using the commercially available insulin NPH (neutral protamine Hagedorn) (Hagedorn et al., 1936) and Ultralente (Hallas-Møller et al., 1952) suspensions as model systems, two practical methods for preparing crystals can be described. In one approach, crystals are grown from a solution containing all of the ingredients at the proper concentrations that comprise the final formulation. NPH insulin is an example where this strategy is employed. The other method involves preparing a concentrated crystal suspension that is then diluted with a suitable, aqueous suspension vehicle to produce the final formulation. This type of process is used for preparing Ultralente insulin. Details of the crystallization processes used for NPH and Ultralente insulin preparations have been thoroughly reviewed elsewhere (Brange, 1987; Schlichtkrull, 1961), and the information is only summarized here to illustrate the concepts that can be applied to other peptides or proteins. Regardless of the method used, it is important to devise a process that is both reproducible and produces crystalline material having appropriate physical and chemical properties for a pharmaceutical suspension.

A schematic representation of the NPH insulin crystallization process is shown in Figure 9.1. NPH suspension is actually prepared by cocrystallizing insulin with the highly basic (pI about 13) peptide protamine. Two solutions are used for the crystallization. One solution contains insulin and protamine dissolved in water at

---

* Time limitations are necessary because of concerns over inadvertent microbiological contamination. For solutions prior to sterilization by filtration, there is also potential for accumulation of endotoxins. No minimum time is suggested here; however, such processing times should be accomplished as expediently as possible.

acidic pH, and the other contains dibasic sodium phosphate adjusted to slightly basic conditions. Both solutions also contain the additional ingredients (i.e., excipients) necessary to complete the crystallization and produce the final formulation. Greater details regarding the nature of the additional ingredients are provided in Section 9.4 on excipient selection. Precipitation is initiated by combining the solutions in a 1:1 ratio causing a rapid change to neutral pH conditions. An amorphous material forms immediately, which then transforms over time (approximately 24 hours) to form rod-shaped crystals approximately 3–6 µm long and 1–1.5 µm wide.

Krayenbühl and Rosenberg (1946) have reported systematic studies evaluating the requirements for producing insulin NPH crystals. These conditions can generally be applied to prepare NPH suspensions of human, bovine, or porcine insulin. NPH-type protamine cocrystalline suspensions have been developed for monomeric insulin analogs LysB28, ProB29-human insulin (DeFelippis, 1995), and AspB28 insulin (Balschmidt, 1996) for the purpose of preparing biphasic mixtures (see Section 9.3.1.4). In both cases, modifications to the standard crystallization procedure were necessary to produce appropriate crystals. LysB28 and ProB29-human insulin–protamine crystallization is complete within 24 hours, but requires temperature control (15°C) as well as additional modifications to the precipitation procedure. These changes include the preparation of separate insulin analog and protamine solutions at neutral pH that are then combined, and maintaining the neutral pH of this mixture during initial precipitation and throughout the entire crystallization process. In the case of AspB28 analog, crystals are reported to form only after extended storage (6 days) at refrigerated conditions. These examples highlight that even when dealing with known processes and related proteins, optimization of appropriate crystallization conditions is often necessary. The modifications to the NPH procedure necessary to prepare LysB28 and ProB29-human insulin–protamine crystals have been argued to result from the altered self-association properties of the analog (DeFelippis et al., 1997). Pharmacological studies have been reported for both analog suspensions (DeFelippis et al., 1996; Radziuk et al., 1996; Weyer et al., 1997) and the time–action profiles were shown to be similar to that of protamine suspensions prepared with other insulin species.

Protamine cocrystallization may be an option for preparing other peptide and protein suspensions since the complex formation is driven by electrostatic interactions and most proteins will not be as positively charged as protamine. Binding can be expected to result in precipitation, but the extent to which crystal growth is achieved, if at all, is dependent on the system under investigation. Another reason for considering protamine cocrystallization as an option is the fact that protamine sulfate has been proven to be an acceptable excipient in marketed preparations although immunogenicity is a consideration (see discussion in Section 9.4). Examples of other peptides and proteins for which protamine suspensions have been described include alpha interferon (Yim, 1988), insulinotropin (glucagon-like-peptide-1-(7-37)) (Kim and Rose, 1995), somatostatin (Brazeau et al., 1974; Martin et al., 1974; Tannenbaum and Colle, 1980), glucagon (Buch and Buch, 1983; Naets and Guns, 1980), and gonadotropins (Donini, 1974). The aim for the preparation of these protein–protamine complexes was to achieve an extended activity; however, testing was only limited to investigational studies. Table 9.2 provides examples of other peptide and protein suspensions proposed in the patent and scientific literature.

TABLE 9.2

Examples of Peptide and Protein Suspensions[a]

Peptide or Protein	Suspension Characteristics	Status
Human OB protein	Crystalline or amorphous	Patent literature
Infliximab	Crystalline	Scientific literature
Insulinotropin	Crystalline	Scientific literature
Interleukin-4	Crystalline	Patent literature
Rituximab	Crystalline	Scientific literature
Trastuzumab	Crystalline	Scientific literature
Zinc-interferon alpha-2B	Crystalline	Scientific literature
Alpha interferon	Protamine complex	Patent literature
Glucagon	Protamine complex	Scientific literature
Gonadotropins	Protamine complex	Patent literature
Insulinotropin	Protamine complex	Scientific literature
Octanoyl-N^ε-LysB29-human insulin; human insulin; protamine cocrystals	Protamine complex	Scientific literature
Somatostatin	Protamine complex	Scientific literature
ACTH	Oleaginous suspension	Scientific literature
Bovine somatotropin	Oleaginous suspension	Patent literature
Growth hormone releasing hormones	Oleaginous suspension	Patent literature
Porcine somatotropin	Oleaginous suspension	Patent literature

[a] The information provided in the table is not to be considered exhaustive. Readers should consult relevant electronic databases of scientific and patent literature for other examples of peptide or protein suspensions not listed.

A novel hybrid insulin NPH-type cocrystal concept has been proposed for controlled release delivery (Brader et al., 2002). The process for preparing crystalline particles involved using human insulin in combination with a less soluble insulin derivative, octanoyl-N^ε-LysB29-human insulin (C8-HI). Solutions containing various ratios of the two insulins were mixed with a solution of protamine to produce suspensions having analogous physicochemical properties to NPH. However, the dissolution rate of the cocrystals was slower with increasing proportion of the lipophilically modified insulin. Pharmacokinetic and glucodynamic studies in a dog model showed that the 75% C8-HI cocrystal suspension displayed a release pattern flatter and longer than human NPH. The authors suggested that pharmacological properties could be tailored by adjusting the ratio of the native and chemically modified proteins. Whether this approach can be generalized to other proteins depends on the ability to modify the target protein without altering its biological activity or crystallization behavior.

In contrast to the NPH crystallization procedure, Ultralente suspensions are prepared without using protamine sulfate (Figure 9.2). Precipitation is initiated by adjusting the conditions to the isoelectric point of insulin in the presence of zinc ions,

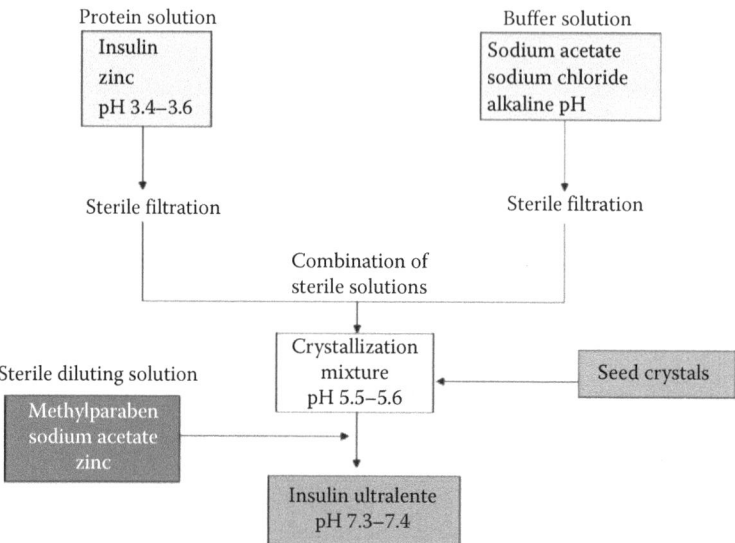

FIGURE 9.2   Insulin Ultralente pharmaceutical process. The procedure involves preparing a concentrated suspension of insulin crystals that is then diluted with vehicle containing additional excipients (e.g., preservative) required for the final suspension preparation.

sodium chloride, and sodium acetate, which is accomplished by mixing an acidic solution of insulin with a buffer such that the appropriate pH is achieved. Most of the ingredients required for the final preparation are present during crystallization except antimicrobial preservative. Concentrations for insulin and other ingredients are tenfold higher during crystal growth and a diluent containing preservative is used to dilute the concentrated suspension to produce the final preparation. Because a monodisperse particle size distribution is desired for the final preparation, predetermined amounts of seed crystals are added during the crystallization phase. Seeding also effectively eliminates self-nucleation. For more detailed information regarding the preparation of insulin seed crystals see Schlichtkrull (1957). Commercial Ultralente preparations contain rhombohedral crystals in the approximate size range of 20–30 μm.

The method described for making Ultralente crystals is specific to insulin and results from its unique self-association and ligand binding properties; therefore, it cannot be expected that the exact procedure will be generally applicable to other peptides or proteins. However, this example demonstrates viable possibilities for other molecules. In cases where crystallization can only be accomplished under defined conditions and cannot include all required excipients for a pharmaceutically acceptable preparation, a separate crystal suspension can be manufactured first followed by dilution with vehicle containing any additional required ingredients. The addition of seed crystals is also a possibility to initiate crystallization or control particle size distribution. There are important considerations in either case. Studies evaluating the compatibility between crystals and ingredients (excipients) in the suspending vehicle are necessary to ensure that degradation of the particles or other changes in morphology do not occur. A good example of the importance of this particular requirement is the use of methylparaben, a somewhat inferior antimicrobial agent, for the Ultralente suspension since studies have shown that phenol or phenolic-derivatives negatively influence the physical stability of the preparation (Brange, 1987). Incorporation of seed crystals into a pharmaceutical process necessitates establishing a suitable aseptic method of manufacture, developing assays for characterization, implementing routine testing plans and stability assessments, and defining a material storage program for the seed preparation. Validation of the manufacturing process to produce the seed crystals and their use in the crystallization unit operation for preparation of the pharmaceutical suspension product are additional requirements.

Crystallization experiments reported for insulinotropin (glucagon-like-peptide-1-(7-37)) are illustrative of the types of studies that are necessary to devise a process for preparing a pharmaceutically acceptable crystalline suspension and further substantiate one of the considerations mentioned earlier (Kim and Haren, 1995). Analogous to Ultralente insulin, a microcrystalline suspension of insulinotropin was desired to achieve a sustained release preparation. Kim and Haren (1995) demonstrated that insulinotropin can be crystallized from a salt solution adjusted to a pH near its isoelectric point, but these crystals transform into amorphous material when mixed with m-cresol. However, the crystals remain intact if they are first treated with a zinc soaking procedure prior to the addition of phenolic preservative. Note the similarity to the Ultralente insulin process of controlled addition of the necessary ingredients without compromising crystalline morphology.

Other examples of suspensions containing crystalline or precipitated peptides or proteins have been reported (see Table 9.2). A crystalline form of human interleukin-4 has been described and proposed to have potential application as a slow release pharmaceutical preparation (Hammond, 1991). The introduction of seed crystals, grown by techniques typically used to prepare crystals for x-ray crystallography, was suggested as a method for accelerating large scale crystallizations. It was further implied that the pharmaceutical preparation would require complexation of the protein with metal ions although no method for introducing preservatives was identified. Reichert et al. (1996) proposed a potential controlled release, crystalline suspension of zinc-interferon alpha-2B. In this example, crystals grown in microgravity were formulated as a suspension and evaluated in primates. The pharmacological data on the crystalline suspension showed a four- to sixfold increase in the serum half-life of the protein compared to a soluble formulation. Donini (1974) described long-acting suspensions containing gonadotropins precipitated by the addition of divalent metal ions and appropriate pH adjustments. A crystalline suspension of human OB protein was demonstrated by Brems et al. (2002). In this case, crystals were grown from a specific solvent system, harvested, and then suspended in a more physiologically acceptable vehicle.

Crystalline suspensions have been put forth as potential approaches for overcoming certain challenges associated with protein pharmaceutical delivery. For example, therapeutic monoclonal antibodies typically require large amounts of protein per dose. Solubility constraints may necessitate administering higher volumes of the preparation that can only be delivered by intravenous infusion. Even if highly concentrated solution preparations were possible, unmanageable viscosity and/or poor physicochemical stability could be limiting factors. Yang et al. (2003) discussed crystallization as a broadly applicable method for achieving high-concentration preparations of monoclonal antibodies that can be delivered in small volumes suitable for subcutaneous injection. Using batch process methods, crystals were prepared for three commercially available monoclonal antibodies (rituximab, trastuzumab, and infliximab) to exemplify the concept. These authors examined the pharmacological properties of trastuzumab and infliximab suspensions in animal models. An extended serum pharmacokinetic profile was demonstrated for infliximab following subcutaneous injection compared to solutions of the monoclonal antibodies delivered subcutaneously or intravenously. Higher bioavailability may have also been observed with the infliximab suspension, although the result was not definitive. The suspension of trastuzumab injected subcutaneously was shown to be efficacious, with the magnitude of the effects comparable to those for a solution of the monoclonal antibody administered intravenously.

Protein crystallization of hGH has been suggested as an alternative to encapsulation with biodegradable polymers for preparing extended release formulations of the protein (Govardhan et al., 2005). See Section 9.10 for details concerning the challenges related to encapsulation strategies. Crystals of hGH were first prepared in batch processes. Harvested crystals from the mother liquor then underwent a coating step that involved complexation of polyelectrolytes (polyarginine or protamine sulfate) onto the surfaces. The authors argue that the molecular coating with polycations provides additional control over dissolution of the protein. Studies in animal models evaluating a suspension of polyarginine-coated hGH showed a pharmacokinetic release profile lasting over several days, and comparable efficacy obtained from once weekly subcutaneous injection relative to seven daily injections of a solution preparation.

9.3.1.2   Amorphous Suspensions

Semilente insulin is an example of a suspension composed of amorphous particles. The suspension is prepared by performing the Ultralente crystallization under less than optimum pH conditions. Physical stability has been correlated to the degree of flocculation (Brange, 1987) suggesting that optimization of a final preparation will be necessary even if amorphous precipitation is easily accomplished. While it is difficult to predict if amorphous suspensions of a given peptide or protein will have the desired pharmacological properties and necessary stability, Semilente insulin preparations highlight the fact that development of a completely crystalline suspension may not be an absolute requirement. For example, an amorphous suspension formulation of the human OB protein has been described in the patent literature (Brems et al., 2002). Particles were formed by precipitating the protein in the presence of zinc chloride at neutral pH. This preparation was claimed to have improved stability at higher concentrations and physiological pH compared to a solution of the protein. The suspension further demonstrated a sustained release effect in mice.

9.3.1.3   Crystalline and Amorphous Mixtures

The Lente insulin preparation is an example of a suspension containing a mixture of particles having different morphologies derived by combining Semilente and Ultralente to produce a 30:70 mixture of amorphous to crystalline material. It should be recognized that the suspension is intentionally prepared in this manner and does not result from incomplete crystallization. There was a specific therapeutic rationale related to injection frequency for which this insulin preparation was developed. Whether a similar justification can be identified for other potential peptides or protein suspensions is clearly dependent upon the system under investigation. If a suspension composed of a specific ratio of amorphous to crystalline material is desired, conditions for both precipitation and crystallization need to be elucidated along with establishing necessary compatibility. The Lente insulin preparation also demonstrates that crystalline suspensions containing a certain defined level of amorphous material are possible alternatives. For example, this might be a useful option in cases where complete crystallization is not possible. Key requirements for a suspension of this type include a batch-to-batch reproducible degree of crystallinity and demonstration that the crystalline to amorphous ratio remains constant over the shelf-life and during the intended use period.

9.3.1.4 Crystalline and Solution Mixtures

Biphasic preparations containing mixtures of solution and suspension phases of insulin provide examples of this category. Once again, the impetus for developing such formulations was based on optimization of injection therapy. The soluble component of the preparation provides the prandial insulin effect to address meal-related insulin requirements while the suspension phase provides protracted activity (basal) all in a single injection. Insulin biphasic mixtures containing either Ultralente or NPH as the suspension component have been described.

A mixture containing Ultralente crystals (described by Brange, 1987) is a special case whereby the solubility differences of two insulin types (porcine and bovine) was exploited. Rapitard® is composed of 75% bovine insulin Ultralente and 25% soluble porcine insulin. Because bovine insulin is less soluble than porcine insulin, the ratio of suspension to soluble insulin is relatively stable with only a small portion of bovine insulin dissolved in solution. This particular preparation is no longer marketed.

Biphasic mixtures composed of human insulin NPH and soluble regular human insulin in numerous ratios of 10:90 through 50:50 (solution:suspension) have been commercialized. Solution:suspension mixtures of the rapid acting insulin analogs LysB28, ProB29-human insulin, and AspB28 insulin are also currently available (see Beals et al., 2008 for details). Similar to the NPH and soluble insulin mixtures, these preparations combine the analog protamine crystalline suspensions with solutions of the respective analog. These preparations must contain the same insulin species in the suspension and the solution phases (i.e., homogeneous mixture) as exchange* can occur in heterogeneous mixtures over long-term storage (Edwards et al., 1996). If appropriate physicochemical stability can be demonstrated, however, it may be feasible to prepare biphasic suspensions composed of two completely different proteins in the solid and solution phases (i.e., heterogeneous mixture). This type of preparation can be achieved, for example, by exploiting crystallization conditions for one protein that are totally inadequate for the other. Differences between

---

* The term exchange refers to a process whereby peptide or protein molecules transfer between the solution and solid phases or vice versa.

the molecular properties of the two proteins would bias against exchange between the two phases over a practical period of time. Examples of this type of biphasic mixture have been described in the patent literature for insulin solution and glucagon-like peptide-1 (GLP-1) suspension preparations (DeFelippis et al., 2007b) and GLP-1 solutions combined with insulin suspensions (DeFelippis et al., 2007a).

Depending on the treatment regimen intended for a specific peptide or protein, such a formulation approach offered by biphasic mixtures might be a useful option. As an alternative to deliberately designing a biphasic mixture, one could also envision a suspension that by the nature of the crystallization procedure is not completely crystalline and contains a known and constant concentration of soluble protein or peptide. This situation may occur in cases where crystal growth is not 100% complete. For example, Yim (1988) has indicated that the conditions for producing an insoluble complex of alpha interferon and protamine can be adjusted to allow for a specific portion of soluble interferon for immediate effect. To be commercially viable, however, the solid to solution ratio in such a suspension must be reproducible and stable, and the preparation should have some practical pharmacological purpose especially if a significant amount of soluble material exists.

9.3.2 COMBINATION OF PARTICLES AND VEHICLE

In this method of suspension manufacture, the drug is prepared as a dry solid independently from the vehicle (Figure 9.3). As with small molecule processing procedures, there are at least four methods for producing the sterile powder. These include crystallization, lyophilization, spray drying, and supercritical fluid particle formation (SCFPF).

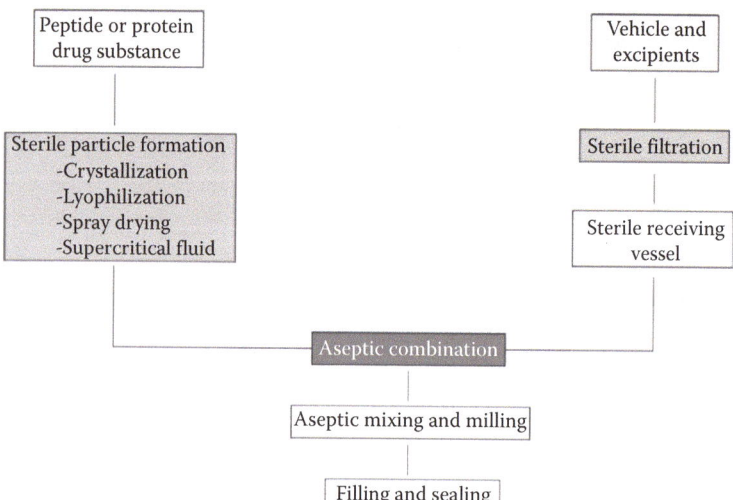

FIGURE 9.3 Diagram illustrating a suspension process involving the combination of particles and vehicle. Dry sterile particles are prepared independently and then aseptically combined with sterilized vehicle.

9.3.2.1   Preparation of Particles

Crystallization: Many of the concepts relating to crystallization previously described can be applied to produce crystals for this formulation approach. A key difference involves harvesting the crystals from the mother liquor and, if necessary, further processing them (e.g., washing and/or drying).

Lyophilization: The proper application of this technique to peptides and proteins has been reviewed elsewhere (Wang, 2000; Nail et al., 2002; see also Chapter 10). Bulking or stabilizing agents can be incorporated into the process as needed.

Spray drying: The active ingredient is dissolved in the solvent and sprayed into a drying chamber. Rapid solvent evaporation is accomplished using a hot stream of sterile gas resulting in the formation of uniform spherical particles (Sollohub and Cal, 2010).

SCFPF: Supercritical carbon dioxide is used as either solvent or antisolvent to achieve supersaturation and subsequent generation of particles. Based on the equipment configuration and methodology employed, three SCFPF techniques have been described. These include Rapid Expansion of Supercritical Solution (RESS), Gas Antisolvent (GAS), and Precipitation with Compressed Antisolvents (PCA). Detailed descriptions and comprehensive reviews comparing the various SCFPF techniques can be found in the studies of Yeo et al. (1993) and Pasquali and Bettini (2008).

Given the propensity of peptides and proteins to denature under high stress conditions, crystallization and lyophilization are more generally applicable to this class of compounds. The ability of these two methods to maintain the sterility of the dried material is also proven. The spray drying process might result in denaturation at the liquid–air interface resulting from the atomization of the protein solution and the high temperature required to evaporate the solvent (Yeo et al., 1993); however, the technique may be appropriate for small peptides that lack higher-order structure or selected proteins with higher-order structure. For example, Mumenthaler et al. (1994) have observed that spray drying caused denaturation of hGH, but a similar process did not adversely affect tissue-type plasminogen activator. Increasing attention has been focused on the potential of SCFPF for pharmaceutical processing in particular its suitability to peptide and protein particle formation. For example, a process for preparing a microparticulate insulin powder (<5 μm) by GAS expansion of dimethylsulfoxide was reported by Yeo et al. (1993). Snavely et al. (2002) reported a PCA technique to obtain particles of insulin by precipitating the protein from solutions of 1,1,1,3,3,3-hexafluoro-2-propanol using supercritical carbon dioxide as antisolvent.

Regardless of the method employed to produce particles, some additional biophysical characterization studies should be performed to confirm that particle processing procedures do not adversely affect the properties of the molecule. Particle size reduction of the solid material by milling and sieving may be required after drying depending on the technique employed and the size of the particles initially produced. The impact of these additional operations on the integrity of the peptide or protein structure needs to be assessed.

### 9.3.2.2	Preparation of Vehicle and Combination

An aqueous or nonaqueous (oleaginous) vehicle containing any necessary excipients is prepared separately from particle formation. Examples of nonaqueous vehicles include any highly purified natural or synthetic oils such as sesame, peanut, or other vegetable oils. Depending on the solubility of the constituents and the overall viscosity of the vehicle, sterilization can be accomplished by either filtration or autoclaving. The sterile combination approach offers more flexibility in the choice of vehicle (aqueous or nonaqueous) since particle growth is accomplished independently.

Once processing of each section is completed, the dry particles and vehicle are aseptically combined. Some form of agitation is required to achieve a homogenous dispersion of particles. In the case of peptides or proteins, appropriate controls should be in place to ensure that the dispersion process does not result in denaturation or other physical changes.

### 9.3.2.3	Peptide and Protein Suspensions Prepared by Combination of Particles and Vehicle

An example of a protein suspension prepared by the combination of particles and vehicle is bovine somatotropin (bST) (Bramley et al., 1989; Ferguson, 1997; Ferguson et al., 1990; Mitchell, 1991). This highly viscous suspension product is marketed under the trademark Prosilac® and is used for increasing milk production in dairy cows. The thick suspension results in protracted release of bST thereby minimizing injection frequency. The vehicle in this case is sesame oil. The oil is thickened by incorporating a wax such as beeswax although a variety of other agents can serve this purpose. Dry powders of the growth hormone are prepared by lyophilization and then dispersed into the vehicle. Mitchell (1991) described a similar suspension, but the growth hormone is complexed with metal ions.

Preparations closely matching the bST suspension design have been described for other growth hormones (Martin, 1986) and growth hormone releasing factors (Brooks and Needham, 1994). A long-acting parenteral suspension of adrenocorticotropic hormone (ACTH) in sesame oil gelled with aluminum monostearate has also been reported (Chien, 1981). It should be recognized that the highly viscous nature of suspensions such as the one described for bST will require the use of large gauge (14–16 G) needles making such preparations generally unattractive for parenteral administration to humans.

## 9.4	EXCIPIENT SELECTION

It would be difficult to identify all possible examples of excipients or auxiliary substances used in parenteral suspensions as selection is highly dependent on the properties of the active ingredient and the desired final preparation. Some excipients are integral to the production of particles and their presence is essential for maintaining specific properties. This situation is especially true for the insulin preparations described earlier. Other excipients, such as preservatives (antimicrobial agents), are included to meet various regulatory requirements for multiuse parenteral pharmaceutical products for injection. Thus, the choice and concentration of excipients is a major consideration in the design of suspension preparations. Not only should these ingredients perform their intended functions, but also optimization studies along with thorough evaluations for compatibility with the other constituents of the preparation and the container-closure system are essential. The quality of each excipient and consistency batch-to-batch are other important considerations. Any trace levels of metal ions, salts, organic agents or other impurities could compromise particle generation and/or impact stability. Therefore, defined specifications and an incoming raw material testing plan for excipients should be part of the overall control strategy for the suspension manufacturing process. These requirements impose limitations on excipient choices in addition to the prerequisite that they be acceptable for use in pharmaceutical products for injection.

The various excipients that are used in parenteral suspensions are categorized as buffering agents, isotonicity modifiers, preservatives, stabilizers, complexing agents, or other auxiliary agents. In some cases, these ingredients may have dual functions as described in Sections 9.4.1 through 9.4.6.

9.4.1 Buffering Agents

Physiologically tolerated buffers are added to maintain the pH in a desired range and some examples include sodium phosphate, sodium bicarbonate, sodium citrate, and sodium acetate. The addition of a buffer is not absolutely necessary if it can be demonstrated that the formulation maintains the desired target pH range. In certain cases, these agents are present as a result of the process for achieving particle formation, yet have no significant buffering capacity at the pH of the final preparation. Ultralente insulin provides an example of such a situation. Sodium acetate is present during crystal growth at pH 5.5, but the final suspension is adjusted to neutral pH conditions where the buffering capacity is minimal. Potential interactions between buffers and metal ions must be considered, as reaction products can lead to compromised stability.

9.4.2 Isotonicity Modifiers

These agents are added to minimize pain that can result from cell damage due to osmotic pressure differences at the injection depot. Glycerin and sodium chloride are examples used in insulin suspensions. Effective concentrations can be determined by osmometry using an assumed osmolality of 285 mOsmol/kg for serum as a guide (Siegel, 1990). Typical concentrations of 7 and 16 mg/mL are used for sodium chloride and glycerin, respectively. The agent to be chosen may be dictated by the need to have a particular ingredient present during particle formation as is the case for sodium chloride in the Lente insulin preparations. The two examples of isotonicity modifiers differ in ionic strength (sodium chloride:high ionic strength; glycerin:low ionic strength) and these properties might influence the choice of one over the other depending upon compatibility and stability considerations.

9.4.3 Preservatives (Antimicrobial Agents)

Multidose parenteral preparations require the addition of preservatives at sufficient concentration to minimize the risk of patients becoming infected upon injection. The testing approaches and requirements for demonstrating antimicrobial effectiveness are described in various pharmaceutical compendia (e.g., USP and Ph. Eur.). Another important point regarding antimicrobial effectiveness testing concerns the use of oleaginous suspension vehicles that may complicate test methodology because of the immiscibility with aqueous microbiological media (Workman and Clayton, 1996).

Typical preservatives for parenteral suspensions include *m*-cresol, phenol, methylparaben, ethylparaben, propylparaben, butylparaben, chlorobutanol, benzyl alcohol, phenylmercuric nitrate, thimerosol, sorbic acid, potassium sorbate, benzoic acid, chlorocresol, and benzalkonium chloride (Boyett and Davis, 1989; Denyer and Wallhaeusser, 1990; Nash, 1988). Toxicology issues will likely impose limitations on the use of other chemicals especially for chronic use applications.

The type of preservative and concentration chosen may also be influenced by factors related to crystal growth, maintaining acceptable suspension stability, or compatibility with already grown crystals in addition to achieving necessary antimicrobial effectiveness. For example, insulin Ultralente cannot be formulated with phenol as the crystal morphology is destroyed over time (Brange, 1987), but methylparaben does not exhibit this effect. The insulinotropin crystallization is also conducted in the absence of preservative, and the formed crystals will transform into amorphous material if *m*-cresol is added before a zinc soaking treatment (Kim and Haren, 1995). In contrast, insulin NPH crystals require a phenolic preservative for crystal growth (Krayenbühl and Rosenberg, 1946), and a mixture of *m*-cresol and phenol in a defined ratio is present in commercial preparations.

9.4.4 STABILIZERS

Stabilizers include a variety of agents that impart stability to particles or the entire suspension. General categories include metal ions (zinc, calcium, etc.), salts used to produce crystals, and organic molecules. Divalent metal ions play a pivotal role in insulin self-assembly and bringing about crystallization. It has been demonstrated that the addition of excess zinc ions to human insulin Ultralente suspensions extends the duration of activity (Hoffmann, 1994). Zinc ions were shown to impact the physical stability of NPH insulin (Massey and Sheliga, 1988; Massey et al., 1988), and were found to be critical for stabilizing insulinotropin crystals (Kim and Haren, 1995). The various salts necessary to achieve crystal growth may also serve as stabilizers in the final suspension. Insulin Ultralente suspensions cleverly exploit the requirement for sodium chloride for crystal growth by also using the ingredient as a tonicity modifier. Organic ligands, such as phenolic preservatives, in addition to serving a role as antimicrobial agents may additionally function as stabilizers.

9.4.5 COMPLEXING AGENTS

Protamine sulfate is an example of a complexing agent used to prepare suspensions. As excess protamine is undesirable from an immunogenicity standpoint (Ellerhorst et al., 1990; Horrow, 1985; Nell and Thomas, 1988) and may impact the stability of biphasic (solution:suspension) mixtures by complexing some of the soluble component, the exact ratio required to completely complex all of the available peptide or protein needs to be determined. Under appropriate conditions, no detectable free protamine or peptide/protein remains in the supernatant. This condition is defined as the isophane ratio (Hagedorn et al., 1936). Originally described using a turbidimetric analysis, high-performance liquid chromatography (HPLC) was used to evaluate the isophane ratio for an insulinotropin–protamine complex (Kim and Rose, 1995).

9.4.6 OTHER AUXILIARY AGENTS

Depending on the type of suspension and its properties and intended use, various auxiliary agents might be included in the preparation. The types of ingredients that might also be added can be classified as wetting agents, flocculating/suspending agents (surfactants, hydrophobic colloids, or electrolytes), viscosity modifiers, antibiotics, and antioxidants. Specific examples of each class of agent can be found in the scientific literature Boyett and Davis (1989) and Nash (1988). Acids and bases such as hydrochloric acid and sodium hydroxide are auxiliary agents necessary for pH adjustment during particle formation and of the final suspension. The choice of auxiliary agent and its concentration may be highly dependent on whether the suspension is for acute or chronic applications. Furthermore, the compatibility of auxiliary agents with drug particles must be demonstrated.

It should be apparent to a pharmaceutical scientist involved in development of suspensions that the formulation and process are integrally related especially in the case of in situ particle growth. Therefore, excipient selection cannot be considered independent of development aspects relating to overall design of the suspension and its manufacturing process.

9.5   GENERAL REQUIREMENTS FOR SUSPENSION PRODUCTS

In addition to demonstrating appropriate chemical, physical, and microbiological stability over its shelf-life and during its intended in-use period, a well-formulated suspension should have the following characteristics (Akers et al., 1987):

1. Resuspension of particles is accomplished with reasonable agitation.
2. Rapid settling of dispersed particles does not occur.
3. Particles can be homogeneously dispersed such that consistent doses are obtained repeatedly.
4. Interaction forces between particles are optimized to prevent the formation of tightly packed agglomerates that are difficult to disperse.
5. The suspension manufacturing process reproducibly produces particles with properties that are maintained batch to batch and during the preparation's intended shelf-life and in-use periods.
6. The suspension product must be easily drawn up into a syringe through a 20–25 gauge (or smaller) needle and readily expelled.

Optimization of these characteristics is an essential part of the development process for suspension products.

Characteristics 1–5 are concerned with special requirements for suspensions relating to elegance, physical attributes, and stability. These items are addressed in greater detail in Section 9.6.2. Requirement 5 is especially important for suspensions intended to have specific delayed or controlled release profiles as the properties of the particles will govern pharmacology.

Characteristic 6 relates to suspension properties essential for administration of proper doses. Needles for parenteral injections have become increasingly smaller in diameter in an attempt to minimize pain and increase patient compliance. While 20–25 gauge needles are acceptable, certain insulin injector devices are currently utilizing 28 or 31 gauge needles. Suspensions composed of large particles could clog these narrow gauge needles affecting the ability to withdraw a proper dose from a container into a syringe (syringeability) and/or the ability to inject the dose into the patient (injectability). In addition to these particle size considerations, suspensions that are composed of particles that tend to agglomerate or employ highly viscous vehicles can also affect syringing and injection operations or patient acceptability. Exceptionally viscous suspensions are not necessarily out of the question as the bST product demonstrates; however, the preparation requires 14–16 gauge needles for subcutaneous administration. Clearly, such a suspension is of limited practicality for human health products.

During development of a suspension, both syringeablity and injectability need to be examined with appropriate in vitro studies. Such studies may involve withdrawals and expulsions using the intended needles and syringes and/or delivery devices. Dose potency of the active ingredient is then confirmed by a suitable analytical assay. Chien et al. (1981) have devised an apparatus that allows quantitative measurements of syringeability. The apparatus was shown to be appropriate for parenteral solutions and suspensions, especially those that are nonaqueous.

9.6   TESTING AND OPTIMIZATION OF CHEMICAL, PHYSICAL, AND MICROBIOLOGICAL PROPERTIES

As with any pharmaceutical product, there is a need to both optimize and test the chemical, physical, and microbiological properties of the preparation and demonstrate appropriate stability over the shelf-life and during the intended in-use period. Suspensions are somewhat more complicated in this regard because optimization of physical properties is extremely challenging and additional testing is usually required. However, many of the concepts relating to other dosage forms also apply to suspensions so that in some cases only a general overview is provided. The reader is referred to other chapters in this book for additional details. "The International Conference on Harmonisation (ICH) guidelines Q1A(R2), Q5C, Q1B, and Q6B" should also be consulted for useful information related to testing of pharmaceutical and biopharmaceutical products.*

---

* *Stability testing of new drug substances and products.* ICH, Q1A(R2), February 2003; *Stability testing of biotechnological products.* ICH, Q5C, November 1995; *Photostability testing of new drug substances and products.* ICH, Q1B, November 1996; *Specifications: Test procedures and acceptance criteria for biotechnological/biological products.* ICH, Q6B, March 1999. These guidance documents are available at www.ich.org.

9.6.1 CHEMICAL PROPERTIES

Peptides and proteins are subject to a variety of chemical modifications resulting in the formation of specific degradation products regardless of the type of pharmaceutical preparation. The kinds of degradation products that can form are dependent on the amino acid sequence of the peptide or protein, formulation conditions, manufacturing operations, and product storage and handling. Since it will be impossible to prevent degradation entirely, the goal of chemical optimization is to identify the degradation products, determine the conditions under which they form most readily, and take steps to minimize their rates of formation.

Reversed-phase, ion exchange, and size-exclusion HPLC assays are particularly useful for peptide and protein products. It should be recognized that such methods must also include an appropriate procedure for solubilizing solid particles. Whenever possible, the most prominent degradation products observed by HPLC should be isolated and subjected to other chemical analyses such as mass spectrometry or sequencing to determine identity and these species may further require activity or toxicology testing. Stability programs should include assays developed to monitor the rates of change of degradation products that can affect product quality and possibly impact the end user. For example, assays for drug concentration and biopotency, other related substances/impurities and higher molecular weight aggregated species are generally good chemical stability indicators. A series of research papers on the chemical stability of insulin (Brange, 1992; Brange and Langkjær, 1992; Brange et al., 1992a, 1992b, 1992c) provide an excellent overview of the experimental approaches to be followed when evaluating protein based pharmaceutical preparations, including suspensions.

Excipients present in the formulation can chemically degrade, interact with various surfaces during manufacturing, interact with the container or closure, or interact with the peptide or protein thereby negatively affecting critical properties of the preparation. Degradation products derived from excipients can also react with the peptide or protein. Therefore, chemical optimization work should also include an evaluation of the excipient properties. The covalent insulin protamine polymer formed in insulin NPH suspensions (Brange, 1992; Brange and Langkjær, 1992; Brange et al., 1992a, 1992b) is an example of a protein/excipient chemical interaction producing an additional degradation product.

Depending on the type of suspension, additional assays may be needed to evaluate the preparation. For example, biphasic mixtures composed of solution and suspension phases may require a method to quantify the proportion of soluble to solid material because the time-action of the two components is different. Simply centrifuging the formulation to separate the phases followed by assay of the supernatant may not be appropriate as the results may be influenced by adsorption of soluble protein onto solid particles. This phenomenon has been observed for the biphasic NPH–soluble insulin mixtures (Brange, 1987; Dodd et al., 1995). To overcome this limitation, Rolim and Bristow (1995) have devised a method whereby treatment of the insulin suspension with a tris buffer effectively desorbs the soluble protein affording estimates for the amount of soluble and solid material.

Other general chemical testing may be undertaken in addition to the routine stability evaluations. Examples include photostability, inverted container stability, packaging compatibility, and aberrant temperature exposure. Photostability testing following conditions described in ICH guideline (Q1B) should be conducted as phenolic preservatives, other excipients, and peptides or proteins (Holt et al., 1977; Kerwin and Remmele, 2007) can degrade when exposed to light for extended periods of time. Inverted container stability is applicable for vial presentations having headspace between the closure and suspension. By including studies on inverted samples, information can be obtained on the potential incompatibility of formulation with the closure. Packaging compatibility with the formulation should be evaluated by performing stability studies in the intended container-closure system for the product. The effect of aberrant temperatures (freezing and heating) on the product should be determined as unpredictable and rather extreme conditions may be encountered during shipping, warehouse storage, or in the hands of the consumer.

9.6.2   Physical Properties

9.6.2.1   General Concepts and Basic Theory: Optimization of Physical Properties of Dispersed Systems

Suspensions are thermodynamically unstable systems and the goal is to design a preparation that is kinetically stable for a sufficient period of time (i.e., shelf-life) so that product performance is not compromised by gross changes in physical properties. Two common physical instability problems are caking and crystal growth. In order to understand the behavior of suspensions and methods for optimization, a basic understanding of theoretical concepts explaining these physical transformations is required. However, a detailed discussion on theory is beyond the scope of this chapter, and the reader is referred to appropriate texts on this subject for additional details.

The Derjaguin, Verwey, Landau, and Overbeek (DVLO) theory was originally devised to explain the stability of colloidal systems, but the principles have also been invoked to explain particle interactions in coarse dispersions such as suspensions (Hiestand, 1964; Kayes, 1977; Martin, 1961; Matthews and Rhodes, 1968, 1970; Schneider et al., 1978). According to DVLO theory, the forces on particles in a dispersion are due to electrostatic repulsion and van der Waals attraction. Note that other forces are usually included to adequately explain interactions in dispersed systems. Potential energy curves for particle interactions are shown in Figure 9.4. The forces at particle surfaces will affect the degree of flocculation and agglomeration observed for suspensions. Thus, DVLO theory provides a framework for understanding the interactions of particles which control physical properties of suspensions.

Referring to the composite curve in Figure 9.4, the collision of particles will not occur if the repulsion energy is high (e.g., low electrolyte concentration in aqueous suspensions). Such a system is referred to as deflocculated. When the particles settle, the energy barrier is overcome and strong attractive forces in the potential well will cause a densely packed sediment to form. Eventually, a hard cake results that is difficult to disperse using normal agitation procedures for resuspension. Such a condition is highly undesirable as a nonuniform dispersion of particles can severely impact dosing reliability.

For coarse dispersions which are flocculated, the potential energy barrier is still too large to be surmounted by approaching particles (Martin, 1961; Schneider et al., 1978).

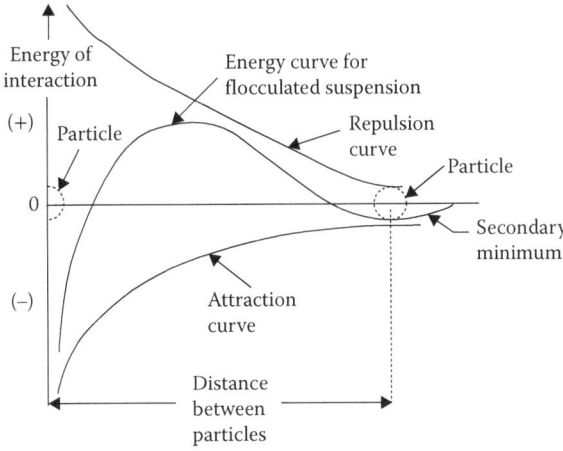

FIGURE 9.4 Potential energy curves for particle interactions in coarse suspensions. Particles in suspension are subject to van der Waals attractive and electrostatic repulsive forces. For coarse dispersions which are flocculated, weaker attractive forces occur at significant interparticle distances at the region referred to as the secondary minimum resulting in the formation of loosely aggregated particles. (Reprinted from Martin, A.N., *J. Pharm. Sci.* 50, 514, 1961. With permission.)

However, weaker attractive forces occur at significant interparticle distances at the region referred to as the secondary minimum in Figure 9.4. Particle interactions in this case result in the formation of loose aggregates (floccules). Flocculation can be induced in a suspension by the addition of a flocculating agent such as an electrolyte. Suspensions that are flocculated are considered pharmaceutically stable because sedimented material is readily redispersed upon typical agitation using procedures such as mild shaking, rotation, and/or inversions of the product container.

The properties of flocculated and deflocculated suspensions are compared in Table 9.3 (Nash, 1988; Zografi et al., 1990). For flocculated suspensions, the

TABLE 9.3

Relative Properties of Deflocculated and Flocculated Particles in Suspension

State	Particle Characteristics	Sedimentation Rate	Sedimentation Volume	Cake	Resuspension
Deflocculated	Exist as separate entities	Slow	Initially suspended, but settles to a small volume	Yes	Compact, difficult to redisperse
Flocculated	Exist as weak aggregates (floccules)	Fast	Settling results in the presence of an obvious, clear vehicle region. Final particle volume is typically large, but may be small	No	Porous, easy to redisperse

sedimentation properties may result in a preparation that appears to contain a majority of clear vehicle upon settling. This condition is not a serious problem provided caking does not occur making the particles difficult to disperse with minor agitation. Ultralente insulin is an example of a suspension that displays a very small sedimentation volume, but the particles are easily resuspended to homogeneity with gentle shaking of the vial. Because Ultralente insulin crystal growth conditions are defined and the suspension has appropriate stability, there is little value in making adjustments to improve its physical appearance upon settling (i.e., large sedimentation volume). Thus, suspension formulation design may require compromises between aesthetic aspects and other desirable physical attributes of the preparation.

Sedimentation volume and zeta potential measurements are useful for optimizing the physical properties of suspensions by providing information on the degree of flocculation. Sedimentation volume is determined by measuring the equilibrium volume of settled particles relative to the total suspension volume after resuspension. The quantity is expressed as a ratio:

$$F = \frac{V_u}{V_o} \tag{9.1}$$

where V_u is the equilibrium volume of sediment and V_o is the total suspension volume.

Zeta potentials are determined to estimate surface charges as discussed in Section 9.7. The relationship between sedimentation volume and zeta potential is illustrated in Figure 9.5 (Martin, 1961). The addition of a flocculating agent causes a progressive reduction in zeta potential and changes in sedimentation volume. There exists a region where the sedimentation volume is maximized (flocculated) and no caking is observed. Note that if too much flocculating agent is added overflocculation and caking can occur. Exposing suspensions to extreme temperature or mechanical stress can also produce this effect (Hem et al., 1986). This example indicates that the zeta potential must be controlled in order to produce a suspension with desirable physical properties.

Most examples of the application of these principles reported in the literature describe small molecule systems (Kayes, 1977; Matthews and Rhodes, 1970). However, proteins carry charge due to ionization of surface groups and specific adsorption of ions (anions or cations), and these concepts are applicable to protein suspensions as well. For example, Kim et al. (1995) described the effect of chloride ions and pH on sedimentation volume and zeta potential of zinc insulin suspensions. Suspensions were prepared at three sodium chloride concentrations and adjusted to different pH values. Sedimentation volume increased while the zeta potential became more negative with increasing pH. At the highest sodium chloride concentration, maximum values for sedimentation volume and zeta potential were observed at pH 6.9 indicating that a flocculated, pharmaceutically stable suspension can be prepared when zinc insulin is precipitated with 120 mM sodium chloride at pH 6.9.

The method for preparing stable suspensions described above is referred to as controlled flocculation. Another method for achieving optimal physical stability of suspensions is termed the structured vehicle approach. In this case, viscosity-imparting suspending agents are added to the vehicle to reduce sedimentation and maintain the particles in suspended state. Depending on the application, structured

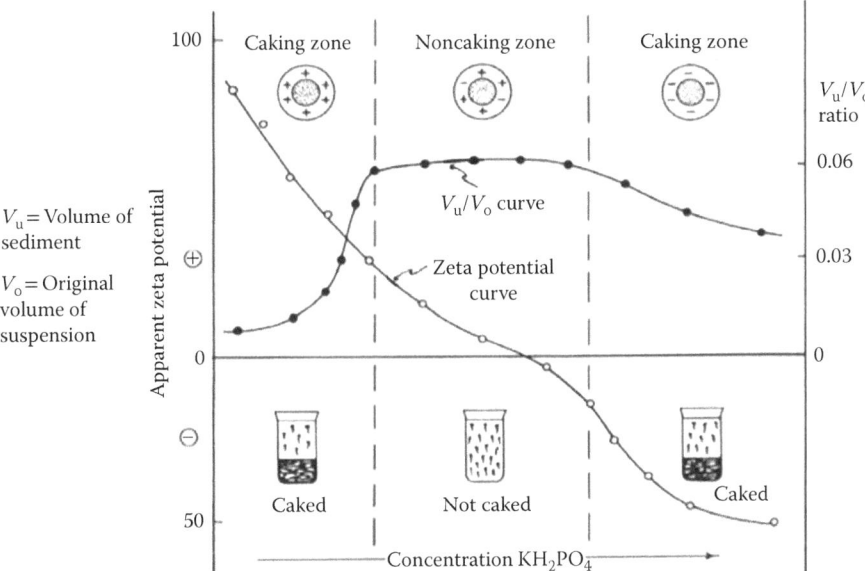

FIGURE 9.5 Relationship between zeta potential and sedimentation volume. The diagram depicts the effect of adding an anionic flocculating agent to a positively charged system. Maximum flocculation, as determined by sedimentation volume measurements, occurs within a narrow range of zeta potential values. The region designated the "no caking zone" defines formulation conditions where caking of the suspension is less likely to occur. (Reprinted from Martin, A.N., *J. Pharm. Sci.* 50, 515, 1961. With permission.)

vehicles might not be appropriate for parenteral preparations because the high viscosity would affect syringeability (Nash, 1988).

Akers et al. (1987) have described the phenomenon of crystal growth in pharmaceutical suspensions. Particle size distribution, dissolution and recrystallization, pH and temperature changes, and polymorphism and solvate formation are factors that can affect crystal growth. The close contact of particles in settled deflocculated suspensions will favor crystal growth by a process referred to as Ostwald ripening (Patel et al., 1986).

One method to retard crystal growth involves the addition of viscosity imparting agents, but again this approach may not be appropriate for parenteral suspensions. The best approaches for minimizing crystal growth in peptide or protein suspensions are to control the particle size distribution, select the correct polymorph (if applicable) or use the controlled flocculation approach. Appropriate testing should also be conducted to determine the impact that pH and temperature changes have on physical stability.

9.6.2.2 Testing of Physical Properties

As highlighted above, the physical properties of suspensions are rather unique compared to other pharmaceutical preparations and much of the testing will be focused in this area. One of the simplest attributes to evaluate is physical appearance. This qualitative assessment is performed in stability programs and after exposing the suspension to various stresses of temperature or mechanical agitation. The preparation must remain "elegant" (no clumping of particles, uniform dispersion of particles upon agitation, particle characteristics remain unchanged, and material does not adhere to container-closure surfaces) after exposure to reasonable conditions of temperature. In instances where the product elegance is compromised, the information can be used to draft instructions for proper storage and use of the product.

Other more quantitative evaluations of physical stability involve measurement of sedimentation volume and rate. The procedure for determining sedimentation volume has been previously described. Sedimentation rate is used to estimate the rate of particle settling, and can be done in conjunction with sedimentation volume determinations by measuring the top boundary of particles progressing downward as a function of time. In addition to determining sedimentation rate for samples exposed to extreme conditions, this important parameter must be evaluated for flocculated suspensions where particles sediment rapidly. After proper resuspension manipulations, the particles must remain dispersed in vehicle for a sufficient time to allow injections of proper doses.

To simulate the various stresses a suspension might encounter during shipping or patient handling, thermal and mechanical testing is routinely performed to evaluate product performance under exaggerated conditions. Methods of this type include some agitation of the suspension in its container induced by either shaking or rotation (e.g., Massey and Sheliga, 1988; Massey et al., 1988). Temperatures are generally elevated during mechanical stress testing as well. Such treatment can result in a rather unsightly appearance of the suspension due to clumping or result in damage to the particles, changes in sedimentation properties, adherence of solids to the container-closure system, and loss or change in concentration depending on the duration of exposure. The same outcomes can result from extended exposure to high temperatures alone as shown for insulin suspensions (Brange, 1987). All of these physicochemical aspects should be evaluated for potential significance to the final product. Shnek et al. (1998) reviewed physical stress testing methods designed to study insulin suspensions and solutions in cartridge containers intended for portable pen-based injector devices. Protocols generally involve exposing product samples to high temperature (e.g., 25 and/or 37°C) combined with mechanical agitation at periodic intervals, and may include simulated dosing following resuspension over the course of several days. The extreme conditions employed with these procedures far exceed those encountered during typical patient use. Nevertheless, regulatory authorities may request data from such experiments to support recommended in-use periods for the product.

In the same manner stress studies are performed at elevated temperatures to understand the impact on physical attributes, the effects of extreme cold temperature exposures must be similarly evaluated. Graham and Pomeroy (1978) have studied the effects of freezing and thawing on insulin suspensions. Freezing and thawing caused an increase in sedimentation rate attributed to aggregation of the crystals. Microscopy also revealed some crystal damage; however, bioactivity was unaffected. Studies exploring extreme temperature excursions (low and high) are also useful for understanding the impact of potential aberrant shipping conditions since refrigeration is typically required for peptide and protein products to maintain appropriate stability.

As highlighted earlier, syringing studies are generally required for suspension preparations stored in vials to ensure that the preparation can be used reliably. Syringing can also be performed to evaluate physical stability. Procedures involve performing daily resuspension manipulations and withdrawal of doses from product containers under conditions that mimic the intended use profile. Evaluations are routinely conducted at room temperature and continue for a time period corresponding to the proposed in-use dating period. The physical appearance of the suspension is examined throughout and a thorough chemical analysis of the material remaining at the end of test is conducted. A common occurrence for suspensions evaluated in this manner is the adherence of solids to the wall of the vial container primarily in the region closest to the stopper. This material forms as the suspension flows downward upon settling. Vehicle drains leaving solid particles remaining on the interior glass surface. Dosing studies with suspensions contained in cartridge containers are also necessary. The approaches are similar to those performed for vials except that doses are expelled following resuspension using a delivery device. In addition to evaluating the physicochemical properties of material remaining within the cartridge, concentration determinations are made on expelled samples to ensure suspension homogeneity of the dose. These tests may be performed in combination with physical stress testing protocols used for suspensions in cartridge containers described earlier.

9.6.3 MICROBIOLOGICAL PROPERTIES

While designing a suspension formulation and its manufacturing process, steps must be taken to ensure that the microbiological properties are optimized and maintained in the final product. The process must include appropriate procedures for sterilization taking into consideration the way in which particles are produced. In certain cases, sterile filtration of solutions prior to initiating particle growth is appropriate while other processes involve aseptic combination of preformed sterile particles with a sterilized vehicle. Because of the sensitivity of peptides and proteins to extreme conditions, careful thought must be given to the manner in which sterilization is accomplished. For example, terminal sterilization is unlikely to be tolerated by peptides and proteins.

Many of the issues associated with identifying the appropriate antimicrobial preservative for a suspension have been described in Section 9.4; however, there are additional points to be considered regarding optimization. Antimicrobial effectiveness testing should be conducted once all of the ingredients and concentrations are established, as the results will reflect the overall composition of the preparation including final pH conditions. Therefore, it should not be assumed that a given concentration of preservative providing sufficient antimicrobial efficacy for one preparation will be suitable for another. Furthermore, the efficacy of preservatives varies, with some being more potent than others. The reports of Allwood (1982), O'Neill and Mead (1982), and Denyer and Wallhaeusser (1990) compare the properties and relative antimicrobial effectiveness of some commonly used preservatives.

Phenolic preservative concentrations can be decreased due to absorption into tubing during mixing and recirculation operations. Permeation through the closure and chemical degradation of these agents is also possible. To account for possible degradation over the shelf-life, formulations varying in preservative concentration should be subjected to antimicrobial effectiveness testing to define the effective range wherein efficacy is maintained and to establish specifications. In addition, preservative excesses may be necessary to account for loss during manufacturing and to achieve final target concentrations.

9.7 TECHNIQUES FOR CHARACTERIZING AND OPTIMIZING SUSPENSIONS

A number of techniques can be applied throughout the development of peptide and protein suspensions to achieve the optimal properties described in Sections 9.5 and 9.6. Characterization and optimization efforts can be directed towards the particle formation process, the particles themselves, or the suspension properties of the preparation. This section will introduce the various techniques available, but our intent is to provide a general insight into what is available and its application rather than to provide detailed descriptions for each method. The reader is referred to appropriate and readily available textbooks for more detailed information on a given technique. This survey is by no means comprehensive either as it focuses primarily on solid state applications. It should be emphasized that an understanding of the solution state properties of the peptide and protein being formulated can be extremely valuable, if not essential to suspension development efforts. Furthermore, chemical analyses on the solid peptide or protein particles should be considered to establish the composition of the material.

9.7.1 MICROSCOPY

Perhaps one of the most basic techniques, the importance of microscopy cannot be overstated. White light microscopy can be used to visualize particles in the size range of 0.4–100 μm (Martin, 1993), and the technique is particularly useful for characterizing nonspherical or amorphous particles. Microscopy can also be used to monitor particle growth over time.

9.7.2 PARTICLE SIZING

Information on the particle size is essential throughout all stages of suspension development. In addition, particle size can be an important property for evaluating product stability and ensuring process control. A number of techniques are available to determine particle size and size distributions each having associated pros and cons for a given application (Kelly and Lerke, 2005). Regardless of the methodology employed in the various commercial instruments, measured parameters are usually reported in terms of the diameter of spherical particles. In many cases, particles may not be spherical or could be irregularly shaped so that sizes are approximate and will vary amongst techniques. Particle size distributions, usually included as part of the analysis, may be more useful as they provide information on the population of species present in the sample. The choice of technique will depend on the nature of the sample to be measured and the appropriateness of the methodology.

Orr and Spence (1977) reported on the use of the Coulter counter for a variety of pharmaceutical applications. Of particular interest, they describe a quantitative assay for determining the amount of amorphous material in crystalline insulin suspensions. In similar studies, laser-light diffraction spectroscopy was used to estimate the particle sizes and distributions of commercially available insulin suspensions (Komatsu et al., 1996). Komatsu et al. (1996) also demonstrated stress-induced (sonication) reduction of particle size in certain cases presumably due to crystal damage. This observation highlights how particle size analysis can be useful for examining the effects of physical stress on suspension products such as mixing and recirculation typically used in manufacture, and agitation caused by shipping and consumer use.

9.7.3 ELECTROPHORETIC LIGHT SCATTERING

Electrophoretic light scattering (ELS) provides a direct measure of the velocity of particles moving in an electric field. Velocity is obtained by measuring the Doppler shift of laser light scattered from a moving particle electrophoresing in liquid. The electrophoretic mobility (U), which is proportional to the surface charge density of the particle, is determined using the following equation:

$$U = \frac{V}{E} \tag{9.2}$$

where V is the electrophoretic velocity and E is the applied electric field.

The zeta potential (ζ) is derived from U using the relationship

$$U = \frac{\zeta \varepsilon}{\eta} \tag{9.3}$$

where ζ is the zeta potential, η is the viscosity, $\varepsilon = \varepsilon_{o}D$, D is the dielectric constant, and ε_{o} is the permittivity of free space. Zeta potentials are usually expressed in millivolts.

As surface charge governs particle interactions, determining zeta potentials is useful during development of pharmaceutical suspensions as the quantities can be correlated with physical attributes and stability of coarse dispersions (see Section 9.6.2.2). The studies of Martin (1961), Kayes (1977), Schneider et al. (1978), and Kim et al. (1995) provide detailed information on the application of ELS for controlling the physical properties of suspensions. In another relevant application of Doppler ELS, the technique has been used to study the pH-dependent surface charge of NPH insulin (Dodd et al., 1995). The results were used to explain the adsorption of soluble insulin onto the NPH crystals in biphasic mixtures.

9.7.4 DYNAMIC LIGHT SCATTERING

Although primarily applicable to solution studies, the technique can be employed in suspension development as a tool to study the potential for systems to form certain particles. It is widely recognized that protein crystallization begins with aggregation of individual molecules in solution. Aggregates formed during the early prenucle-ation phase will determine the potential of the system to form crystals. The goal of dynamic light scattering experiments is to determine whether precipitation or crystallization will occur based on aggregation behavior in solution. Such informa-tion can assist with screening activities for appropriate crystallization conditions. Kadima et al. (1991) have shown that differences in aggregate size can be used to define conditions where solubility is maintained, precipitation occurs, or crystals form.

9.7.5 CALORIMETRY

Differential scanning calorimetry (DSC) has been applied to study the crystal growth mechanism of a protein suspension as demonstrated by Ooshima et al. (1997). DSC curves were interpreted in terms of population of species differing in the degree of self-association. Such information might be useful for predicting crystallization behavior. In another example of the application of calorimetry, isothermal titrating calorimetry has been used to study adsorption of soluble insulin onto NPH crystals and obtain estimates for the thermodynamic parameters associated with this process (Dodd et al., 1995).

9.7.6 SCANNING PROBE MICROSCOPY

The scanning probe microscopy techniques of scanning tunneling microscopy and atomic force microscopy (AFM) with specific applications to pharmaceutical sys-tems have been reviewed (Santos and Castanho, 2004; Shakesheff et al., 1996). AFM has many advantages for the study of biological molecules including no requirement for sample treatment allowing imaging in the natural state, high resolution (nanome-ter) of surface topography, and the ability to image in liquid. Related to the subject matter of this review, the AFM technique is appropriate for characterizing protein crystal packing and growth mechanisms. For example, Yip and Ward (1996) have used tapping-mode atomic force microscopy (TMAFM) to identify polymorphs of bovine insulin. The in situ imaging capabilities of the technique allowed the direct visualization of the effects of additives and other parameters as crystal growth occurs. Real-time TMAFM was used to study crystal growth characteristics of $Lys^{B28}Pro^{B29}$ insulin revealing that subtle changes in molecular conformation of the analog influ-ence interhexamer interactions in the solid state (Yip et al., 1998a). The technique was also applied to characterize Ultralente crystals prepared from human, porcine, and bovine insulin (Yip et al., 1998b). Differences in hexamer molecular packing of bovine insulin Ultralente provided a possible explanation for slower dissolution of these crystals relative to those prepared with human or porcine insulin. Low tem-perature TMAFM performed in solution allowed assessment of interfacial struc-ture, morphology, and growth characteristics associated with $Lys^{B28}Pro^{B29}$-protamine crystals (Yip et al., 2000). This application demonstrates the utility of AFM for exploring the effects of temperature on crystallization behavior. While AFM has been shown to be a useful tool for crystal characterization, it should be emphasized that the molecular resolution achieved in the studies reported for insulin did require larger crystals than those typical of the microcrystalline suspensions for pharmaceutical use. AFM can also be used to examine surface morphology of polymeric materials for drug delivery applications.

9.7.7 INFRARED AND RAMAN SPECTROSCOPY

Vibrational spectroscopy techniques are particularly useful for obtaining structural information on solid peptide or protein particle samples produced by the various methods discussed in Section 9.3.2.1. Because some of these processing procedures can result in denaturation, characterization studies should be performed on the material to confirm structural integrity. Examples of the application of these techniques include the use of infrared spectroscopy to detect denaturation induced by lyophilization (Prestrelski et al., 1993a, 1993b), and application of Raman spectroscopy to characterize various crystal forms of insulin (Tensmeyer and Shields, 1990).

9.7.8 X-RAY CRYSTALLOGRAPHY

As mentioned earlier, x-ray crystallographic data on crystalline particles comprising suspensions are not essential and in most cases not feasible due to the small size or lack of crystal quality. The practical limitation of applying the technique to commercial suspensions is illustrated nicely in the work of Balschmidt et al. (1991). In order to obtain the crystal structure of porcine insulin cocrystallized with clupeine-Z (NPH insulin), the addition of 4 M urea to the crystallization mixture was required. While it can be argued that the difference in crystallization methodology may not induce major structural differences, the relevance of the finding to commercial crystals can still be questioned. However, no other technique can provide equivalent detailed information at the molecular level.

9.7.9 X-RAY POWDER DIFFRACTION

This technique has been used for many years to identify organic and inorganic microcrystalline samples. Accurate and reproducible fingerprint information can be obtained by calculating the d-spacing and intensity for each diffraction line, and this method can be used to distinguish between different crystalline polymorphs of the same molecule. The technique has been rarely applied to proteins due to lack of sufficient microcrystalline material and the requirement that the proteins remain immersed in mother liquor. Nevertheless, polymorphs can be identified by changes in d-spacing and intensity data of the resulting powder patterns as demonstrated by Richards et al. (1999). In this work, x-ray powder diffraction was used to elucidate the type of insulin hexamer in the microcrystals of various Ultralente suspensions.

9.7.10 IN VITRO DISSOLUTION

In the development of sustained or controlled-release suspensions it is useful to have an in vitro assay available for quickly approximating dissolution properties. Analogous to dissolution testing for solid dosage forms, the procedure requires some detection method to continuously monitor the release of drug. As an example, Graham and Pomeroy (1984) have developed a continuous-flow spectrophotometric method that can categorize insulin suspension preparations based on clinical time-action classifications. Prabhu et al. (2001) described the use of a spin-filter device to study the factors controlling dissolution of zinc-complexed insulin suspensions.

9.7.11 COMPUTER MODELING AND SIMULATION

Provided a suitable model can be developed, it is possible to apply computational approaches to generate dissolution profiles. Data obtained from actual experimental studies can then be compared to the computer-generated profiles to derive quantitative values for key terms, such as diffusion resistance and surface reaction resistance, useful for predicting dissolution behavior. This approach has been successfully demonstrated for the dissolution of zinc insulin suspensions (Prabhu et al., 2002). The combination of experimental and computational strategies enhances the utility of in vitro studies aimed at optimizing suspension properties relative to the desired pharmacological performance. Applying predictive tools to identify the best candidate formulations may further help to reduce the number of confirmatory tests in animal models to determine in vivo release characteristics.

9.8   CONTAINER-CLOSURE AND DELIVERY SYSTEMS

As with any injectable pharmaceutical preparation, qualification of the container-closure is an essential part of the development program. The general requirements for container-closure selection as set forth in various government regulations and pharmacopoeias (e.g., USP, Ph. Eur., and Japanese) are no different for suspensions; however, some specific considerations merit attention. One particular area of focus relates to the subject of extractables and leachables (Osterberg, 2005). Extractables are chemical substances that are extracted from the container-closure components under exaggerated conditions (e.g., exposure to organic solvents), while leachables are chemical substances that can migrate into the dosage form under typical conditions of storage and use. Studies to determine extractables are used to inform the types of leachables that may be encountered. The primary concern with these impurities is safety due to intrinsic toxicity of the chemical substances, and also the potential impact on the pharmaceutical preparation and properties of the active ingredient. In the case of peptide or protein suspensions, organic or inorganic substances could have deleterious impact on the properties of the particles resulting in physicochemical changes impacting stability or possibly pharmacological properties. There could also be an impact on the physicochemical properties of the peptide or protein since inorganic or organic substances are capable of binding to these molecules. Insulin is a classic example where both metal ions and phenolic derivatives are capable of binding to the protein and inducing conformational changes to the three-dimensional structure (DeFelippis et al., 2001 and the references therein). Therefore, even if a presumably innocuous substance from a toxicological point of view leaches into the product, the potential impact on the properties of the active molecule and the suspension preparation must be thoroughly understood and corrected as appropriate.

There are unique considerations that also must be taken into account for delivery systems intended for suspensions. Multiuse injectors resembling a writing pen have become very popular for delivering insulin products including the NPH and mixture preparations. These devices use a special cartridge container-closure system that is filled with the formulation such that there is essentially no headspace between the seal and plunger. Because suspensions typically require manipulation to disperse settled particles, the injector must be designed to allow various hand manipulations by the patient that are not too difficult or burdensome to perform. Unlike a suspension in a vial that will have a headspace above the aqueous vehicle and requires minimal effort to disperse particles, cartridges generally need more agitation to achieve a homogenous suspension. To assist with the resuspension operation, a single glass bead may be added to the suspension preparations contained in cartridges. Since it is essential to ensure complete suspension homogeneity prior to dosing, the device must provide the health care practitioner and patient a clear viewing window to enable inspection of the suspension.

9.9 SUSPENSION MANUFACTURE

Basic examples of suspension manufacturing processes have been described in Section 9.3. However, it must be appreciated that aseptic manufacturing for injectable products is a highly complex process consisting of multiple unit operations with associated parameters and requiring numerous controls and procedures to ensure reproducible product quality batch-to-batch. A detailed treatment of this subject is beyond the scope of this chapter, but there are some unique aspects for peptide and protein suspensions related to particle formation, scale-up, and filling that warrant special consideration. These topics are discussed in the context of manufacturing process design and creation of appropriate control strategies. Basu et al. (2004) should be consulted for additional information relevant to this subject. This review, focused on protein crystallization of biopharmaceuticals for drug delivery applications, provides an excellent example of a work flow scheme that can be applied to the formulation and manufacturing process development of a peptide or protein suspension.

9.9.1 PARTICLE GENERATION

The starting point for any pharmaceutical manufacturing process is the raw materials that are combined to produce the finished product. For obvious reasons, these chemical agents must be of the highest quality to ensure patient safety. However, consideration must also be given to how material properties may influence particle formation steps of the process. Purity is of utmost importance since trace levels of inorganic or organic agents may foul crystal growth or denature the peptide or protein. To address this concern, additional specification requirements may need to be established that exceed those typically defined in pharmacopoeia monographs. An expanded testing program to confirm appropriate quality of raw materials prior to use may also be necessary.

Throughout the stages of pharmaceutical development, each unit operation associated with the particle generation process needs to be comprehensively studied to determine the parameters that can negatively impact the quality attributes of the active molecule and/or suspension dosage form. As pointed out earlier, peptides and proteins present unique challenges in general due to their inherent molecular fragility, and many operations involved in particle formation can have deleterious effects on structural properties. In addition, the process parameters themselves and the manner by which operations are conducted may impact whether the desired particles are produced.

The order, rate, or location of raw material additions can be extremely important for producing the desired crystalline particle form of a peptide or protein. Temperature can influence particle growth kinetics, or a specific range may be required in order to achieve a particular morphology. Heat gain or loss (rate and overall time) resulting from process conditions can result in denaturation of the active molecule or influence particle generation. Agitation due to mixing (type, rate, intensity, and duration) is another factor requiring thorough exploration and careful control of process parameters when dealing with peptide or protein suspensions. Excessive foaming or entrapped air should be avoided as denaturation of the peptide or protein at the air–liquid interface is possible. If foam is generated and remains at the surface, crystal nucleation can occur resulting in increases in the number of crystals and changes in size distribution as observed by Schlichtkrull (1961). It may not be possible to include wetting agents or surfactants in the suspension so steps must be taken in manufacturing to prevent foaming. The influence of mixing on particle generation, control, and stability has been discussed in a report by Genck (1995).

Process parameter studies can be accomplished using univariate experimentation and the resulting information interpreted to define proven acceptable ranges. However, it is very likely that many of the parameters discussed above will have interaction effects and these can best be examined using multifactorial statistically designed experiments. The linkage between process inputs (i.e., materials and parameters) to quality attributes (molecule and dosage form) can then be used to define a design space* for the particle formation operation. Once a comprehensive understanding of the process parameters influencing particle generation is obtained from development studies, and the performance of the operation confirmed in the manufacturing facility, a control strategy can be devised using a variety of different elements including raw material controls, procedural controls, parameter controls, in-process tests, process monitoring, characterization, and lot release (end product) testing.

Many aspects of the control strategy are more suitably addressed throughout the process rather than reliance on lot release tests. The importance of implementing controls for incoming raw material has already been discussed. Operations associated with the compounding of raw materials can be managed using combinations of procedural, parametric, and/or in-process testing controls. For example, additions can be controlled by the manufacturing procedures listed in the batch record and

---

* The term "design space" is defined as the multidimensional combination and interaction of input variables and process parameters that have been demonstrated to provide assurance of quality. For further details concerning the concept of design space, refer to the following reference: *Pharmaceutical development.* ICH, Q8 (R2), Aug. 2009. This guidance document is available at www.ich.org.

verified by mass determination. If a particular raw material is deemed critical to particle generation and its concentration must be maintained within a narrow range, an in-process test to confirm proper levels might be warranted. Temperature can be controlled via continuous monitoring with probes and recorded with a data collection system. Mixing operations can be managed with parametric controls defined within the manufacturing batch record. Some confirmatory in-process tests for pH and/or osmolality may be conducted to ensure the correct vehicle matrix conditions. Finally, in-process tests for particle morphology, size, and distribution can be incorporated into the control strategy to demonstrate completion of the particle formation operation and proper characteristics prior to forward processing.

9.9.2 SCALE-UP

The scale-up of parenteral suspensions to commercially viable manufacturing processes presents a significant challenge. As pointed out earlier, crystallization of peptides and proteins at small scale is not simple, but the difficulty of the problem is exacerbated by virtue of the large volumes needed and the strict controls required for the preparation of pharmaceutical products. Generally, incremental increases in scale are attempted starting from the bench process and progressing upward in volume to the required batch size. Changes in vessel composition, for example, glass versus stainless steel, and geometry will occur during the transition and this could impact the crystallization. Because of these issues, care must be taken in developing scale-down models and validating their utility for predicting potential outcomes at commercial process scale. In addition, one must consider how certain operations performed with ease in the laboratory such as additions, mixing, fluid transfers, and temperature control will be conducted under the aseptic conditions of a pharmaceutical manufacturing facility.

The other methods of particle generation are not any easier to scale-up (Akers et al., 1987) especially when peptides or proteins are the target compound. Seemingly sound lab or pilot scale procedures can produce undesirable outcomes at larger scale. Therefore, steps must be taken to ensure that particle generation and size reduction operations are accomplished without affecting the structural properties of the molecule or reproducibility of the process. Milling operations for dry peptide or protein particles must be conducted aseptically because no practical means of resterilization is feasible if sterility is compromised. Nonaqueous (oil) vehicles do require special consideration if they are to be sterilized by filtration as filter composition, pore size, and flow rates can impact capture efficiency. Finally, an appropriate strategy needs to be devised for aseptically combining the particles and vehicle.

Following each successive scale-up, it is important to consider comparability* of the product properties relative to known characteristics at the previous scale. Expanded physicochemical evaluation beyond routine testing will need to be employed to ensure, for example, that molecular integrity, particle morphology, suspension/sedimentation

---

* For further details concerning the concept of comparability, refer to the following reference: *Comparability of biotechnological/biological products subject to changes in their manufacturing process.* ICH, Q5E, June 2005. This guidance document is available at www.ich.org.

aspects, dissolution profile, and stability performance (including accelerated and stress conditions) are comparable between scales. Section 9.7 provides examples of the types of characterization techniques that can be employed to assess comparability. Depending upon the stage of clinical development or whether the product is already licensed, it may be necessary to also include nonclinical and/or clinical studies to evaluate in vivo pharmacological performance. Comparability assessments should be considered for other types of changes to the process, such as introducing new raw materials, parameter modifications, or transfer of the process to a different manufacturing facility due to the potential to influence properties of the suspension.

9.9.3 FILLING

Suspension homogeneity must be maintained throughout the filling operation to ensure content uniformity in the finished units. Continuous mixing and recirculation are typically conducted to keep particles homogeneously dispersed. The specific type of agitation required is highly dependent on the sedimentation properties of the particles and nature of the vehicle. Careful examination of parameters, such as mixer configuration, mixing/recirculation speed, and duration, is necessary to determine optimal conditions. Computational modeling approaches may be useful for defining agitation parameters necessary to achieve optimal particle dispersion. The issues associated with mixing peptide and protein suspensions have already been elaborated. While the concerns are similar for recirculation, there are additional considerations. The recirculation operation involves pumping the suspension through tubing and the impact of this agitation on molecular structure and/or particle integrity needs to be assessed. Product interactions with contact surfaces of equipment used for recirculation should be additionally explored since the duration of filling may last several hours. The potential for leachables from recirculation line tubing also exists raising the same concerns described earlier for the container-closure system. Alternatively, adsorption or absorption of certain excipients or the peptide/protein in contact with tubing is possible resulting in diminished concentration. One final consideration for suspension filling involves line stoppages. If this situation does occur, stopping the agitation may be advisable in order to minimize exposure of the product to these physical stress conditions. Sufficient time must be allowed upon restart to ensure homogeneity and some population of the filled units will likely be discarded once filling commences to ensure uniformity has been reestablished.

Since some form of agitation is necessary to properly fill a suspension product, a balance must be achieved so that suspension homogeneity is accomplished without impacting the molecule or the particles. One approach to overcome the filling issues associated with suspensions involves particle formation in individual product containers. In this case, fixed volumes of two solutions may be combined together in the vial initiating particle formation. This filling strategy is limited to suspension products where particle formation in aqueous vehicle is feasible. Furthermore, since a commercial batch size could conceivably yield in excess of ten thousand individual units, a thorough understanding of the particle formation process and the influence of associated parameters is essential. Validation of the process must demonstrate that consistency of suspension properties is achieved for each individual unit.

The control strategy for the filling operation involves confirmation of uniformity, and for peptide or protein based preparations measurement can be achieved by nitrogen determination, colorimetric test, or an HPLC assay. A statistically defined set of samples across the entire filling operation of the batch are typically evaluated due to the destructive nature of testing. Continuous online measurement of optical density is also a possibility, and offers the advantage of nondestructively examining every container for appropriate uniformity. However, process analytical technology approaches require development, validation, and maintenance of measurement equipment and associated computer models.

9.9.4 Control Strategy

Final batch release testing to confirm quality consists of a set of attributes, test methods, and acceptance criteria that comprise the product specification. However, it is important to appreciate that the batch specification only represents one part of the overall control strategy. There are a number of required tests for sterile injectable products including assays for identity, content, purity, extractable volume, sterility, and endotoxin that must be included before a product can be released to the market. Any additional testing that is included at this point depends on the design of the overall control strategy. Many options for implementing various control elements throughout a suspension process have been highlighted in this section, so there may not be a need to repeat certain tests at batch release. For suspension products, it might be appropriate to include in the specification measurement of particle morphology, particle size and distribution, or rheological properties depending on the nature of the final suspension. Multiuse suspension products containing a preservative may include a content determination for this excipient to ensure that the concentration remains in the range effective for antimicrobial effectiveness. Finally, this discussion only considered a subset of the unit operations involved in suspension manufacture and does not represent a complete description of a suitable control strategy for a pharmaceutical product.

9.10 OTHER DISPERSED SYSTEMS

The suspensions described thus far are composed of particles of the peptide or protein dispersed in vehicle containing any necessary excipients. Recall that these formulations were devised with the intention of retarding or controlling release characteristics following injection. While these suspensions display extended activity, the duration of such preparations may still not be long enough to preclude frequent daily injections. For example, the time-action of Ultralente insulin, although variable, is generally greater than 24 hours, but multiple injections of insulin are still required every day. This particular suspension, which was the longest acting insulin preparation at the time, highlights a major shortcoming of the systems described thus far in this chapter. Specifically, is the pharmaceutical effect that is ultimately achieved worth the effort involved in developing such a suspension? The bST suspension is one exception displaying sustained release of bioactive somatotropin for 7–28 days (Ferguson, 1997), although the practical limitations of such preparations have been discussed earlier.

How can the activity of therapeutically useful peptides and proteins be extended to longer durations, and can release of drug be better controlled? One approach involves encapsulation of these molecules into degradable microspheres. This technology has been thoroughly reviewed elsewhere (Bilati et al., 2005; Giteau et al., 2008; Langer, 1990; Li et al., 2008; Pisal et al., 2010; Tracy, 1998) and will only be briefly summarized. These systems represent examples of aqueous injectable suspensions, but in this case with particles having diameters of 1–100 μm that are supplied as dried powders. Prior to injection, the particles are mixed with an appropriate vehicle, dispersed, and administered. Release kinetics is controlled by polymer degradation and diffusion of the drug, and the duration can be adjusted from days to months (Langer, 1996; Pisal et al., 2010).

Microencapsulation involves rather harsh conditions that may involve high shear, organic solvents, or high temperatures (Li et al., 2008). In addition, the encapsulated molecules will be exposed to high body temperature over extended periods of time. As a result of these processing requirements and potential stability issues, the technology was not thought to be appropriate for peptides and proteins. However, examples of encapsulation procedures for peptides and proteins have been described in the literature (Bilati et al., 2005; Giteau et al., 2008; Langer, 1990; Pisal et al., 2010; Tracy, 1998). In general, peptides lacking complex, higher order structure are amenable to the microencapsulation approach. An example of a peptide that has been encapsulated is leuprolide acetate, a synthetic nonapeptide analog of leutenizing hormone-releasing hormone. The microencapsulated peptide is marketed as Lupron® Depot and is used for the treatment of advanced prostatic cancer. Reconstitution of the dried particles with vehicle results in a suspension that is administered intramuscularly at monthly intervals. Table 9.4 lists other examples of microencapsulated peptide preparations approved by regulatory authorities.

The microencapsulation approach can be reliably applied to selected proteins if the unique properties of these molecules are considered as part of process development. This fact is illustrated nicely in the procedure devised for microencapsulating hGH (Johnson et al., 1996). By exploiting the stabilizing effect of zinc ion complexation and using a low temperature method for incorporation during encapsulation, degradable microspheres were prepared containing structurally intact hGH. The

TABLE 9.4
Examples of Commercial Sustained-Release Peptide and Protein Injectable Suspensions

Product Name	Active Ingredient
Lupron Depot®	Luprorelin
Sandostatin LAR®	Octreotide
Somatuline® LA	Lanreotide
Trelstar™ Depot	Triptroelin
Nutropin Depot®a	hGH

a Product is no longer marketed.

dried microspheres were suspended in a vehicle containing carboxymethyl cellulose, polysorbate, and sodium chloride and injected into monkeys to evaluate the pharmacological properties of the preparation. Elevated serum concentrations of hGH were sustained for greater than one month. This pharmacological response was subsequently demonstrated in human clinical trials, and a microsphere preparation of hGH referred to as Nutropin Depot® was successfully commercialized and received regulatory approval. However, it is noted that this product is no longer marketed. The microsphere technology may be applicable to other proteins, with more examples developed in the future as research in the field continues to explore approaches to address the issues of destabilization during processing as well as maintaining stability within the carriers and during release (Bilati et al., 2005; Giteau et al., 2008; Tracy, 1998). Of course, the success of such systems depends on demonstrating the desired pharmacological properties in humans, and the ability to adapt the procedure into a viable commercial manufacturing process. The dose administration requirements relevant to the patient (e.g., required needle gauge for injection and convenience of dosage form preparation) must be additionally considered.

9.11   CONCLUSIONS

Pharmaceutical suspensions are generally considered to be exceedingly complex formulations to develop because of the multitude of factors that must be considered from product design through full-scale manufacturing. The unique chemical and physical properties make peptides and proteins one of the most difficult classes of molecules to develop as therapeutics. Therefore, the combination of these two factors makes developing commercially viable peptide or protein suspensions a significant challenge. This fact is demonstrated by the limited number of peptide or protein suspensions successfully reaching the market, and emerging trends in biotechnology that have taken the approach of reengineering the molecule's properties to achieve desired pharmacological properties when administered as solution preparations (Walsh, 2004). Despite the associated complexities, suspension preparations still remain a feasible option for peptides and proteins. The extended release properties afforded by suspensions are particularly attractive from the standpoint of patient compliance and convenience. Furthermore, suspensions of peptides or proteins can provide a means to achieve a specific pharmacological effect without the need to create a new molecular entity thereby creating a more streamlined path to product lifecycle management through line extensions.

Building upon the concepts devised for small molecule suspensions and the years of accumulated knowledge obtained for insulin suspensions dating back to the 1930s, we have summarized approaches for designing, developing, and manufacturing suspensions of other peptides and proteins. Methods for producing particles and suitable vehicle have been described, as well as strategies that can be used to prepare peptide and protein suspensions having optimal chemical, physical, and microbiological properties. Techniques and procedures for evaluating these characteristics to meet the various pharmaceutical requirements for suspension preparations have also been addressed. Finally, we have briefly touched upon the issues and challenges associated with the manufacture of parenteral suspensions, and describe how a suitable control strategy can be devised. The numerous examples and practical information provided herein will assist formulation scientists in the development of high quality suspension preparations of peptides or proteins.

REFERENCES

Akers, M.J., Fites, A.L., and Robison, R.L. 1987. Formulation design and development of parenteral suspensions. *J. Parenter. Sci. Technol.* 41:88–96.

Allwood, M.C. 1982. The effectiveness of preservatives in insulin injections. *Pharm. J.* 229:340.

Balschmidt, P. 1996. AspB28 insulin crystals. United States Patent. Patent Number: 5,547,930.

Balschmidt, P., Hansen, F.B., Dodson, E.J., Dodson, G.G., and Korber, F. 1991. Structure of porcine insulin cocrystallized with clupeine-Z. *Acta Cryst.* B47:975–986.

Basu, S.K., Govardhan, C.P., Jung, C.W., and Margolin, A.L. 2004. Protein crystals for the delivery of biopharmaceuticals. *Expert Opin. Biol. Ther.* 4:301–317.

Beals, J.M., DeFelippis, M.R., and Kovach, P.M. 2008. Insulin. In *Pharmaceutical Biotechnology Fundamentals and Applications*, 3rd edn, eds. D.J.A. Crommelin, R.D. Sindelar, and B. Meibohm, 265–280. New York: Informa Healthcare USA, Inc.

Bilati, U., Allémann, E., and Doelker, E. 2005. Strategic approaches for overcoming peptide and protein instability within biodegradable nano- and microparticles. *Eur. J. Pharm. Biopharm.* 59:375–388.

Boyett, J.B., and Davis, C.W. 1989. Injectable emulsions and suspensions. In *Pharmaceutical Dosage Forms Disperse Systems*, eds. H.A. Lieberman, M.M. Rieger, and G.S. Banker, Vol. 2, 379–416. New York: Marcel Dekker, Inc.

Brader, M.L., Sukumar, M., Pekar, A.H., McClellan, D.S., Chance, R.E., Flora, D.B., Cox, A.L., Irwin, L., and Myers, S.R. 2002. Hybrid insulin cocrystals for controlled release. *Nat. Biotechnol.* 20:800–804.

Bramley, M.R., Carter, A.B., and Dunwell, D.W. 1989. Somatotropin formulations. European Patent Application. Publication Number: 0, 314, 421.

Brange, J. 1987. *Galenics of Insulin: The Physico-Chemical and Pharmaceutical Aspects of Insulin and Insulin Preparations.* Berlin Heidelberg: Springer-Verlag.

Brange, J. 1992. Chemical stability of insulin 4. Mechanisms and kinetics of chemical transformations in pharmaceutical formulation. *Acta Pharm. Nord.* 4:209–222.

Brange, J. 1994. *Stability of Insulin Studies on the Physical and Chemical Stability of Insulin in Pharmaceutical Formulation.* Dordrecht, The Netherlands: Kluwer Academic Publishers.

Brange, J., Hallund, O., and Sørensen, E. 1992a. Chemical stability of insulin 5. Isolation, characterization and identification of insulin transformation products. *Acta Pharm. Nord.* 4:223–232.

Brange, J., Havelund, S., and Hougaard, P. 1992b. Chemical stability of insulin. 2. Formation of higher molecular weight transformation products during storage of pharmaceutical preparations. *Pharm. Res.* 9:727–734.

Brange, J. and Langkjær, L. 1992. Chemical stability of insulin 3. Influence of excipients, formulation, and pH. *Acta Pharm. Nord.* 4:149–158.

Brange, J., Langkjær, L., Havelund, S., and Vølund, A. 1992c. Chemical stability of insulin. 1. Hydrolytic degradation during storage of pharmaceutical preparations. *Pharm. Res.* 9:715–726.

Brazeau, P., Rivier, J., Vale, W., and Guillemin, R. 1974. Inhibition of growth hormone secretion in the rat by synthetic somatostatin. *Endocrinology* 94:184–187.

Brems, D.N., French, D.L., and Speed, M.A. 2002. Stable, active, human OB protein compositions and methods. United States Patent Application. Publication Number: US 2002/0019352 A1.

Brooks, N.D., and Needham, G.F. 1994. Injectable extended release formulations and methods. United States Patent. Patent Number: 5,352,662.

Buch, J., and Buch, A. 1983. Sustained effect of zinc-protamin-glucagon in hyperlipidaemic patients. *Acta Pharmacol. Toxicol.* 53:188–192.

Carter, C.W., Jr. 1990. Efficient factorial designs and the analysis of macromolecular crystal growth conditions. In *Methods: A Companion to Methods in Enzymology Protein and Nucleic Acid Crystallization*, ed. C.W. Carter, Jr., Vol. 1, 12–24. Duluth, MN: Academic Press, Inc.

Chien, Y.W. 1981. Long-acting parenteral drug formulations, *J. Parenter. Sci. Technol.* 35:106–139.

Chien, Y.W., Przybyszewski, P., and Shami, E.G. 1981. Syringeability of nonaqueous parenteral formulations-development and evaluation of a testing apparatus. *J. Parenter. Sci. Technol.* 35:281–284.

DeFelippis, M.R. 1995. Monomeric insulin analog formulations. United States Patent. Patent Number: 5,461,031.

DeFelippis, M.R., Bakaysa, D.L., Bell, M.A., Heady, M.A., Li, S., Pye, S., Youngman, K.M., Radziuk, J., and Frank, B.H. 1997. Preparation and characterization of a cocrystalline suspension of [Lys^{B28}, Pro^{B29}]-human insulin analogue. *J. Pharm. Sci.* 87:170–176.

DeFelippis, M.R., Bakaysa, D.L., Youngman, K.M., Radziuk, J., and Frank, B.H. 1996. Preparation and characterization of neutral protamine lispro (NPL) suspension. *Diabetes* 45(Suppl 2):74A.

DeFelippis, M.R., Chance, R.E., and Frank, B.H. 2001. Insulin self-association and the relationship to pharmacokinetics and pharmacodynamics. *Crit. Rev. Ther. Drug Carrier Syst.* 18:201–264.

DeFelippis, M.R., DiMarchi, R.D., and Ng, K. 2007a. Pre-mixtures of GLP-1 and basal insulin. United States Patent. Patent Number: 7,238,663 B2.

DeFelippis, M.R., DiMarchi, R.D., Ng, K., and Trautman, M.E. 2007b. Biphasic mixtures of GLP-1 and insulin. United States Patent. Patent Number: 7,179,788 B2.

Denyer, S.P., and Wallhaeusser, K-H. 1990. Antimicrobial preservatives and their properties. In *Guide to Microbiological Control in Pharmaceuticals*, eds. S.P. Denyer and R.M. Baird, 251–273. England: Ellis Horwood Limited.

Dodd, S.W., Havel, H.A., Kovach, P.M., Lakshminarayan, C., Redmon, M.P., Sargeant, C.M., Sullivan, G.R., and Beals, J.M. 1995. Reversible adsorption of soluble hexameric insulin onto the surface of insulin crystals cocrystallized with protamine: An electrostatic interaction. *Pharm. Res.* 12:60–68.

Donini, P. 1974. Long-active gonadotropins. United States Patent. Patent Number: 3,852,422.

Drenth, J. 1994. *Principles of Protein X-ray Crystallography*. New York: Springer-Verlag.

Ducruix, A.F., and Ries-Kautt, M.M. 1990. Solubility diagram analysis and the relative effectiveness of different ions on protein crystal growth. In *Methods: A Companion to Methods in Enzymology Protein and Nucleic Acid Crystallization*, ed. C.W. Carter, Jr., Vol. 1, 25–30. Duluth, MN: Academic Press, Inc.

Durbin, S.D., and Feher, G. 1996. Protein crystallization. *Annu. Rev. Phys. Chem.* 47:171–204.

Edwards, S.L., DeFelippis M.R., Frank, B.H., Kilcomons, M.A., Sheliga, T.A., Stickelmeyer, M.P., Youngman, K.M., and Havel, H.A. 1996. *Assessment of the stability of insulin lispro mixtures with human insulin NPH*. Poster presentation at the 211th National Meeting of the American Chemical Society, March 24–28, New Orleans, LA.

Ellerhorst, J.A., Comstock, J.P., and Nell, L.J. 1990. Protamine antibody production in diabetic subjects treated with NPH insulin. *Am. J. Med. Sci.* 299:298–301.

Ferguson, T.H. 1997. Peptide and protein delivery for animal health applications. In *Controlled Drug Delivery Challenges and Strategies*, ed. K. Park, 289–308. Washington, DC: American Chemical Society.

Ferguson, T.H., Harrison, R.G., and Moore, D.L. 1990. Injectable sustained release formulation. United States Patent. Patent Number: 4,977,140.

Frokjaer, S., and Otzen, D.E. 2005. Protein drug stability: A formulation challenge. *Nat. Rev. Drug Discov.* 4:298–306.

Genck, W.J. 1995. Mixing's influence on precipitations. *Chem. Process.* January: 46–50.

Gilliland, G.L., and Bickham, D.M. 1990. The biological macromolecule crystallization database: A tool for developing crystallization strategies. In *Methods: A Companion to Methods in Enzymology Protein and Nucleic Acid Crystallization*, ed. C.W. Carter, Jr., Vol. 1, 6–11. Duluth, MN: Academic Press, Inc.

Giteau, A., Venier-Julienne, M.C., Aubert-Pouëssel, A., and Benoit, J.P. 2008. How to achieve sustained and complete protein release from PLGA-based microparticles? *Int. J. Pharm.* 350:14–26.

Govardhan, C., Khalaf, N., Jung, C.W., Simeone, B., Higbie, A., Qu, S., Chemmalil, L., Pechenov, S., Basu, S.K., and Margolin, A.L. 2005. Novel long-acting crystal formulation of human growth hormone. *Pharm. Res.* 22:1461–1470.

Graham, D.T., and Pomeroy, A.R. 1978. The effects of freezing on commercial insulin suspensions. *Int. J. Pharm.* 1:315–322.

Graham, D.T., and Pomeroy, A.R. 1984. An in-vitro test for the duration of action of insulin suspensions. *J. Pharm. Pharmacol.* 36:427–430.

Hagedorn, H.C., Jensen, B.N., Krarup, N.B., and Wodstrup, I. 1936. Protamine insulinate. *J. Am. Med. Assoc.* 106:177–180.

Hallas-Møller, K., Petersen, K., and Schlichtkrull, J. 1952. Crystalline and amorphous insulin-zinc compounds with prolonged action. *Science* 116:394–399.

Hammond, G. 1991. Crystalline interleukin-4. International Patent Application. Publication Number: WO 91/19742.

Hem, S.L., Feldkamp, J.R., and White, J.L. 1986. Basic chemical principles related to emulsions and suspension dosage forms. In *The Theory and Practice of Industrial Pharmacy*, eds. L. Lachman, H.A. Lieberman, and J.L. Kanig, 100–122. Philadelphia: Lea & Febiger.

Hiestand, E.N. 1964. Theory of coarse suspension formulation. *J. Pharm. Sci.* 53:1–18.

Hoffmann, J.A. 1994. Insulin formulation. European Patent Application. Publication Number: 0, 646, 379 A1.

Holt, L.A., Milligan, B., Rivett, D.E., and Stewart, F.H.C. 1977. The photodecomposition of tryptophan peptides. *Biochim. Biophys. Acta* 499:131–138.

Horrow, J.C. 1985. Protamine: A review of its toxicity. *Anesth. Analg.* 64:348–361.

Johnson, O.L., Cleland, J.L., Lee, H.J., Charnis, M., Duenas, E., Jaworowicz, W., Shepard, D., Shahzamani, A., Jones, A.J.S., and Putney, S.D. 1996. A month-long effect from a single injection of microencapsulated human growth hormone. *Nature Med.* 2:795–799.

Kadima, W., McPherson, A., Dunn, M.F., and Jurnak, F. 1991. Precrystallization aggregation of insulin by dynamic light scattering and comparison with canavalin. *J. Cryst. Growth* 110:188–194.

Kayes, J.B. 1977. Pharmaceutical suspensions: Relation between zeta potential, sedimentation volume and suspension stability. *J. Pharm. Pharmacol.* 29:199–204.

Kelly, R.N., and Lerke, S.A. 2005. Particle size measurement technique selection within method development in the pharmaceutical industry. *Am. Pharm. Rev.* 8:72–81.

Kerwin, B.A., and Remmele, R.L., Jr. 2007. Protect from light: Photodegradation and protein biologics. *J. Pharm. Sci.* 96:1468–1479.

Kim, Y., Cuff, G.W., and Morris, R.M. 1995. Effect of chloride ion on the sedimentation volume and zeta potential of zinc insulin suspensions in neutral pH range. *J. Pharm. Sci.* 84:755–759.

Kim, Y., and Haren, A.M. 1995. The application of crystal soaking technique to study the effect of zinc and cresol on insulinotropin crystals grown from a saline solution. *Pharm. Res.* 12:1664–1670.

Kim, Y., and Rose, C.A. 1995. Precipitation of insulinotropin in the presence of protamine: Effect of phenol and zinc on the isophane ratio and the insulinotropin concentration in the supernatant. *Pharm. Res.* 12:1284–1288.

Komatsu, H., Kitajima, A., and Okada, S. 1996. Estimation of average particle sizes and size distributions of commercially available human-insulin aqueous suspensions using laser-light diffraction spectroscopy. *Chem. Pharm. Bull.* 44:1966–1969.

Krayenbühl, C., and Rosenberg, T. 1946. Crystalline protamine insulin. *Rep. Steno Mem. Hosp. Nord. Insulin Lab* 1:60–73.

Langer, R. 1990. New methods of drug delivery. *Science* 249:1527–1533.

Langer, R. 1996. Controlled release of a therapeutic protein. *Nat. Med.* 2:742–743.

Li, M., Rouaud, O., and Poncelet, D. 2008. Microencapsulation by solvent exchange: State of the art for process engineering approaches. *Int. J. Pharm.* 363:26–39.

Martin, A. 1993. *Physical Pharmacy*, 4th Edn. Philadelphia: Lea & Febiger.

Martin, A.N. 1961. Physical chemical approach to the formulation of pharmaceutical suspensions. *J. Pharm. Sci.* 50:513–517.

Martin, J.L. 1986. Prolonged release of growth promoting hormones. Australian Patent Application. Application Number: AU-A-61092/86.

Martin, J.B., Renaud, L.P., and Brazeau, P., Jr. 1974. Pulsatile growth hormone secretion: Suppression by hypothalamic ventromedial lesions and by long-acting somatostatin. *Science* 186:538–540.

Massey, E.H., and Sheliga, T.A. 1988. Human insulin (HI) isophane suspension (NPH) with improved physical stability. *Pharm. Res.* 5:S-34.

Massey, E.H., Tensmeyer, L.G., and Sheliga, T.A. 1988. *Aggregation of human insulin (HI) in isophane suspension formulations (NPH).* Presentation at the 196th ACS National Meeting, September 25–30, Los Angeles, CA.

Matthews, B.A., and Rhodes, C.T. 1968. Some studies of flocculation phenomena in pharmaceutical suspensions. *J. Pharm. Sci.* 57:569–573.

Matthews, B.A., and Rhodes, C.T. 1970. Use of the Derjaguin, Landau, Verwey, and Overbeck theory to interpret pharmaceutical suspension stability. *J. Pharm. Sci.* 59:521–525.

McPherson. A. 1982. *Preparation and Analysis of Protein Crystals.* New York: John Wiley & Sons, Inc.

Mitchell, J.W. 1991. Prolonged release of biologically active somatotropin. United States Patent. Patent Number: 5,013,713.

Mumenthaler, M., Hsu, C.C., and Pearlman, R. 1994. Feasibility study on spray-drying protein pharmaceuticals: Recombinant human growth hormone and tissue-type plasminogen activator. *Pharm. Res.* 11:12–20.

Naets, J.P., and Guns, M. 1980. Inhibitory effects of glucagon on erythropoiesis. *Blood* 55:997–1002.

Nail, S.L., Jiang, S., Chongprasert, S., and Knopp, S.A. 2002. Fundamentals of freeze-drying. *Pharm. Biotechnol.* 14:281–360.

Nash, R.A. 1988. Pharmaceutical suspensions. In *Pharmaceutical Dosage Forms Disperse Systems,* eds. H.A. Lieberman, M.M. Rieger, and G.S Banker, Vol. 1, 151–198. New York: Marcel Dekker, Inc.

Nell, L.J., and Thomas, J.W. 1988. Frequency and specificity of protamine antibodies in diabetic and control subjects. *Diabetes* 37:172–176.

O'Neill, J.J., and Mead, C.A. 1982. The parabens: Bacterial adaptation and preservative capacity. *J. Soc. Cosmet. Chem.* 33:75–84.

Ooshima, H., Urabe, S., Igarashi, K., Azuma, M., and Kato, J. 1997. Mechanism of crystal growth of protein: Differential scanning calorimetry of thermolysin crystal suspension. In *Separation and Purification by Crystallization,* eds. G.D. Botsaris and K. Toyokura, 18–27. Washington, DC: American Chemical Society.

Orr, N.A., and Spence, J. 1977. Applications of particle size analysis in the pharmaceutical industry. *Analyst* 102:466–472.

Osterberg, R.E. 2005. Potential toxicity of extractables and leachables in drug products. *Am. Pharm. Rev.* 8:64–67.

Pasquali, I., and Bettini, R. 2008. Are pharmaceutics really going supercritical? *Int. J. Pharm.* 364:176–187.

Patel, N.K., Kennon, L., and Levinson, R.S. 1986. Pharmaceutical suspensions. In *The Theory and Practice of Industrial Pharmacy*, eds. L. Lachman, H.A. Lieberman, and J.L. Kanig, 479–501. Philadelphia: Lea & Febiger.

Pisal, D.S., Kosloski, M.P., and Balu-Iyer, S.V. 2010. Delivery of therapeutic proteins. *J. Pharm. Sci.* 99:2557–2575.

Prabhu, S., Jacknowitz, A.I., and Stout, P.J. 2001. A study of factors controlling dissolution kinetics of zinc complexed protein suspensions in various ionic species. *Int. J. Pharm.* 217:71–78.

Prabhu, S., Jacknowitz, A.I., and Stout, P.J. 2002. Evaluating the dissolution behavior of zinc-complexed protein suspensions by computer modeling and simulation. *Drug Dev. Indust. Pharm.* 28:703–709.

Prestrelski, S.J., Arakawa, T., and Carpenter, J.F. 1993a. Separation of freezing- and drying-induced denaturation of lyophilized proteins using stress-specific stabilization. *Arch. Biochem. Biophys.* 303:465–473.

Prestrelski, S.J., Tedeschi, N., Arakawa, T., and Carpenter, J.F. 1993b. Dehydration-induced conformational transitions in proteins and their inhibition by stabilizers. *Biophys. J.* 65:661–671.

Radziuk, J., Bradley, B., Welsh, L., DeFelippis, M.R., and Roach, P. 1996. Profiles of biological activity after subcutaneous administration of mixtures of Lys^{B28}-Pro^{B29} human insulin (lispro) in soluble and neutral protamine formulations. *Diabetes* 45(Suppl 2):218A.

Reichert, P., Nagabhushan, T.L., Long, M.M., Bugg, C.E., and DeLucas, L.J. 1996. Macroscale production and analysis of crystalline interferon alpha-2B in microgravity on STS-52. In *Space Technology & Applications International Forum (STAIF-96)*, ed. M.S. El-Genk, Vol. 361, Pt. 1, 139–148. Woodbury, New York: American Institute of Physics.

Richards, J.P., Stickelmeyer, M.P., Frank, B.H., Pye, S., Barbeau, M., Radziuk, J., Smith, G.D., and DeFelippis, M.R. 1999. Preparation of a microcrystalline suspension formulation of $Lys^{B28}Pro^{B29}$-human insulin with ultralente properties. *J. Pharm. Sci.* 88:861–867.

Rolim, C., and Bristow, A.F. 1995. Determination of insulin-in-solution in biphasic isophane insulin formulations. *Pharmeuropa* 7:22–25.

Santos, N.C., and Castanho, M.A.R.B. 2004. An overview of the biophysical applications of atomic force microscopy. *Biophys. Chem.* 107:133–149.

Schlichtkrull, J. 1957. Insulin crystals IV. The preparation of nuclei, seeds and monodisperse insulin crystal suspensions. *Acta Chem. Scand.* 11:299–302.

Schlichtkrull, J. 1961. *Insulin Crystals.* Copenhagen, Denmark: Novo.

Schneider, W., Stavchansky, S., and Martin, A. 1978. Pharmaceutical suspensions and the DVLO theory. *Am. J. Pharm. Educ.* 42:280–289.

Shakesheff, K.M., Davies, M.C., Roberts, C.J., Tendler, S.J.B., and Williams, P.M. 1996. The role of scanning probe microscopy in drug delivery research. *Crit. Rev. Ther. Drug Carrier Syst.* 13:225–256.

Shnek, D.R., Hostettler, D.L., Bell, M.A., Olinger, J.M., and Frank, B.H. 1998. Physical stress testing of insulin suspensions and solutions. *J. Pharm. Sci.* 87:1459–1465.

Siegel, F.P. 1990. Tonicity, osmoticity, osmolality and osmolarity. In *Remington's Pharmaceutical Sciences*, 18th Edn, ed. A.R. Gennaro, 1481–1498. Easton, PA: Mack Publishing Co.

Snavely, W.K., Subramaniam, B., Rajewski, R.A., and DeFelippis, M.R. 2002. Micronization of insulin from halogenated alcohol solution using supercritical carbon dioxide as an antisolvent. *J. Pharm. Sci.* 91:2026–2039.

Sollohub, K., and Cal, K. 2010. Spray drying technique: II. Current applications in pharmaceutical technology. *J. Pharm. Sci.* 99:587–597.

Tannenbaum, G.S., and Colle, E. 1980. Ineffectiveness of protamine zinc somatostatin as a long-acting inhibitor of insulin and growth hormone secretion. *Can. J. Physiol. Pharmacol.* 58:951–955.

Tensmeyer, L.G., and Shields, J.E. 1990. The Raman spectra of crystalline 4Zn, 2Zn, and Na insulin. In *Raman and Luminescence Spectroscopies in Technology II*, eds. F. Adar and J.E. Griffiths, Vol. 1336, 222–234. Bellingham, Washington: Proc. SPIE.

Tracy, M.A. 1998. Development and scale-up of a microsphere protein delivery system. *Biotechnol. Prog.* 14:108–115.

Tung, M., and Gallagher, D.T. 2009. The biomolecular crystallization database version 4: Expanded content and new features. *Acta Cryst.* D65:18–23.

Walsh, G. 2004. Second-generation biopharmaceuticals. *Eur. J. Pharm. Biopharm.* 58:185–196.

Walsh G. 2006. Biopharmaceutical benchmarks. *Nat. Biotechnol.* 24:769–776.

Wang, W. 2000. Lyophilization and development of solid protein pharmaceuticals. *Int. J. Pharm.* 203:1–60.

Weyer, C., Heise, T., and Heinemann, L. 1997. Premixed formulation of B28Asp and NPH-insulin: Pharmacodynamic properties of a 30/70-stable mixture. *Diabetologia* 40(Suppl 1): A350.

Workman, W.E., and Clayton, R.A. 1996. Microbial sterility testing of oil-formulated bovine somatotropin using Tween® 80 dispersion. *J. Pharm. Biomed. Anal.* 15:193–200.

Yang, M.X., Shenoy, B., Disttler, M., Patel, R., McGrath, M., Pechenov, S., and Margolin, A.L. 2003. Crystalline monoclonal antibodies for subcutaneous delivery. *Proc. Natl. Acad. Sci.* 100:6934–6939.

Yeo, S-D., Lim, G-B., Debenedetti, P.G., and Bernstein, H. 1993. Formation of microparticulate protein powders using a supercritical fluid antisolvent. *Biotech. Bioeng.* 41:341–346.

Yim, Z. 1988. Stable interferon complexes. European Patent Application. Publication Number: 0, 281, 299.

Yip, C.M., Brader, M.L., DeFelippis, M.R., and Ward, M.D. 1998a. Atomic force microscopy of crystalline insulins: The influence of sequence variation on crystallization and interfacial structure. *Biophys. J.* 74:2199–2209.

Yip, C.M., Brader, M.L., Frank, B.H., DeFelippis, M.R., and Ward, M.D. 2000. Structural studies of a crystalline insulin analog complex with protamine by atomic force microscopy. *Biophys. J.* 78:466–473.

Yip, C.M., DeFelippis, M.R., Frank, B.H., Brader, M.L., and Ward, M.D. 1998b. Structural and morphological characterization of ultralente insulin crystals by atomic force microscopy: Evidence for hydrophobically driven assembly. *Biophys. J.* 75:1172–1179.

Yip, C.M., and Ward, M.D. 1996. Atomic force microscopy of insulin single crystals: Direct visualization of molecules and crystal growth. *Biophys. J.* 71:1071–1078.

Zografi, G., Schott, H., and Swarbrick, J. 1990. Disperse systems. In *Remington's Pharmaceutical Sciences*, 18th Edn, ed. A.R. Gennaro, 257–309. Easton, PA: Mack Publishing Co.

10 Rational Design of Solid Protein Formulations

Bingquan (Stuart) Wang and Michael J. Pikal

CONTENTS

10.1 INTRODUCTION

Therapeutic proteins are important healthcare products for the treatment of many critical diseases, such as diabetes, hemophilia, and cancer. However, the inadequate stability of proteins in an aqueous solution often limits their distribution and use (Carpenter et al., 1994). Thus, the protein products are commonly dried to increase stability, with freeze-drying (lyophilization) and spray-drying being the major drying methods employed (Pikal, 1994). The storage stability of proteins in the glassy solids is of great significance to the pharmaceutical industry, and stabilization of labile biomolecules is becoming increasingly important as more biopharmaceuticals are introduced into the marketplace (Wang, 2000).

Due to the drying stresses associated with freeze-drying and spray-drying processes, pure protein can experience significant damage upon drying (Pikal, 1994). Thus, stabilizers are often required in a protein formulation to protect protein stability, with sugars such as sucrose and trehalose being the most common stabilizers (Carpenter et al., 1997; Pikal, 1994). Excipients typically interact with the protein by forming hydrogen bonds, and the extent of protein–excipient interaction is found to be important for controlling the protein native structure (Carpenter et al., 1994). Protein secondary structure preservation during freeze-drying, as measured by Fourier Transform Infrared (FTIR), has been reported to play an important role in protein stability during storage (Carpenter et al., 1998). Therefore, protein formulation through rational selection of excipients can greatly impact protein stability.

In addition to formulation factors, the drying process also plays a significant role in the stability of dried proteins. The quality of dried product depends on the process used to prepare the material, with thermal history being one critical factor (Abdul-Fattah et al., 2007b; Wang et al., 2010). Molecular motions of the protein and potential reactants in the glassy matrix are necessary for protein degradation (Pikal, 1994), and such motions can be either highly cooperative global motion coupled to viscosity (α-relaxation) or local noncooperative motions with a short timescale (β-relaxation) (Wang et al., 2010). Modulation/reduction of these motions through formulation and/or processing can potentially lead to protein stabilization. Therefore, it is very important to understand the role of molecular mobility in controlling protein stability (Pikal, 2004).

The concept of Quality by Design (QbD) is proposed in the ICH guidelines Q8, Q9, and Q10. Basically, QbD means that the quality should be achieved by rational design of the product and process rather than by testing representative samples of the finished product and rejecting defective units. QbD requires a thorough investigation of the relationship between critical process parameters and the product's critical quality attributes. To practice QbD from the beginning of the development process, the interplay between formulation and drying process and the resulting effect on protein stability needs to be understood. While many important generalizations have been developed in the area of freeze-drying and spray-drying in the past two decades, several critical stability-related issues remain, including the role of specific protein–excipient interactions in determining pharmaceutical stability, the relationship between global and local mobility and their roles in long-term stability, and the linkage between the current understanding of protein formulation and industry formulation practice.

This chapter is an attempt to discuss the role of both thermodynamic and kinetic factors in determining stability of biopharmaceuticals during processing and storage in the solid state. The characterization of protein–excipient interactions in dried solids is summarized, and the correlations of stability with both global motion with a long timescale and local motion with a short timescale are presented. These mechanisms of protein stabilization are then put into practice, and practical formulation guidelines are outlined for solid protein formulations using common excipients. Finally, solid-state characterization of dried powder and biophysical characterization of reconstituted solution that facilitate the rational development of stable protein formulations are discussed. It is our aim to provide a picture of the current understanding of the mechanisms of protein stabilization in the amorphous solid such that this knowledge could be easily used to maximize pharmaceutical stability through a science-driven formulation optimization strategy.

10.2 SPRAY-DRYING AND FREEZE-DRYING PROCESSES AND DENATURATION STRESSES

10.2.1 SPRAY-DRYING PROCESS

Spray-drying is a continuous drying process that is widely used to produce powders for pulmonary delivery, and it consists of three major steps (Cal and Sollohub, 2010). The first step is the atomization process, which involves feeding the solution through an atomizer nozzle placed inside the drying chamber. When the solution flows out of the nozzle orifice, the liquid stream breaks up into fine droplets in the drying chamber with the aid of an atomizing gas such as air or nitrogen (Abdul-Fattah et al., 2007a). Different nozzle designs such as the usage of rotary or ultrasound nozzles allow for control of droplet size and distribution. The second step is the drying step, which involves the interaction of atomized droplets with the drying gas at a high temperature. The solvent contained within the fine droplets is vaporized, which results in the formation of solid product particles. Self-cooling effect by the active evaporation results in a much lower product temperature than the gas temperature. During the third step, the dried particles are separated from the drying gas by a cyclone separator and then collected in a receptacle tank.

The spray-drying process parameters play important roles in determining the product quality attributes, such as the particle size, moisture content, surface area, and protein activity (Abdul-Fattah et al., 2007b). Process parameters that can be controlled include the feed rate of the liquid stream, the pressure of atomizing gas, the inlet temperature, relative humidity, and flow rate of the drying gas (Cal and Sollohub, 2010). Outlet temperature (T_{outlet}), drying efficiency, and product quality are dependent on all of these controllable parameters. Due to self-cooling effects, both product temperature and temperature inside the drying chamber correspond more closely to T_{outlet}. Protein instability arising from variations in surface area and composition may also be important, but are outside the scope of this review, and the interested reader is directed to a recent review published elsewhere (Abdul-Fattah et al., 2007a).

The first atomization step results in the creation of a large air–water interface, which may be extremely damaging to the protein due to adsorption to the interface, unfolding, and aggregation. Proteins are amphiphilic polyelectrolytes and therefore can adsorb to liquid–air and liquid–solid interfaces (Chang et al., 1996b). Proteins usually assume native conformations in solution through hydrophobic interaction, and thus nonpolar amino acids in the protein core are buried inside to minimize contact with water, while more hydrophilic amino acids are exposed to the aqueous environment. When a protein is exposed to a hydrophobic air–water interface by means of atomization, the protein may unfold and expose more hydrophobic groups to the nonaqueous phase (Abdul-Fattah et al., 2007a; Kreilgaard et al., 1998).

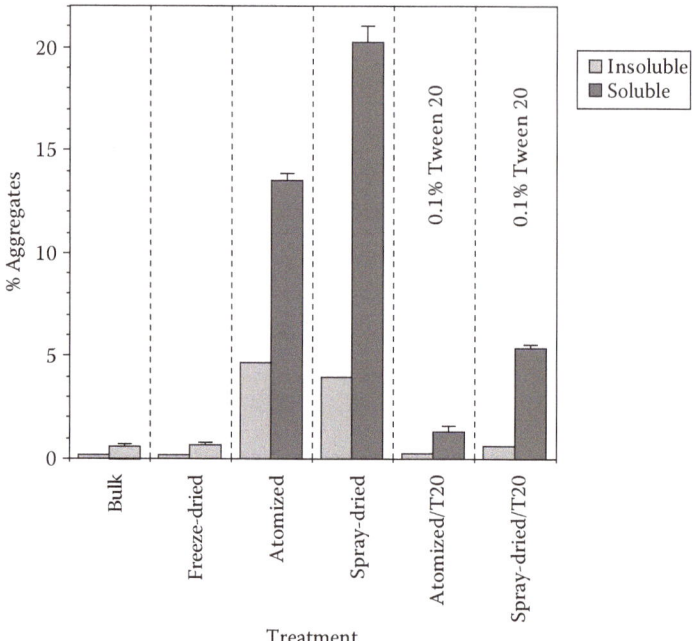

FIGURE 10.1   In-process stability of human growth hormone (hGH) upon exposure to different processes. The formulation is 2 mg/mL hGH, 88 mM mannitol, and 5 mM sodium phosphate (pH 7.8). Soluble aggregate formation was studied by size exclusion chromatography. (Reproduced from Abdul-Fattah, A.M., et al., *J. Pharm. Sci.* 96, 1886, 2007a; with data from Mumenthaler et al., *Pharm. Res.* 11, 1994.)

Figure 10.1 shows the in-process stability for human growth hormone (hGH) given different treatments (Abdul-Fattah et al., 2007a). The bulk material used for the spray-drying has a low soluble aggregate of about 0.5%. Using hot air as the drying medium, we can dry hGH formulated with mannitol to a residual moisture content of ≤4%. However, the spray-drying process results in formation of about 20% soluble aggregates and 5% insoluble aggregates (Mumenthaler et al., 1994), suggesting the presence of severe denaturation stress for the protein during the spray-drying process. To identify further the major stress during the spray-drying, samples after the atomization step are analyzed. About 19% of aggregates are produced upon atomization, suggesting that surface denaturation at the air–liquid interface of the droplets plays a major role in the degradation of the protein (Mumenthaler et al., 1994). Similar observations have been made on atomization of excipient-free recombinant hGH (2 mg/mL) (Maa et al., 1998), which resulted in 25% insoluble aggregates and 17.2% soluble aggregates. However, freeze-drying the same formulation results in negligible change in terms of total aggregates (see Figure 10.1), suggesting that lyophilization results in much smaller stress than does spray-drying, at least for hGH.

To minimize the aggregation caused by spray-drying, surfactant is added into the protein formulation (Abdul-Fattah et al., 2007a; Adler et al., 2000; Chang et al., 1996b; Yu et al., 2006). The addition of 0.1% (w/v) Tween 20 into the hGH formulation has reduced the formation of soluble and insoluble aggregates by approximately 90% during atomization (see Figure 10.1). Addition of surfactant has reduced the formation of soluble and insoluble aggregates upon spray-drying to 5% and 1%, respectively. Similarly, spray-drying with Tween 80 also has led to the improvement of in-process stability of other proteins such as lactose dehydrogenase (LDH) and bovine serum albumin (BSA) (Adler et al., 2000). These studies clearly show that the presence of nonionic surfactants such as Tween 20 and 80 could protect a protein during spray-drying. Note, however, that the increase in aggregation during the drying stage (i.e., difference between spray-dried and atomized) is not significantly reduced by addition of surfactant (Figure 10.1). Thus, while addition of surfactant is very effective in limiting aggregation during atomization, its impact on aggregation inhibition during the drying stage is relatively small.

10.2.2 FREEZE-DRYING PROCESS

Freeze-drying (or lyophilization) is a drying method that involves freezing the solvent, usually water, and subliming the crystalline solvent to convert all solutes into a dry solid. Thus, lyophilization is unique among drying methods in that the material is dried from a frozen solid, which is generally preferable for sensitive biologics from a stability point of view. As illustrated in Figure 10.1 for hGH, freeze-drying generally results in a smaller stress to the protein than does spray-drying and better stability. However, it is a longer, more energy-intensive, and therefore much more expensive process than spray-drying. Although more costly, freeze-drying remains the most common drying method for producing powders for parenteral administration, especially for biological products (Carpenter et al., 1997; Crowe et al., 1998; Pikal, 1994, 2004; Wang, 2000). Therefore, this chapter focuses on the stabilization of protein product produced by freeze-drying, although many of the principles and formulation generalizations also apply to spray-dried products.

Freeze-drying consists of three main stages: freezing, primary drying, and secondary drying. Freezing is actually a desiccation step, during which most of the water is converted to ice and separates from solute. The solute phase becomes highly concentrated as more and more water is crystallized. At the end of the freezing stage, the concentrated solute phase usually contains roughly 20% of water (w/w), or less than 1% of the water in the original filled solution (i.e., before ice formation) (Tang and Pikal, 2004). The freezing stage typically takes several hours to complete and typically involves temperatures around −40°C or lower.

Primary drying is the step where water is removed through ice sublimation. This step begins after the chamber pressure is reduced to below the vapor pressure of ice. The chamber pressure would impact both heat and mass transfer and thus is an important parameter in the freeze-drying process. The chamber vacuum should be maintained well below the ice vapor pressure at the product temperature to allow a high sublimation rate. During primary drying, the shelf temperature is raised to supply heat to compensate for heat removed by ice sublimation, and the removed water vapor is subsequently condensed onto the cold coils/plates (less than −50°C) in the condenser. The primary drying stage is typically the longest stage of freeze-drying, and may last as long as a week, depending on the nature of the formulation, solute concentration, and fill volume. The product temperature plays a significant role in controlling the product quality and cycle time. For a system composed of nearly all crystalline material, primary drying should be performed below the eutectic temperature (T_{eu}) to prevent melt-back. In contrast, for a system dominated by the amorphous phase, primary drying should be performed 2–5°C below the collapse temperature (T_c) or glass temperature of the freeze concentrate (T_g') to avoid product collapse and to maintain the elegant cake structure (Tang and Pikal, 2004). A small safety margin (≤ 1°C below T_g') can result in collapse in a subgroup of product vials or microcollapse, whereas a big safety margin (≥ 5°C below T_g') would prolong primary drying unnecessarily, and hence extend the whole cycle time.

After primary drying is complete, an amorphous product still contains a relatively high amount of residual water (5–20% on a dried solid basis, depending on the formulation). This high amount of moisture will reduce the glass transition temperature of the product (T_g) and result in instability issues. Therefore, a secondary drying step is needed to reduce further the residual moisture, normally to about 1% or less. In this step, unfrozen water is removed by desorption from the solute phase (Pikal, 2004). The shelf temperature in secondary drying is generally much higher than that used for primary drying to facilitate the diffusion-controlled desorption process. However, ramping from primary drying to secondary drying temperature should be slow to avoid product collapse arising from a product temperature higher than T_g. For the final portion of secondary drying, it is usually better to run a high shelf temperature for a short time (4–6 hours) than a low temperature for a long period (Pikal et al., 1990). Studies indicate that the water desorption rate decreases dramatically after 4–6 hours at a given temperature.

The freezing and drying processes create many destabilizing stresses for biological materials. The removal of bulk water from the protein phase reduces or eliminates the hydrophobic interactions that are generally believed to be critical for the thermodynamic stabilization of tertiary structure. Variation in freezing rate can present variable stresses to the protein, depending on the nature of the protein. Slow freezing rates may increase protein damage in systems prone to phase separation, since phase separation is often kinetically controlled and time consuming. On the other hand, a high freezing rate can result in a larger degree of supercooling and the formation of more/smaller ice crystals (i.e., larger specific surface area). A relatively large ice–liquid interface often correlates directly with high loss of secondary and tertiary structure during freezing. When solutions of several proteins are quench-frozen in liquid nitrogen as opposed to slow cooling, an increased level of aggregation has been observed for a number of proteins, including phosphofructokinase (PFK), LDH, glutamate dehydrogenase (GDH), tumor necrosis factor (TNF), and interleukin-1 (IL-1) receptor antagonist (Chang et al., 1996b). Therefore, it is generally not advisable to freeze in liquid nitrogen or load vials onto precooled shelves for lyophilization of protein products.

Freeze concentration during ice crystallization presents another stress since the highly concentrated solution results in much higher protein and salt concentrations (Pikal, 1994, 2001; Tang and Pikal, 2004). Protein–protein interactions increase with increasing concentration and can lead to aggregation. Since second-order reactions increase in rate as concentration increases, the rates of intermolecular aggregation during freezing may be faster in a freeze concentrate in spite of the lower temperature during the freezing stage (Pikal, 1994). A high ionic strength can also promote degradation of proteins, as demonstrated by the detrimental impact of NaCl on hGH stability (Pikal et al., 1991). In addition, a large pH shift can occur when one buffer component crystallizes during freezing. For example, the sodium phosphate buffer system can have a dramatic decrease in pH (about 4 pH units) due to the crystallization of Na_2HPO_4 (Pikal, 2001). Such shifts can obviously be detrimental to proteins sensitive to pH shifts, and thus it is important to select an appropriate buffer during the formulation stage.

Dehydration is the ultimate result of any drying method, and drying presents a major stress to protein molecules. A fully hydrated protein has a monolayer of water covering the protein surface, which is generally thought to be critical for protein functionality. The removal of this hydration shell during the secondary drying stage results in a reduction in the total number of protein–water hydrogen bonds, and it can disrupt the native state of a protein and cause denaturation and aggregation (Wang et al., 2009b). In addition, the dehydration process may cause uneven moisture distribution throughout a product cake. This moisture heterogeneity may lead to partial overdrying in some locations and result in different extent of microcollapse and protein damage (Pikal and Shah, 1997).

10.3 MECHANISM OF PROTEIN STABILIZATION IN DRIED SOLID BY EXCIPIENTS

Proteins can experience a variety of chemical degradations (e.g., deamidation, oxidation, hydrolysis, and covalent aggregation) as well as physical aggregation (Carpenter et al., 1997; Lai and Topp, 1999). Due to the various stresses mentioned in the preceding section, protein molecules tend to degrade and lose activity during the lyophilization process (Pikal, 2004). It is very important to understand the chemical and physical degradation pathways of each protein, so that appropriate stabilization strategies may be developed, including optimization of formulation pH, moisture, disaccharide, and surfactant. Use of radical scavenging excipients and antioxidants for protein stabilization is highly specific to each protein (Costantino et al., 1994; Wang, 2000). Here, we focus on the more general physical stabilization mechanisms.

10.3.1 Stabilization by Surfactant

As mentioned in the previous section, addition of surfactants such as Tween 20 or 80 into the formulation can significantly reduce the formation of protein aggregates during spray-drying or lyophilization (Abdul-Fattah et al., 2007a; Adler et al., 2000; Chang et al., 1996b; Yu et al., 2006). Several possible mechanisms can explain the protection of proteins from aggregation by nonionic surfactants. First, surfactant molecules displace protein molecules from the air–water or water–container interface and thus less protein will be susceptible to unfolding and aggregation arising from interaction with these interfaces. Because surfactant molecules diffuse faster than protein due to their smaller molecular size (Mumenthaler et al., 1994), they will quickly occupy these interfaces. Additionally, it is possible that less damage would occur to the protein even if some protein molecules do adsorb at these interfaces due to the lower surface tension after surfactant adsorption (Randolph and Jones, 2002).

The second mechanism of protein stabilization by surfactant involves its binding to certain hydrophobic sites on protein molecules, and hence a decrease in protein intermolecular interactions that would lead to aggregation (Chang et al., 1996b; Randolph and Jones, 2002). If surfactant binds to the native folded protein, then stabilization can be explained by the increased free energy of protein unfolding (i.e., the free energy of the native state is lowered by the binding interaction). If surfactants bind transiently with partially unfolded protein molecules, they might act as chemical chaperones and sterically hinder protein–protein interactions that would otherwise lead to aggregation.

Another mechanism has been proposed by which a surfactant can decrease protein aggregation. The surfactant can slow the dissolution rate of lyophilized powder, which helps to decrease protein aggregation (Webb et al., 2002). Thus, the stage at which a surfactant is presented in the formulation (i.e., during lyophilization or during reconstitution) may influence its ability to stabilize the protein.

10.3.2   Stabilization by Saccharides and Polyols

Saccharides, particularly disaccharides, and polyols not only can stabilize proteins against aggregation and denaturation (Chang et al., 1993), but also may protect products from chemical degradation (Pikal et al., 1991). For an excipient to function as a protein stabilizer, the excipient needs to be in the same amorphous phase as the protein; otherwise, the potential protection function will be completely lost (Carpenter et al., 1997).

In general, three mechanisms of protein stabilization by excipients have been proposed for stabilization in solids. First, if the reactive moiety is diluted in an inert matrix, molecular interactions (reactions) between the reactive molecules are greatly reduced because the number of nearest neighbors that are capable of reacting is reduced. In contrast to the simple dilution effect, which does not require any chemical interaction between drug and excipients beyond preventing phase separation, the leading thermodynamic stabilization mechanism for rationalization of stabilization during drying, often called the water substitute hypothesis, requires a specific interaction between protein and stabilizer, at least at some point during the formation of the solid, such that the native conformation is favored (Carpenter and Crowe, 1994). In addition, there is a pure kinetic stabilization mechanism, which we denote the "glass dynamics hypothesis" (Franks, 1994).

10.3.2.1   Dilution Effect on Intermolecular Reaction

One likely mechanism for the stabilization of a protein by excipient is the dilution effect. Since the drug molecules are diluted in a solid matrix, the probability of forming close contacts between two reactive molecules (i.e., two protein molecules) decreases, and thus a dimerization reaction such as protein aggregation slows down (Chang et al., 2005). This dilution effect does not demand any particular interaction between protein and excipient beyond maintaining a single amorphous phase, and thus any excipients that would be miscible may be used to stabilize the protein. This mechanism impacts only an intermolecular reaction such as protein aggregation in the solid state, and does not apply to stabilization against degradation where an intramolecular reaction is rate limiting.

To estimate quantitatively the impact of the dilution effect, a simplified dilution model has been developed (Chang et al., 2005). In this model, based on random distribution of solutes according to molecular volume, the reaction rate in protein–excipient formulations should be roughly proportional to the volume (or mass) fraction of protein in the formulation. However, the stabilization effect arising from dispersing the protein in a disaccharide matrix is much greater than that can be accounted by the dilution effect alone (Chang et al., 2005; Wang et al., 2009a). Thus, other mechanisms are more important for the stabilization of protein molecules, at least in these systems.

10.3.2.2 Thermodynamic Stabilization Based on Protein–Excipient Interaction

A thermodynamic stabilization mechanism generally requires that a stabilizer functions by increasing the free energy of protein denaturation, and stability is conferred by shifting the equilibrium between native and unfolded conformation toward the more stable native state (Arakawa et al., 2001). The water substitute hypothesis is one example of a thermodynamic stabilization mechanism. This hypothesis, also called a specific interaction mechanism, proposes that sugar molecules form hydrogen bonds at specific sites on the surface of the protein and thus substitute for water–protein hydrogen bonds that stabilize a protein in an aqueous solution (Carpenter et al., 1992). When water is removed during drying, the loss of water–protein hydrogen bonds results in a perturbation to the native structure of protein, which may lead to protein degradation since the unfolded or partially unfolded conformation is generally more reactive. When a stabilizer such as a disaccharide or polyol is added to the protein formulation, the hydroxyl groups in the stabilizer can form hydrogen bonds to the water binding sites of the protein, and thus substitute for the water–protein hydrogen bonds that are believed critical to protein conformational stability. The characterization of specific interactions between protein and sugar in the solid state has been developed over the past two decades. The protein–sugar interactions in the solid state have been mainly investigated using the following methods: FTIR, gravimetric water sorption, and solution calorimetry.

10.3.2.2.1 FTIR Measurement of Protein Structure

Solid-state FTIR spectroscopy is the first technique widely used for the characterization of protein–sugar interactions in dried protein solids (Prestrelski et al., 1993). The sensitivity of vibrational spectroscopy to variations in both geometric arrangements of atoms and hydrogen bonding enables infrared spectroscopy to discriminate between the various secondary structures. The majority of protein structural information can be obtained from the amide I band, which is located from 1600 cm^{-1} to 1700 cm^{-1} (Dong et al., 1995). This band originates primarily from the amide C=O stretching vibration, and is highly impacted by hydrogen bonding (Prestrelski et al., 1993). Since different secondary structures, such as α-helix, β-sheet, and random coil, have their characteristic band positions, the change in each peak reflects variations in secondary structure of the protein (Dong et al., 1995). FTIR techniques based on the KBr pellet transmission mode have been the most popular method for protein structure characterization in dry samples. Other techniques such as attenuated total reflection, diffuse reflectance FTIR, and Raman spectroscopy have also been used to study potential structural changes after drying (Sane et al., 2004; Souillac et al., 2002a).

IR studies suggest the presence of an interaction between sugar and protein, which is quite similar to the water–protein interaction. A peak at 1580 cm⁻¹, which is ascribed to the H-bond interaction with the carboxylate group, is observed in lysozyme dried with sugar, whereas such a peak was not observed in the dried protein without sugar (Carpenter et al., 1998). Figure 10.2 shows the second derivative amide I spectra of several proteins before and after freeze-drying. When the proteins are lyophilized alone, the structures of α-lactalbumin and lactate dehydrogenase are significantly perturbed, while granulocyte colony stimulating factor (GCSF) is minimally impacted. However, when α-lactalbumin is freeze-dried with 200 mM sucrose, the structure of the freeze-dried protein is much closer to the structure in the corresponding aqueous solution (Carpenter et al., 1998). These observations demonstrate the normally observed strong correlation between sugar–protein H-bonding and native structure preservation and therefore suggest that H-bonding between sugar and protein is a major cause of native structure preservation during lyophilization. The extent of protein secondary structure perturbation during freeze-drying was found to be very sensitive to the formulation, with the largest alteration in the structure of dry protein generally occurring in formulations without sugar stabilizers (Chang et al., 2005). The preservation of native structure during processing may also play a significant role in the storage stability in the dried solid. However, one must recognize that while many studies show the correlation between storage stability and native structure preservation, correlations do not necessarily mean a cause-and-effect relationship exists. In addition, there are also reports of failure to correlate stability with the degree of retention of the native structure (Costantino et al., 1996; Schule et al., 2007).

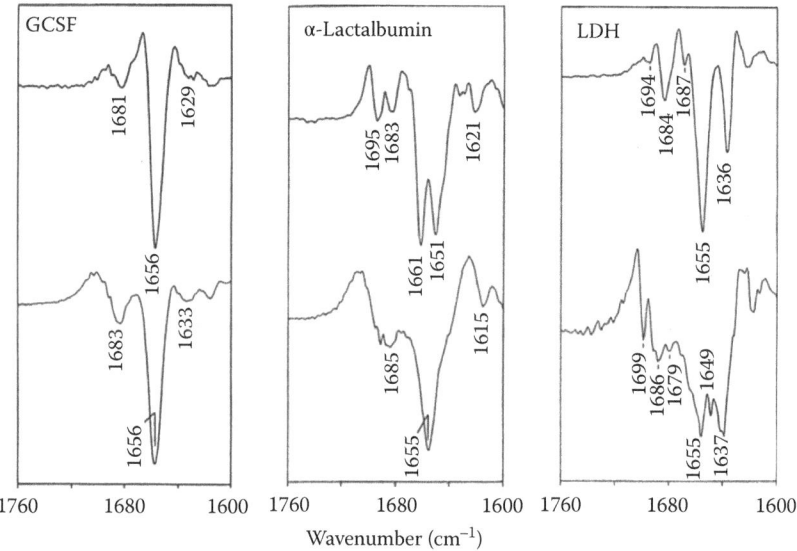

FIGURE 10.2 Second-derivative amide I spectra of granulocyte colony-stimulating factor (GCSF), α-lactalbumin, and lactate dehydrogenase (LDH) in an aqueous solution (upper spectra) and freeze-dried solid (lower spectra). (Reproduced from Carpenter, J.F., et al., *Eur. J. Pharm. Biopharm.* 45, 231, 1998.)

10.3.2.2.2   Gravimetric Water Sorption Analysis

Gravimetric sorption analysis, producing water sorption isotherms, is another method that may be used to study sugar–protein interaction in dried solids. Protein reactivity generally increases with increasing water content due to the ability of water to enhance conformational flexibility, increase molecular mobility, and participate as a reactant in some degradation pathways (Shamblin, 2004). Thus, it is important to investigate the water sorption behavior of a protein sample simply from a pragmatic viewpoint, and water sorption isotherms are therefore commonly obtained to determine the amount of water absorbed at given relative humidities. Water sorption isotherms can be obtained by the simple traditional method, by directly measuring the water content after incubation at various relative humidities, or by using an automatic gravimetric sorption analyzer, which consists of a microbalance within a computer-controlled humidified environment.

The water monolayer coverage, which is related to the number of strong water-binding sites in the protein, may be relevant to the interaction between proteins and excipients and therefore to the pharmaceutical stabilization of proteins. As a dried protein is rehydrated, the adsorbed water interacts initially with strong water-binding sites, followed by adding to weakly binding sites, up to the point where further water exhibits essentially bulk properties (Costantino et al., 1998). Water monolayer coverage can be evaluated using several models, including the Brunauer–Emmett–Teller (BET) equation (Costantino et al., 1998). The monolayer coverage for an amorphous solid has a different meaning from the monolayer formation on a molecularly impervious solid surface. For an amorphous system, the water is absorbed, and the monolayer coverage is related to the number of strong binding sites and not to the geometric surface area. For proteins colyophilized with sugar, which remain amorphous after freeze-drying, the water monolayer coverage is found to be lower than that expected based on additive contributions of the pure protein and pure sugar (Lopez-Diez and Bone, 2000; Salnikova et al., 2008; Wang et al., 2009b). This nonideal behavior suggests that amorphous sugars and proteins interact in such a way to reduce the availability of strong water-binding sites; that is, the sugar occupies some of the water binding sites on the protein, presumably by hydrogen bonding to the polar sites on the protein as does water. Thus, BET analysis of water sorption isotherms provides information on specific interactions between the excipient and the protein, which may help elucidate stabilization mechanisms.

10.3.2.2.3   Solution Calorimetry for Enthalpy of Dissolution Measurement

Solution calorimetry may also be used to evaluate protein–carbohydrate interactions in the dried protein formulation (Souillac et al., 2002b). The enthalpies of solution for pure protein, pure excipient, and colyophilized mixture can be measured with heat of solution calorimetry, and the difference in enthalpies of solution between colyophilized and physical mixtures is a reflection of protein–excipient interaction in the solid state in much the same way as water monolayer coverage data (Souillac et al., 2002b). As shown in Figure 10.3, the expected linear relationship between enthalpy of solution and the mass fraction of protein in the formulations is observed for physical mixtures of protein and sugar, where there are no protein–excipient interactions. In contrast, a nonlinear relationship between the enthalpy of solution

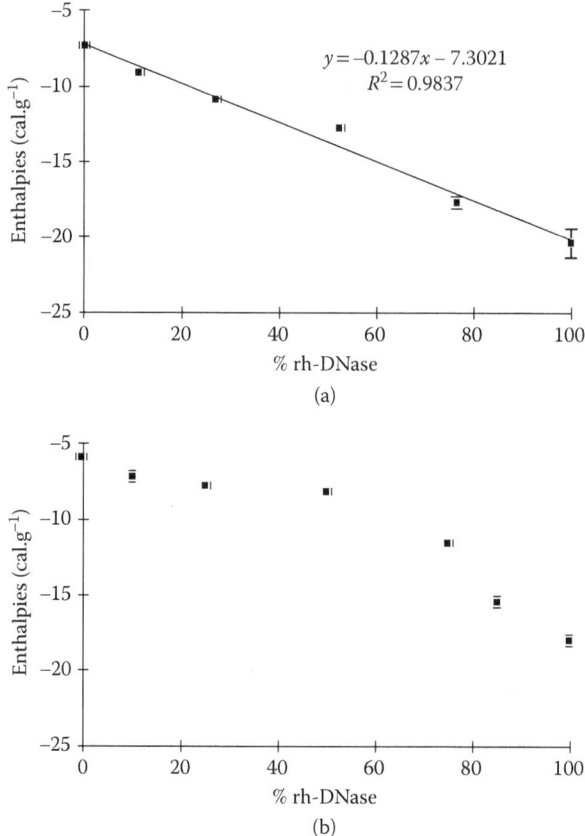

FIGURE 10.3 Enthalpies of solution in water for rh-DNase–sucrose mixtures containing different weight percentages of protein. (a) Physical mixtures; (b) freeze-dried samples. Enthalpy of solution was measured using an isoperibol calorimeter. (Reproduced from Souillac, P., et al., *Int. J. Pharm.* 235, 207, 2002a.)

and protein content is observed for the colyophilized formulations. Positive deviations from linearity are detected for a number of colyophilized protein–sugar samples, which included four different proteins and the two common stabilizers, sucrose and trehalose (Souillac et al., 2002b). This nonadditive behavior provides strong evidence for the presence of interactions between protein and disaccharide in the lyophilized protein samples. The extent of deviation from linearity is found to be strongly dependent on the protein mass fraction in the protein formulations, with the largest protein–sugar interaction occurring at about protein–sugar mass ratio of 1:1 (see Figure 10.3). This technique has been further extended to study protein–amino acid systems (Tian et al., 2006) and in another study to compare the amplitude of specific interaction between hGH and different sugars (Salnikova et al., 2008).

The presence of protein–sugar interaction is undoubtedly important for protein stability in dried solids; for example, significant protein–stabilizer interactions are essential to maintain a single-phase amorphous system of protein and stabilizer.

However, such interactions do not rule out the importance of kinetic factors in stabilizing protein in a glass. Proteins do degrade during storage in a glassy solid, which indicates that there is a thermodynamic driving force for protein degradation (i.e., free-energy change must be <0 or the degradation would never occur). However, if due to kinetic constraints the degradation is negligible on the timescale of importance, the system is still stable even though the degradation is thermodynamically favored (Pikal, 2004). Therefore, the decrease in free energy is a necessary but not a sufficient condition for protein degradation in a dried glass. Moreover, the magnitude of change in free energy does not necessarily correlate with degradation in the glassy solid, and the kinetic factors can play a dominant role in controlling pharmaceutical stability of a protein in a glass (Pikal, 2004; Wang et al., 2009a, 2010).

10.3.2.3 Kinetics Stabilization Based on Reduction in Molecular Mobility

The glass dynamics theory, also called the vitrification hypothesis, is a pure kinetic mechanism (Franks, 1994; Franks et al., 1991). It proposes that the sugar forms a rigid, inert matrix in which the protein is molecularly dispersed, and the limited mobility in the glassy matrix greatly slows the protein mobility necessary for protein degradation (Franks et al., 1991). Thus, if the motion of reactants and protein in the solid is sufficiently slow on the timescale of experiment, the protein will not significantly degrade regardless of what the free energy of denaturation or chemical degradation might become (Pikal, 2004).

Figure 10.4 shows a schematic of enthalpy as a function of temperature. As the liquid is cooled rapidly, the system passes the melting temperature without crystallization and becomes a super-cooled liquid. Here, the molecules are still continuously undergoing rotational and translational motion and are free to access all possible states of the system, and the system is still in thermodynamic equilibrium except for

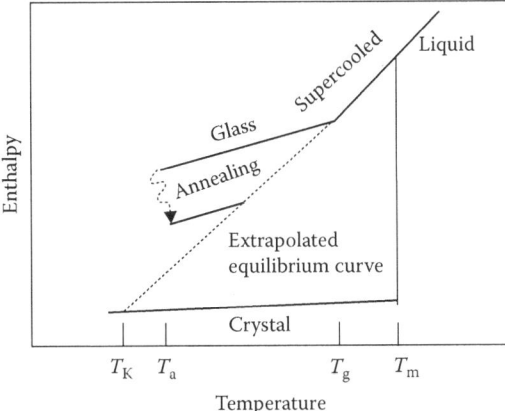

FIGURE 10.4 Schematic representation of enthalpy as a function of temperature showing the annealing behavior in a glass. T_K: Kauzman temperature, where configurational entropy approaches zero, and below which processes requiring molecular rearrangement will theoretically cease; T_a: aging or annealing temperature; T_g: glass transition temperature; T_m: melting temperature.

the lack of crystallization (Angell, 1988). As the system is further cooled, viscosity increases progressively and molecules move more and more slowly until a certain point where they do not have a chance to rearrange configurations to maintain structural equilibrium (Hodge, 1994). This temperature point is called the glass transition temperature, which is a transition between an equilibrium super-cooled liquid and a glassy solid (Angell, 1988; Hodge, 1994). Essentially, a glass is a supercooled liquid that on the timescale of interest is no longer in structural equilibrium due to lack of mobility. Below T_g, material properties (e.g., volume and enthalpy) deviate from those of the extrapolated equilibrium curve, and the resulting glass is in a nonequilibrium state, whose exact properties depend on the process details and thermal history involved in preparation (Angell, 1988; Hodge, 1994).

An amorphous solid is similar to a crystalline material in that the molecular mobility is low in these solids compared to the liquid state. However, some configurational mobility does exist in the glass, even though the timescale can be extremely long (Shamblin et al., 1999). Moreover, as noted earlier, a glass is also different from a liquid in that a glass is not in equilibrium and the configurations frozen in at T_g can change with time (Angell, 1988; Hodge, 1994). Due to its higher enthalpy and free energy, a glass tends to relax toward the extrapolated supercooled equilibrium curve and release heat during this relaxation process. This process is known as structural relaxation, physical aging, or annealing. Thus, molecular mobility in a glass is not zero even at temperatures well below T_g, and thus amorphous pharmaceuticals still have instability issues.

The motions responsible for the glass transition require highly cooperative motions of the neighboring molecules and groups of atoms and are directly coupled to viscosity (Angell, 1988; Hodge, 1994; Shamblin et al., 1999). These global motions have a timescale of typically over 100 seconds near the glass transition temperature and timescales of months or years well below T_g. In addition to this primary or global relaxation (i.e., α-relaxation), glasses also exhibit secondary relaxations such as β-relaxation, Johari–Goldstein relaxations, NMR relaxation, and fast dynamics as measured by neutron scattering (Cicerone and Soles, 2004; Yoshioka et al., 2006). The rich glass dynamics are illustrated in Figure 10.5, where the broad α and β, fast process, and boson peak are shown with different time (or frequency) range (Lunkenheimer et al., 2000). β-relaxation is believed to be a local, largely noncooperative process in the sense that interactions between the moving entity and the neighboring molecules are weak. The timescale of these local motions is within a broad range from nanoseconds to seconds, which is much shorter than global motion at similar temperatures (Lunkenheimer et al., 2000). Also, the activation energy for local motion is typically about 10 to 70 kJ/mol, which is much lower than the activation energy for global cooperative motion (about 200 to 500 kJ/mol) (Johari, 1973). The potential importance of both α-motions and β-relaxation is discussed below.

10.3.2.3.1   Global Motions (α-Relaxation)

The timescale of global motions is long compared to other classes of motion, ranging from minutes near the glass transition temperature to many years well below T_g. The temperature dependence of global mobility above T_g can generally be analyzed by the Williams–Landel–Ferry model (Yoshioka and Aso, 2007). On the basis of the

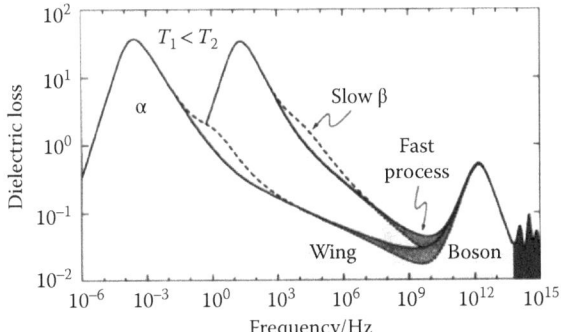

FIGURE 10.5 Schematic view of the frequency-dependent dielectric loss in amorphous glasses at two different temperatures. Different characteristic features, not necessarily all present in a single glass former, are shown in the order of increasing frequency: the α-relaxation peak, a slow β-relaxation peak, the Johari–Goldstein relaxation wing, the minimum with possible contribution of a fast process, and the boson peak. (Reproduced from Lunkenheimer, P., et al., *Contemp. Phys.* 41, 15, 2000.)

model, it is frequently assumed that in a series of samples, increased T_g would lead to better stability. While the chemical degradation rate seems to correlate with $T–T_g$ for several systems (Yoshioka and Aso, 2007), T_g often fails to correlate with protein stability at temperatures well below T_g (Chang, 2003; Chang et al., 1996a). For example, a good glass former with high T_g, such as dextran, does not necessarily provide better stability than glass with a low T_g (Chang et al., 1996a). In addition, although trehalose has a higher T_g than sucrose, it is not necessarily a better stabilizer (Pikal et al., 2008). Thus, even though T_g is an important parameter and the protein–excipient formulation needs to remain in the glassy state to ensure product elegance, the T_g of the glass is not a good predictor of the molecular mobility relevant to stability at temperatures below the glass transition region.

The structural relaxation time, which is theoretically directly proportional to the viscosity, is found to be a better indicator of global dynamics than $T–T_g$ (Pikal, 2002, 2004). When a glass is kept at a constant temperature below T_g, it will relax and release heat (Hodge, 1994). The structural relaxation processes are nonexponential—an observation normally attributed to the spatial heterogeneity of independently relaxing substates; that is, there are contributions from different substates of various sizes and configurational entropies. The kinetics of the relaxation process is often described by the empirical Kohlrausch–Williams–Watt (KWW) equation (Pikal, 2004):

$$\Phi(t) = \exp\left[-\left(\frac{t}{\tau}\right)^{\beta}\right] \tag{10.1}$$

where $\Phi(t)$ is the relaxation function at time t, τ is the mean structural relaxation time, and β ($0 < \beta \le 1$) is a measure of the width of distribution of relaxation times, with small values of β indicating a broad distribution.

Structural relaxation time in a pharmaceutical glass is commonly measured using differential scanning calorimetry (DSC) or the thermal activity monitor (TAM). The enthalpy lost during the relaxation process will be recovered during a DSC heating scan, and can be measured by integration of the T_g-overshoot peak (Luthra et al., 2008a; Shamblin et al., 1999). This enthalpy loss due to relaxation below T_g is measured as a function of aging time, and the kinetics are analyzed by a fit of the KWW equation to calculate τ and β. The TAM can directly measure the rate of enthalpy relaxation (i.e., power) at a constant temperature, and the derivative version of the KWW equation may be fitted to the power–time data to obtain structural relaxation time (Luthra et al., 2008b; Wang and Pikal, 2010). The KWW equation assumes that τ and β are constant throughout the measurement process. However, aging does occur during the calorimetric experiment, and τ increases continuously. Analysis using the KWW function gives a τ value that is larger than that in the initial state and a value of β that is too small. However, theoretical modeling shows that τ^β is an accurate representation of the initial state of the sample, and thus the τ^β value is a more robust parameter for representing the mean structural relaxation time in the initial sample (Kawakami and Pikal, 2005).

There is growing experimental evidence for a correlation between drug stability and structural relaxation time of the glassy system (Duddu et al., 1997; Pikal, 2002; Yoshioka and Aso, 2007). For example, a good correlation is noticed between aggregation of an IgG1 antibody with the reduced time (t/τ) below T_g (Duddu et al., 1997). Also, a strong coupling exists between the rate of protein degradation and structural relaxation time in human growth hormone systems colyophilized with several sugars (Figure 10.6) (Pikal et al., 2008). The coupling coefficient (i.e., the slope in the graph) for aggregation is found to be greater than that for chemical degradation, including methionine oxidation and deamidation. It is obvious that sucrose provides better stability than trehalose at equivalent structural relaxation time. Actually, stability of a sucrose formulation at a given temperature is about a factor of two better than the corresponding trehalose formulation even though the trehalose formulations give a much longer structural relaxation time (Figure 10.6). The better stability in sucrose formulation cannot be explained by differences in protein secondary structure and global mobility; however, local mobility differences may be critical in determining stability trends.

10.3.2.3.2   Local Motions (β-Relaxations)

In addition to the global mobility described above, polymer or protein molecules can experience faster motions involving only part of the molecule, such as movement of some functional groups, terminal arm motion, rotations of side chains, hinge bending, and segmental motion of the polymer backbone (Hill et al., 2005). Furthermore, it is found that secondary relaxation processes also exist in rigid small molecules (like disaccharides) where there is no little or no conformational flexibility. Johari has attributed β-relaxations occurring in the glassy state to islands of mobility (Johari, 1973), that is, the noncooperative hindered motion of some molecules in large cages of other molecules. These noncooperative local motions have a timescale between nanoseconds to seconds, which is much shorter than the global motions related to viscosity.

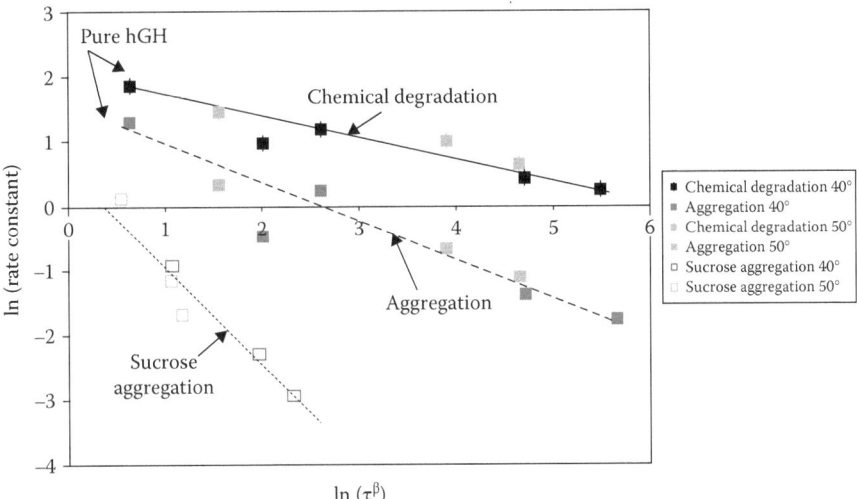

FIGURE 10.6 **(See color insert.)** Correlation between stability and structural relaxation time constant (τ^{β}) of hGH formulations. Formulations studied include pure hGH, stachyose (1:1), and trehalose (1:1, 1:3, and 1:6). These ratios are weight ratios of protein–saccharide. The triangles represent aggregation data in sucrose formulations. The rate constant is from square root of time (months) kinetics, and τ is in hours. Note that sucrose formulations are more stable than expected based upon the structural relaxation time constant. (Reproduced from Pikal, M.J., et al., *J. Pharm. Sci.* 97, 5106, 2008.)

Proteins may also undergo internal motions such as binding and release of a substrate from the interior of an enzyme; such motions are required for protein function (Cicerone and Soles, 2004; Hill et al., 2005). The response of a protein's internal motions to temperature change is similar to that of a small-molecule glass and a polymer glass in that different types of motions freeze out at different temperatures. The internal motions of proteins also undergo glass-like behavior upon cooling to near a dynamical transition temperature (T_d), below which the protein activity is lost, and the unfolding is inhibited (Hill et al., 2005). If these internal motions are linked to α-relaxations in the matrix via interactions with the matrix, protein motion becomes coupled with the structural rigidity of the matrix, thereby damping internal motion in the protein and stabilizing the protein.

Numerous studies have presented evidence that there is a correlation of stability with local motions at temperatures well below T_g (Cicerone and Soles, 2004; Cicerone et al., 2003; Wang et al., 2009a). The presence of small amounts of plasticizer (about 5% w/w relative to disaccharide), such as glycerol and sorbitol, stabilizes enzymes and IgG1 protein in sugar glasses, even though the global mobility increased (i.e., T_g decreased, and structural relaxation time decreased) upon addition of such plasticizers (Cicerone and Soles, 2004; Cicerone et al., 2003). A neutron backscattering spectrometer with high energy resolution is used to study high-frequency dynamics of the system, where only those motions on the nanosecond timescale can be measured (Cicerone and Soles, 2004). The scattering intensity at different scattering angles can be recorded while the sample is heated from as low as 10 K to above T_g.

The hydrogen-weighted mean-square atomic displacement $<u^2>$ can be obtained by fitting a harmonic oscillator model to the data. The larger is the mean square amplitude of motion ($<u^2>$), the higher is the local mobility.

As seen from Figure 10.7 (inset), a neutron scattering study showed that the fast local dynamics in trehalose glasses is hindered upon addition of 5% plasticizer (glycerol). Thus, although a low-molecular-weight excipient, such as glycerol, plasticizes the global motions, the fast dynamics or local motions are anti-plasticized (Cicerone and Soles, 2004). This fast dynamics study is consistent with results from a molecular dynamics simulation that shows the strongest H-bonding network as well as the greatest suppression of fast dynamics for the 5% glycerol formulation (Dirama et al., 2005). The role of local dynamics in determining the stability of an IgG1 antibody prepared with different drying methods and formulations is also investigated (Abdul-Fattah et al., 2007b). Degradation rate variations with formulation correlates with both global mobility and local dynamics as measured by $<u^2>$, but when the stability variable is the method of drying, the correlation is best with $<u^2>$.

Solid-state NMR has also been used to study the relationship between storage stability and local mobility of proteins (Yoshioka et al., 2006, 2007). NMR measures predominantly rotational motions, and these local motions have a timescale from microseconds to seconds. The degradation rate of insulin freeze-dried with dextran is greater than insulin freeze-dried with trehalose at low relative humidity (12%), an observation that is correlated to the inhibition of β-relaxations by trehalose as evidenced by higher $T_{1\rho}$ (rotating frame spin lattice relaxation time) values (Yoshioka and Aso, 2007; Yoshioka et al., 2006). Also, aggregation stability of β-galactosidase (GA) correlates with local mobility in a series of lyophilized GA

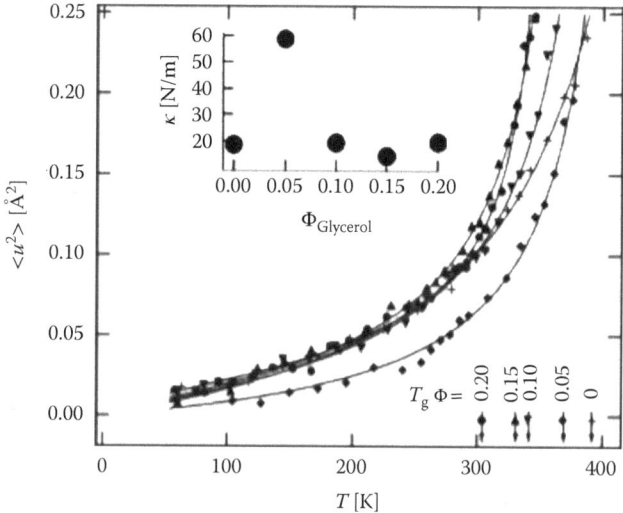

FIGURE 10.7 Debye–Waller factors from trehalose glasses with different amounts of glycerol: 0 (+), 0.05 (♦), 0.10 (▼), 0.15 (▲), and 0.20 (●). (Inset) Spring constants of glasses. The x-axis also shows the variation of T_g of the matrix as a function of glycerol fraction. (Reproduced from Cicerone, M.T., and Soles, C.L., *Biophys. J.* 86, 3836, 2004.)

formulations containing trehalose, sucrose, or stachyose (Yoshioka et al., 2007). The rank order of GA aggregation rate at temperatures below T_g is sucrose < trehalose < stachyose. No significant differences in the second-derivative FTIR spectra of the protein formulations are found. The rank order of the ability of excipients in decreasing the local mobility of GA appears to be the same as the rank order of their ability to decrease aggregation rate, suggesting that local mobility measured by NMR $T_{1\rho}$ is a primary factor affecting the stability of these GA formulations.

Stability well below T_g for the above systems is found to be better predicted by local mobility measured by either neutron backscattering or NMR than global mobility. However, relaxation time measured by DSC and other techniques (such as isothermal calorimetry) appears to be a better explanation of stability, at least for some systems (Bakaltcheva et al., 2007; Pikal, 2002, 2004). Thus, the mobility most predictive of stability may be global mobility and/or some type of local mobility. Although there has been some progress in the understanding of the relationship between global and local mobility (Cicerone and Soles, 2004; Wang and Pikal, 2010; Wang et al., 2009a, 2010), a number of important questions remain: Why do local motions with very short timescales control long-term stability with a timescale of months? Which type of mobility is most critical for protein stability under a given set of circumstances, global or local motion? To what extent do the mobilities measured with different techniques correlate with each other? More systematic studies are needed to have a better understanding in the field.

10.4   PRACTICAL PROTEIN FORMULATION GUIDELINES

The traditional formulation of a solid protein is usually a largely empirical process involving much trial and error. This practice might be defended because each protein is somewhat unique in both physical and chemical behavior and tends to exhibit a unique stability profile. Our view is that although the specific characteristics of each protein must be appropriately recognized and some empirical studies are necessary, one may also make use of general principles that reflect the past decade or two of research in protein stabilization. Here, we attempt to provide such guidelines in an effort to facilitate QbD efforts.

10.4.1   STABILITY ENHANCEMENT THROUGH EXCIPIENTS

The presence of each excipient in the formulation must be justified, and two critical steps are needed to support a rational formulation development process. The first step is to evaluate the scope of the stability problem during processing and storage during the early phase of a formulation project (Pikal, 1990, 1994). To assess potential degradation pathways caused by lyophilization process, samples generated from freeze–thaw, lyophilization, and reconstitution are assayed and compared to the stability of the starting solutions. Also, accelerated stability studies are used to assess the degradation during high-temperature storage. The freeze-dried samples are generally subjected to short accelerated stability for about 4 weeks at high temperatures (about 40°C or higher, but at least 10°C below T_g). To obtain a moderate amount of degradation to assist the analytical method development and to survey the scope of the instability problem, these studies generally use the pure drug or a simple drug formulation with some bulking agent simply to prevent blow-out* during lyophilization.

The second critical step in the formulation development process is to add stabilizers to minimize the stability issues identified from the processing and storage studies. A surfactant may be used to prevent protein loss from the ice–water or air–water interface if degradation occurs during freeze–thaw (or atomization process in spray-drying and freeze-drying). In addition, disaccharide and amino acids may be added as excluded solvents to stabilize thermodynamically the native conformation of the protein (Carpenter et al., 1997). If the drying process results in significant protein degradation, nonreducing disaccharide should be included to protect the protein (Pikal, 1990). If there is severe storage instability, addition of a high level of stabilizer and a tight control of moisture level are needed (Pikal, 1994). In the following section, the use of such common excipients as buffer, saccharide, crystalline bulking agent, and surfactant for solid protein formulations is discussed in detail.

10.4.1.1 Buffer Type and Concentration

Many proteins are stable only in a narrow pH range, and buffer is generally used to control the target pH. At extreme pH values, proteins tend to undergo structural unfolding or denaturation, which can greatly impact protein aggregation and/or bioactivity. Moreover, the solution pH can significantly affect the rate of chemical degradations such as deamidation and fragmentation (Wang, 1999). For example, the storage stability of an IgG1 antibody in the solid state is sensitive to the pH of protein solution before lyophilization (Chang et al., 2005), especially at a low level of sugar stabilizer. Therefore, it is essential to optimize the solution pH used to prepare a solid formulation.

Buffer type is an important factor in determining the potential for pH change upon freezing, and it may also impact stability in several ways. For example, selective crystallization of buffer during freezing may cause large pH changes, and acetate buffer may evaporate during drying and cause a major pH shift in the reconstituted solid. Histidine, on the other hand, may change to a yellowish color upon long-term storage. It is prudent to screen different buffer types and identify potential problems early in the formulation stage.

Freeze concentration may selectively crystallize one buffer species, causing a massive pH change. For example, the pH of sodium phosphate can drop from 7.5 to 4.5 upon selective crystallization of Na_2HPO_4 (Anchordoquy and Carpenter, 1996), which may denature pH-sensitive proteins. This problem can be minimized by use of only a low concentration of buffer, such that the buffer concentration (wt%) is much lower than the concentration of other solutes. Under such conditions, the buffer normally does not crystallize and remains in the amorphous phase, thus avoiding the large pH shift that could otherwise occur (Pikal, 1990). In general, high buffer capacity is not required in a freeze-dried product, so use of a low concentration of buffer does not compromise pH control in solutions.

---

* The rapid flow of water vapor from the subliming ice may carry away the product from the vial and create a container with minuscule amounts of product sitting at the bottom. This phenomenon is called blow-out, and it generally occurs when the concentration of the product is relatively low.

10.4.1.2 Stabilizer

Saccharides have been commonly used as stabilizers in lyophilized formulations. Nonreducing saccharides such as sucrose, trehalose, and raffinose are chemically inert, and tend to resist crystallization during lyophilization. These stabilizers can hydrogen bond with the protein to stabilize the protein native structure, and also provide a glassy matrix for the protein to dampen the molecular motion necessary for degradation.

Even though sucrose has been extensively used in commercial formulations, trehalose provides several advantages over sucrose, such as higher T_g and stronger resistance to acid hydrolysis at low pH. At low pH (pH below 5), sucrose will hydrolyze into reducing sugars and then react with an amine group of the protein via a Maillard reaction. Also, moisture uptake from ambient can be a serious problem for sucrose-based products designed for inhalation delivery. In this case, trehalose is a better option than sucrose due to its relatively higher T_g (Pikal et al., 1991). Even after some exposure to ambient conditions, the T_g of a trehalose formulation is still likely to be much higher than room temperature. However, if the pH is greater than 5 and the product can be well sealed in the final container, sucrose may provide better stabilization at the storage temperatures of interest, at least for many proteins, as demonstrated for hGH (Pikal et al., 2008).

For the appropriate level of sugar stabilizer, a higher stabilization effect is generally achieved with a higher weight ratio of stabilizer to protein (Wang et al., 2009a). However, a high total solid content (i.e., > 10%) may slow the drying due to the high product resistance to the transfer of water vapor, and should generally be avoided. Thus, a weight ratio of stabilizer to protein ranging from 1:1 to 10:1 is generally recommended. Since maximum stabilization is not always needed, sufficient stabilization may be achieved with weight ratio of only 0.5:1, as is the case for a monoclonal antibody (Chang et al., 2005) and an interleukin (Wang et al., 2009a).

10.4.1.3 Crystalline Material

A bulking agent is commonly used to provide mass and mechanical strength to the dried cake. When the total drug concentration is less than 1%, a freeze-dried product may blow out because of the momentum transfer from flowing vapor to the dried cake (Pikal, 2001). This blowing out event can be avoided by using a bulking agent at a sufficient concentration (>3% w/w). Mannitol or glycine is commonly used as a bulking agent due to their favorable thermal behavior (Pikal, 1990). Both compounds have relatively high eutectic temperatures (slightly lower than 0°C), and can readily crystallize during the freezing step to provide mechanical strength to the cake. Thus, it is relatively easy to design a process that produces an elegant product in a short time. In addition, crystalline material can be used to facilitate the freeze-drying process for formulations with disaccharide stabilizers. Due to the relatively low collapse temperature, sugar stabilizers need to be dried at low temperatures, and the lyophilization process may be relatively long. The presence of a bulking agent makes it possible to dry the formulations at relatively high temperatures. Mixtures of amorphous stabilizer (i.e., sucrose) and crystalline bulking agent can be very useful in cases of low-dose drug products since adequate stabilization can be obtained in a process that is fast and robust. However, it is critical to ensure essentially complete crystallization of the bulking agent because the noncrystallized portion will give a low T_g' that complicates primary drying and may also lead to storage stability issues due to the low T_g in the dried product (Pikal, 2004). The weight ratio of bulking agent to amorphous components, including protein and stabilizer, is critical for controlling the crystallization behavior of mannitol or glycine (Liao et al., 2007).

Proteins do not crystallize at all during lyophilization and thus stay in the amorphous phase. A protein can also inhibit the crystallization of a bulking agent due to protein–excipient interaction. To achieve maximum crystallization of mannitol or glycine, the bulking agent needs to be present in significant excess (Pikal, 1990). A weight ratio of about 3:1 of bulking agent to all other solid components is generally recommended. The degree of crystallinity can be measured with an x-ray diffractometer, and the degree of crystallinity in a system can be influenced by the nature and the amount of the additives. In addition, the process-related parameters such as cooling rate, annealing temperature, and annealing duration also have significant impact on crystallinity (Liao et al., 2007).

However, the above generalizations are general guidelines, and in some cases, it may be necessary to formulate outside these guidelines. We also emphasize that, for low-dose proteins, one can effectively employ a mixture of stabilizer and bulking agent. Here, for a low protein amount at about 0.1 mg/mL, the sugar level may be as high as 10 mg/mL (i.e., sugar–protein mass ratio of 100), with a mannitol level of 50 mg/mL. Such a system will not show collapse due to the crystalline fraction and may provide a stable environment for the protein even if the primary drying temperature is well above the normal collapse temperature of the sugar–protein system. Even though cake collapse is not acceptable for esthetic reasons, drying above the normal collapse temperature or Tg′ may not result in adverse impact on the product quality. For example, the impact of drying above the collapse temperature has been studied for recombinant factor VIII (rFVIII) and α-amylase in sucrose–glycine formulation (Wang et al., 2004). For rFVIII, there is no difference in the biological activity for product freeze-dried under different drying conditions when stored at 5°C or 25°C. At 40°C storage, however, the stability of the collapsed product appears to be even better than that of normal product freeze-dried with a conservative cycle. For α-amylase, there is no impact of process temperature on stability at any of these three storage temperatures. Similarly, it has been recently reported that collapse during primary drying (using an aggressive cycle) did not have a negative effect on either the in-process or long-term storage stability of an IgG1 in a sucrose formulation (Schersch et al., 2010).

10.4.1.4 Surfactant

Surfactants are often used in protein liquid formulations to avoid product loss due to adsorption to the filling line and containers, particularly when the protein concentration is low (Manning et al., 2010). In addition, surfactants can reduce protein aggregation at ice–water interfaces during freezing (Chang et al., 1996b). Therefore, surfactants can perform a stability function even though they are not often categorized as stabilizers.

Tween 20 and 80 are the most commonly used nonionic surfactants for protein stabilization. Low concentrations of these surfactants are often sufficient to serve this purpose due to their relatively low critical micelle concentrations (Chi et al., 2003; Chou et al., 2005). For example, the humanized monoclonal antibody (IgG1) Avastin® contains 0.4 mg/mL Tween 20, and Lucentis® contains 0.01% Tween 20. Tween 80 at a concentration of 0.01% protects both LDH and GDH from denaturation during freezing and thawing (Chang et al., 1996b). Tween 80 at concentrations from 0.005% to 0.01% also protects several other proteins from freezing denaturation, including IL-1, TNF, GDH, and PFK (Chang et al., 1996b; Wang, 2000). Moreover, the addition of high amount of Tween in a formulation should be approached with caution because these materials often contain residual peroxides that can destabilize the protein system via oxidation reactions.

10.4.2   CHARACTERIZATION TO IMPROVE STABILITY

Following the general formulation rules described above will provide some reasonable formulation candidates. It is also important to use various characterization tools to evaluate both the molecular mobility of the solid and conformational changes upon drying. The QbD concept stresses the scientific understanding of both the formulation and manufacturing process, and these characterization tools facilitate the rational design of protein formulation with acceptable stability profiles.

Molecular mobility should be tested in solid formulations. DSC should be used to characterize molecular mobility via measurement of T_g and/or estimation of relaxation times through measurement of the time dependence of enthalpy recovery (Shamblin et al., 1999). The product T_g should be well above the anticipated storage temperature, even with the moisture increase in the product during long-term storage arising from moisture uptake from stopper.

Structural information of protein and crystalline excipients should also be obtained. The secondary structure of the protein in the dried powder has been found to be very useful in predicting long-term stability of protein formulations. Solid-state FTIR is quick for protein structure characterization, and is highly recommended for use in the development of dried protein products (Carpenter et al., 1998). For formulations with easily crystallizable components, analysis of the dried cake by powder x-ray diffraction is very important to ensure product quality. Depending on the formulation and freeze-drying cycle, mannitol or glycine can exist in several different polymorphic forms, and the crystallization pattern can impact the reconstitution time, moisture, and stability (Liao et al., 2007; Pikal, 1999). Moreover, at least a semiquantitative measure of degree of crystallinity can be obtained by x-ray diffraction, which is often critical to product performance. For example, the release of water from mannitol hemi-hydrate during long-term storage can adversely impact the stability. Therefore, the qualitative analysis of crystalline phase composition and quantitative analysis of degree of crystallinity are important for these formulations.

In addition to the above solid-state characterizations, characterization of the reconstituted solution is also critical. The use of CD, FTIR, or fluorescence allows the quick screening of protein formulations, and helps to understand better the stability of a lead protein formulation through investigation of the thermal folding behavior. In addition, these spectroscopic tools can be used to verify native structure in the protein solution that would be administered. Since the commonly used size exclusion chromatography for aggregate analysis has potential deficiencies due to the possibility of insoluble aggregates being filtered out on the column and the dilution steps involved perhaps changing the fraction of soluble aggregates, some additional methods (often termed orthogonal) have been used to provide additional insights to protein aggregate formation (Gabrielson et al., 2007). Dynamic light scattering is a high-throughput method, and it can provide information about the hydrodynamic radius and polydispersity. The analytical ultracentrifuge is another powerful tool for studying the size and shape of protein molecules in solution, and it is particularly useful in quantifying protein aggregation without disturbing the state of aggregation in the solution being investigated.

Recently, characterization of subvisible and visible protein particles has become a hot topic because protein aggregates/particles are suspected to be a cause of immunogenicity in therapeutic protein dosage forms (Carpenter et al., 2009) (Chapter 8). Since subvisible and visible particles exist in every therapeutic protein product, their sizes and levels can be extremely important product quality attributes. Quantifying these protein particles during stress and real-time stability studies may provide new insights into aggregation pathways and help to assess the potential immune response arising from different types of aggregates. Several new tools, such as Microflow Imaging, Flowcam, Coulter Counter, and Nanoparticle Tracking Analyzer, are being developed to characterize these particles (Wuchner et al., 2010). The improved understanding of particle formation utilizing these characterization methods will facilitate development of stable protein formulations with a minimum of potentially damaging protein aggregates.

10.5 SUMMARY

Knowledge in the field of formulation of lyophilized proteins has been significantly advanced by numerous studies of protein–excipient interactions and the correlation between molecular mobility and stability, which provide a basis for the formulation guidelines discussed in this review. However, the mechanisms of protein stabilization by drying are still not fully understood, and additional studies are needed to understand better the relationships between global and/ or local motion and protein stability during processing and storage. In addition, since protein stability is impacted by formulation and lyophilization processes, development of a stable protein formulation requires a deep understanding of the lyophilization process. Therefore, formulation scientists need to follow best practice for freeze-drying process development, characterize the protein structure and molecular mobility in the candidate formulations, and conduct systematic stability studies at temperatures well below the T_g of these formulations at relevant residual water contents. By taking advantage of the generalities that are established, and systematically investigating the unique properties of the protein of interest, QbD can be accomplished efficiently and product quality optimized without excessive development time and resources.

REFERENCES

Abdul-Fattah, A.M., Kalonia, D.S., and Pikal, M.J., 2007a. The challenge of drying method selection for protein pharmaceuticals: Product quality implications. *J. Pharm. Sci.* 96:1886–1916.

Abdul-Fattah, A.M., Truong-Le, V., Yee, L., Nguyen, L., Kalonia, D.S., Cicerone, M.T., and Pikal, M.J., 2007b. Drying-induced variations in physico-chemical properties of amorphous pharmaceuticals and their impact on stability (I): Stability of a monoclonal antibody. *J. Pharm. Sci.* 96:1983–2008.

Adler, M., Unger, M., and Lee, G., 2000. Surface composition of spray-dried particles of bovine serum albumin/trehalose/surfactant. *Pharm. Res.* 17:863–870.

Anchordoquy, T.J., and Carpenter, J.F., 1996. Polymers protect lactate dehydrogenase during freeze-drying by inhibiting dissociation in the frozen state. *Arch. Biochem. Biophys.* 332:231–238.

Angell, C.A., 1988. Perspective on the glass transition. *J. Phys. Chem. Solids* 49:863–871.

Arakawa, T., Prestrelski, S.J., Kenney, W.C., and Carpenter, J.F., 2001. Factors affecting short-term and long-term stabilities of proteins. *Adv. Drug Deliv. Rev.* 46:307–326.

Bakaltcheva, I., O'Sullivan, A.M., Hmel, P., and Ogbu, H., 2007. Freeze-dried whole plasma: Evaluating sucrose, trehalose, sorbitol, mannitol and glycine as stabilizers. *Thromb. Res.* 120:105–116.

Cal, K., and Sollohub, K., 2010. Spray drying technique. I: Hardware and process parameters. *J. Pharm. Sci.* 99:575–586.

Carpenter, J.F., Arakawa, T., and Crowe, J.H., 1992. Interactions of stabilizing additives with proteins during freeze-thawing and freeze-drying. *Dev. Biol. Stand.* 74:225–238.

Carpenter, J.F., and Crowe, J.H., 1994. An infrared spectroscopic studies of the interaction of trehalose with dried proteins. *Biochemistry* 28:3916–3922.

Carpenter, J.F., Pikal, M.J., Chang, B.S., and Randolph, T.W., 1997. Rational design of stable lyophilized protein formulations: Some practical advice. *Pharm. Res.* 14:969–975.

Carpenter, J.F., Prestrelski, S.J., Anchordoguy, T.J., and Arakawa, T., 1994. Interactions of stabilizers with proteins during freezing and drying. In *Formulation and Delivery of Proteins and Peptides* (ACS Symposium Series, Vol. 567), Eds. Cleland, J.L., and Langer, R., 134–147. Washington, DC: American Chemical Society.

Carpenter, J.F., Prestrelski, S.J., and Dong, A.C., 1998. Application of infrared spectroscopy to development of stable lyophilized protein formulations. *Eur. J. Pharm. Biopharm.* 45:231–238.

Carpenter, J.F., Randolph, T.W., Jiskoot, W., et al., 2009. Overlooking subvisible particles in therapeutic protein products: Gaps that may compromise product quality. *J. Pharm. Sci.* 98:1201–1205.

Chang, B.S., Beauvais, R.M., Dong, A.C., and Carpenter, J.F., 1996a. Physical factors affecting the storage stability of freeze-dried interleukin-1 receptor antagonist: Glass transition and protein conformation. *Arch. Biochem. Biophys.* 331:249–258.

Chang, B.S., Kendrick, B.S., and Carpenter, J.F., 1996b. Surface-induced denaturation of proteins during freezing and its inhibition by surfactants. *J. Pharm. Sci.* 85:1325–1330.

Chang, B.S., Randall, C.S., and Lee, Y.S., 1993. Stabilization of lyophilized porcine pancreatic elastase. *Pharm. Res.* 10:1478–1483.

Chang, L., 2003. Stabilizers in the freeze drying of proteins: Mechanism of stabilization. PhD diss., University of Connecticut, Storrs, CT.

Chang, L., Shepherd, D., Sun, J., et al., 2005. Mechanism of protein stabilization by sugars during freeze-drying and storage: Native structure preservation, specific interaction, and/or immobilization in a glassy matrix? *J. Pharm. Sci.* 94:1427–1444.

Chi, E.Y., Krishnan, S., Randolph, T.W., and Carpenter, J.F., 2003. Physical stability of proteins in aqueous solution: Mechanism and driving forces in nonnative protein aggregation. *Pharm. Res.* 20:1325–1336.

Chou, D.K., Krishnamurthy, R., Randolph, T.W., Carpenter, J.F., and Manning, M.C., 2005. Effects of Tween 20 and Tween 80 on the stability of albutropin during agitation. *J. Pharm. Sci.* 94:1368–1381.

Cicerone, M.T., and Soles, C.L., 2004. Fast dynamics and stabilization of proteins: Binary glasses of trehalose and glycerol. *Biophys. J.* 86:3836–3845.

Cicerone, M.T., Tellington, A., Trost, L., and Sokolov, A., 2003. Substantially improved stability of biological agents in dried form: The role of glassy dynamics in preservation of biopharmaceuticals. *BioProcess Int.* 1:36–38, 40, 42, 44, 46, 47.

Costantino, H.R., Curley, J.G., Wu, S., and Hsu, C.C., 1998. Water sorption behavior of lyophilized protein-sugar systems and implications for solid-state interactions. *Int. J. Pharm.* 166:211–221.

Costantino, H.R., Langer, R., and Klibanov, A.M., 1994. Solid-phase aggregation of proteins under pharmaceutically relevant conditions. *J. Pharm. Sci.* 83:1662–1669.

Costantino, H.R., Schwendeman, S.P., Griebenow, K., Klibanov, A.M., and Langer, R., 1996. The secondary structure and aggregation of lyophilized tetanus toxoid. *J. Pharm. Sci.* 85:1290–1293.

Crowe, J.H., Carpenter, J.F., and Crowe, L.M., 1998. The role of vitrification in anhydrobiosis. *Ann. Rev. Physiol.* 60:73–103.

Dirama, T., Carri, G., and Sokolov, A., 2005. Role of hydrogen bonds in the fast dynamics of binary glasses of trehalose and glycerol: A molecular dynamics simulation study. *J. Chem. Phys.* 122:114501–114508.

Dong, A.C., Prestrelski, S.J., Allison, S.D., and Carpenter, J.F., 1995. Infrared spectroscopic studies of lyophilization-induced and temperature-induced protein aggregation. *J. Pharm. Sci.* 84:415–424.

Duddu, S.P., Zhang, G., and Dal Monte, P.R., 1997. The relationship between protein aggregation and molecular mobility below the glass transition temperature of lyophilized formulations containing a monoclonal antibody. *Pharm. Res.* 14:596–600.

Franks, F., 1994. Long-term stabilization of biologicals. *Biotechnology* 12:253–256.

Franks, F., Hatley, R.H.M., and Mathias, S.F., 1991. Materials science and the production of shelf-stable biologicals. *BioPharm* 4:38, 40–42, 55.

Gabrielson, J.P., Brader, M.L., Pekar, A.H., Mathis, K.B., Winter, G., Carpenter, J.F., and Randolph, T.W., 2007. Quantitation of aggregate levels in a recombinant humanized monoclonal antibody formulation by size-exclusion chromatography, asymmetrical flow field flow fractionation, and sedimentation velocity. *J. Pharm. Sci.* 96:268–279.

Hill, J.J., Shalaev, E.Y., and Zografi, G., 2005. Thermodynamic and dynamic factors involved in the stability of native protein structure in amorphous solids in relation to levels of hydration. *J. Pharm. Sci.* 94:1636–1667.

Hodge, I.M., 1994. Enthalpy relaxation and recovery in amorphous materials. *J. Non-Crystalline Solids* 169:211–266.

Johari, G.P., 1973. Intrinsic mobility of molecular glasses. *J. Chem. Phys.* 58:1766–1770.

Kawakami, K., and Pikal, M.J., 2005. Calorimetric investigation of the structural relaxation of amorphous materials: Evaluating validity of the methodologies. *J. Pharm. Sci.* 94:948–965.

Kreilgaard, L., Jones, L.S., Randolph, T.W., Frokjaer, S., Flink, J.M., Manning, M.C., and Carpenter, J.F., 1998. Effect of Tween 20 on freeze-thawing- and agitation-induced aggregation of recombinant human factor XIII. *J. Pharm. Sci.* 87:1597–1603.

Lai, M.C., and Topp, E.M., 1999. Solid-state chemical stability of proteins and peptides. *J. Pharm. Sci.* 88:489–500.

Liao, X.M., Krishnamurthy, R., and Suryanarayanan, R., 2007. Influence of processing conditions on the physical state of mannitol: Implications in freeze-drying. *Pharm. Res.* 24:370–376.

Lopez-Diez, E.C., and Bone, S., 2000. An investigation of the water-binding properties of protein + sugar systems. *Phys. Med. Biol.* 45:3577–3588.

Lunkenheimer, P., Schneider, U., Brand, R., and Loidl, A., 2000. Glassy dynamics. *Contemp. Phys.* 41:15–36.

Luthra, S.A., Hodge, I.M., and Pikal, M.J., 2008a. Effects of annealing on enthalpy relaxation in lyophilized disaccharide formulations: Mathematical modeling of DSC curves. *J. Pharm. Sci.* 97:3084–3099.

Luthra, S.A., Hodge, I.M., Utz, M., and Pikal, M.J., 2008b. Correlation of annealing with chemical stability in lyophilized pharmaceutical glasses. *J. Pharm. Sci.* 97:5240–5251.

Maa, Y.F., Nguyen, P.A., Andya, J.D., Dasovich, N., Sweeney, T.D., Shire, S.J., and Hsu, C.C., 1998. Effect of spray drying and subsequent processing conditions on residual moisture content and physical/biochemical stability of protein inhalation powders. *Pharm. Res.* 15:768–775.

Manning, M.C., Chou, D.K., Murphy, B.M., Payne, R.W., and Katayama, D.S., 2010. Stability of protein pharmaceuticals: An update. *Pharm. Res.* 27:544–575.

Mumenthaler, M., Hsu, C.C., and Pearlman, R., 1994. Feasibility study on spray drying protein pharmaceuticals: Recombinant human growth hormone and tissue-type plasminogen activator. *Pharm. Res.* 11:12–20.

Pikal, M.J., 1990. Freeze-drying of proteins part II: Formulation selection. *BioPharm* 3:26–30.

Pikal, M.J., 1994. Freeze-drying of proteins: Process, formulation, and stability. In *Formulation and Delivery of Proteins and Peptides* (ACS Symposium Series, Vol. 567), Eds. Cleland, J.L., and Langer, R., 120–133. Washington, DC: American Chemical Society.

Pikal, M.J., 1999. Impact of polymorphism on the quality of lyophilized products. In *Polymorphism in Pharmaceutical Solids* (Drugs and the Pharmaceutical Sciences, Vol. 95), Ed. Brittain, H.G., 395–419. New York: Marcel Dekker.

Pikal, M.J., 2001. Freeze-drying. In *Encyclopedia of Pharmaceutical Technology*, Vol. 6, Eds. Boylan, J., and Swarbrick, J., 275–303. New York: Marcel Dekker.

Pikal, M.J., 2002. Chemistry in solid amorphous matrices: Implication for biostabilization. In *Amorphous Food and Pharmaceutical Systems*, Ed. Levine, H., 257–277. Cambridge, UK: The Royal Society of Chemistry.

Pikal, M.J., 2004. Mechanisms of protein stabilization during freeze-drying and storage: The relative importance of thermodynamic stabilization and glassy state relaxation dynamics. In *Freeze-Drying/Lyophilization of Pharmaceutical and Biological Products* (Drugs and the Pharmaceutical Sciences, Vol. 137), Eds. Rey, L., and May, J.C., 63–107. New York: Marcel Dekker.

Pikal, M.J., Dellerman, K.M., Roy, M.L., and Riggin, R.M., 1991. The effects of formulation variables on the stability of freeze-dried human growth hormone. *Pharm. Res.* 8:427–436.

Pikal, M.J., Rigsbee, D., Roy, M.L., Galreath, D., Kovach, K.J., Wang, B.Q., Carpenter, J.F., and Cicerone, M.T., 2008. Solid state chemistry of proteins: II. The correlation of storage stability of freeze-dried human growth hormone (hGH) with structure and dynamics in the glassy solid. *J. Pharm. Sci.* 97:5106–5121.

Pikal, M.J., and Shah, S., 1997. Intravial distribution of moisture during the secondary drying stage of freeze drying. *PDA J. Pharm. Sci. Technol.* 51:17–24.

Pikal, M.J., Shah, S., Roy, M.L., and Putman, R., 1990. The secondary drying stage of freeze drying: Drying kinetics as a function of temperature and chamber pressure. *Int. J. Pharm.* 60:203–217.

Prestrelski, S.J., Tedeschi, N., Arakawa, T., and Carpenter, J.F., 1993. Dehydration-induced conformational transitions in proteins and their inhibition by stabilizers. *Biophys. J.* 65:661–671.

Randolph, T.W., and Jones, L.S., 2002. Surfactant-protein interactions. *Pharm. Biotechnol.* 13:159–175.

Salnikova, M.S., Middaugh, C.R., and Rytting, J.H., 2008. Stability of lyophilized human growth hormone. *Int. J. Pharm.* 358:108–113.

Sane, S.U., Wong, R., and Hsu, C.C., 2004. Raman spectroscopic characterization of drying-induced structural changes in a therapeutic antibody: Correlating structural changes with long-term stability. *J. Pharm. Sci.* 93:1005–1018.

Schersch, K., Betz, O., Garidel, P., Muehlau, S., Bassarab, S., and Winter, G., 2010. Systematic investigation of the effect of lyophilizate collapse on pharmaceutically relevant proteins. i: Stability after freeze-drying. *J. Pharm. Sci.* 99:2256–2278.

Schule, S., Friess, W., Bechtold-Peters, K., and Garidel, P., 2007. Conformational analysis of protein secondary structure during spray drying of antibody/mannitol formulations. *Eur. J. Pharm. Biopharm.* 65:1–9.

Shamblin, S.L., 2004. The role of water in physical transformations in freeze-dried products. In *Lyophilization of Biopharmaceuticals*, Eds. Costantio, H.R., and Pikal, M.J., 229–270. Arlington, VA: AAPS Press.

Shamblin, S.L., Tang, X., Chang, L., Hancock, B.C., and Pikal, M.J., 1999. Characterization of the time scales of molecular motion in pharmaceutically important glasses. *J. Phys. Chem. B* 103:4113–4121.

Souillac, P., Middaugh, C., and Rytting, J., 2002a. Investigation of protein/carbohydrate interactions in the dried state. 2. Diffuse reflectance FTIR studies. *Int. J. Pharm.* 235:207–218.

Souillac, P.O., Costantino, H.R., Middaugh, C.R., and Rytting, J.H., 2002b. Investigation of protein/carbohydrate interactions in the dried state. 1. Calorimetric studies. *J. Pharm. Sci.* 91:206–216.

Tang, X., and Pikal, M.J., 2004. Design of freeze-drying processes for pharmaceuticals: Practical advice. *Pharm. Res.* 21:191–200.

Tian, F., Sane, S., and Rytting, J., 2006. Calorimetric investigation of protein/amino acid interactions in the solid state. *Int. J. Pharm.* 310:175–186.

Wang, B., Cicerone, M.T., Aso, Y., and Pikal, M.J., 2010. The impact of thermal treatment on the stability of freeze-dried amorphous pharmaceuticals: II. Aggregation in an IgG1 fusion protein. *J. Pharm. Sci.* 99:683–700.

Wang, B.Q., and Pikal, M.J., 2010. The impact of thermal treatment on the stability of freeze dried amorphous pharmaceuticals: I. Dimer formation in sodium ethacrynate. *J. Pharm. Sci.* 99:663–682.

Wang, B.Q., Tchessalov, S., Cicerone, M.T., Warne, N.W., and Pikal, M.J., 2009a. Impact of sucrose level on storage stability of proteins in freeze-dried solids: II. Correlation of aggregation rate with protein structure and molecular mobility. *J. Pharm. Sci.* 98:3145–3166.

Wang, B.Q., Tchessalov, S., Warne, N.W., and Pikal, M.J., 2009b. Impact of sucrose level on storage stability of proteins in freeze-dried solids: I. Correlation of protein-sugar interaction with native structure preservation. *J. Pharm. Sci.* 98:3131–3144.

Wang, D.Q., Hey, J.M., and Nail, S.L., 2004. Effect of collapse on the stability of freeze-dried recombinant factor VIII and alpha-amylase. *J. Pharm. Sci.* 93:1253–1263.

Wang, W., 1999. Instability, stabilization, and formulation of liquid protein pharmaceuticals. *Int. J. Pharm.* 185:129–188.

Wang, W., 2000. Lyophilization and development of solid protein pharmaceuticals. *Int. J. Pharm.* 203:1–60.

Webb, S.D., Cleland, J.L., Carpenter, J.F., and Randolph, T.W., 2002. A new mechanism for decreasing aggregation of recombinant human interferon-gamma by a surfactant: Slowed dissolution of lyophilized formulations in a solution containing 0.03% polysorbate 20. *J. Pharm. Sci.* 91:543–558.

Wuchner, K., Buchler, J., Spycher, R., Dalmonte, P., and Volkin, D.B., 2010. Development of a microflow digital imaging assay to characterize protein particulates during storage of a high concentration IgG1monoclonal antibody formulation. *J. Pharm. Sci.* 99:3343–3361.

Yoshioka, S., and Aso, Y., 2007. Correlations between molecular mobility and chemical stability during storage of amorphous pharmaceuticals. *J. Pharm. Sci.* 96:960–981.

Yoshioka, S., Miyazaki, T., and Aso, Y., 2006. Beta-relaxation of insulin molecule in lyophilized formulations containing trehalose or dextran as a determinant of chemical reactivity. *Pharm. Res.* 23:961–966.

Yoshioka, S., Miyazaki, T., Aso, Y., and Kawanishi, T., 2007. Significance of local mobility in aggregation of beta-galactosidase lyophilized with trehalose, sucrose or stachyose. *Pharm. Res.* 24:1660–1667.

Yu, Z., Johnston, K.P., and Williams III, R.O., 2006. Spray freezing into liquid versus spray-freeze drying: Influence of atomization on protein aggregation and biological activity. *Eur. J. Pharm. Sci.* 27:9–18.

11 Peptide and Protein Drug Delivery Systems for Nonparenteral Routes of Administration

Ulrik Lytt Rahbek, František Hubálek, and Simon Bjerregaard

CONTENTS

11.1 INTRODUCTION

Although most peptide and protein drugs are dosed by parenteral administration, considerable effort is invested in testing alternative routes of administration, for example, oral, buccal, nasal, pulmonary, rectal, and transdermal. Patient compliance, improved treatment options, and life-cycle extension of patented drugs are the main driving factors. This chapter will describe the difficulties presented by the nonparenteral routes of administration and how they can be overcome. First, the key requirements and considerations of nonparenteral route delivery systems will be presented and, subsequently, the many physical and chemical barriers that need to be overcome will be described. Finally, examples of the delivery strategies utilized today will be presented to give an overview of the current state of the field

11.2 REQUIREMENTS AND CONSIDERATIONS REGARDING CHOICE OF DELIVERY SYSTEMS

During the development of drug delivery systems for alternative administration routes, several factors need to be considered, as highlighted by Mathias and Hussain (2010). Only in relatively few cases can a good match be found between a drug and a drug delivery technology. Some potential considerations are listed below (not in prioritized order).

11.2.1 PHARMACOKINETICS AND PHARMACODYNAMICS

Rate and extent of drug absorption is typically changed by changing administration route. For example, rapid onset of action is typically observed after nasal and pulmonary delivery. This could actually be an advantage in certain acute therapies in which a rapid onset of action is beneficial and potentially improves the therapeutic outcome (Davis and Illum, 2003). Moreover, certain delivery routes might deliver the drug in a more physiologically relevant manner. For example, oral insulin is absorbed from the gastrointestinal (GI) tract into the portal vein (parenteral insulin is absorbed into the systemic circulation), which generates a high portosystemic gradient mimicking the endogenous secretion of insulin. This reduces systemic insulin exposure, as insulin is delivered directly to the liver, which may prevent the excessive weight gain sometimes seen with chronic use of subcutaneous (s.c.) insulin (Mathias and Hussain, 2010). The liver is furthermore a major plasma-glucose–regulating organ; rather than removing glucose from the plasma, it down regulates the glucose output rate. Hence, Arbit and Kidron (2009) suggested that portal insulin delivery may be associated with a reduced risk of hypoglycemia.

A drawback of many alternative drug delivery systems, however, is that they do not offer the same flexibility with respect to dose titration as in the case of s.c. injections. This can be a challenge in therapies where a narrow dose titration is needed.

11.2.2 POTENCY

Low potency drugs, which are dosed in milligram quantities like immunoglobulins, are not likely to be a good match for alternative drug delivery systems. Some administration routes such as nasal and pulmonary delivery have a relative limited dose capacity (<20 mg) (Mathias and Hussain, 2010).

11.2.3 VARIABILITY

Inter- and intrasubject variability with respect to the rate and extent of drug absorption from alternative drug delivery systems is typically higher than that from s.c. administration (Adjei et al., 1992). For example, the nasal bioavailability of salmon calcitonin (Miacalcin Nasal Spray) is 3%, with a range of 0.3–30.6% according to the drug label. Such variability is only acceptable for a drug with a wide therapeutic index. For comparison, s.c. injection of regular insulin under controlled experimental conditions displayed an intraindividual coefficient of variation (CV) of insulin action of 15–25% and an interindividual CV of 20–45% in a study performed by Heinemann (2002).

11.2.4 TOLERABILITY AND SAFETY

Ideally, the effect on local tissue of the drug delivery system is transient and reversible, especially if chronic therapies are considered. Hence, safety of compounds that, for example, alter intestinal permeability will be of some concern, and this subject needs to be addressed during development of an alternative drug delivery system (Aungst, 2000). Little information is generally available on the long-term exposure of functional excipients like permeation enhancers via alternative delivery routes and, consequently, excipients of nonanimal origin that meet monograph specifications (e.g., European Pharmacopoeia or United States Pharmacopoeia/National Formulary) are often preferred. Different excipient/additive compendia or databases can also be an inspiration for a preliminary evaluation of excipient suitability such as the *Handbook of Pharmaceutical Excipients*, the *Handbook of Pharmaceutical Additives*, the FDA Inactive Ingredient List, the WHO Food Additives Series, or the ChemIDplus Advanced. Furthermore, the database SWEDIS (Swedish Drug Information System) also contains a searchable complete composition of all pharmaceutical products on the Swedish market.

11.2.5   Cost of Goods

Most alternative protein and peptide delivery systems are likely to result in a lower bioavailability/biopotency than s.c. administration. Hence, large quantities of pure proteins at reasonable costs should be available even if bioavailabilities of 10–15% in best-case scenarios are achieved.

11.2.6   Delivery Device

Various alternative delivery methods, such as pulmonary, nasal, and transdermal delivery, will require a delivery device, which is the primary interface to the patient. Hence, ease of use will be a determining factor for patient compliance. In one case, patients with diabetes even preferred a pen injection system compared to a pulmonary inhaler (Forst et al., 2009).

11.2.7   Stability

The drug product should have adequate in-use stability, and the development of formulations with good physical and chemical stabilities can be a challenging task (Frokjaer and Otzen, 2005), particularly if the drug product contains reactive compounds/impurities or physical stress is applied during manufacturing or storage. The potential stability and compatibility issues with critical excipients and container/closure systems should therefore ideally be addressed early in pharmaceutical development.

11.3   NONPARENTERAL ROUTES OF DELIVERY

11.3.1   Intranasal Delivery

The nasal cavity is easily accessible and extensively vascularized, and the nasal tissue is highly permeable. Furthermore, the proteolytic activity in the nasal mucosa is relatively limited. Compounds administered via this route are absorbed directly into the systemic circulation, avoiding the hepatic first-pass effect. It is also possible to target the olfactory region from which drugs are able to pass directly into the brain.

Limiting factors for systemic absorption of peptides and proteins from the nasal cavity include the rapid mucociliary clearance and the small absorption surface area of the nasal cavity (approximately 150 cm^2) (Behl et al., 1998). Furthermore, the nasal route is less suitable if permeation enhancers are required due to the risk of nasal irritation. Hence, only permeation enhancers with very high safety profiles would be applicable for nasal drug delivery. Nonetheless, Bentley Pharmaceuticals has attempted to develop of a nasal insulin product (Nasulin) containing a proprietary permeation enhancer, CPE-215 (also known as cyclopentadecalactone). The relative bioavailability of Nasulin compared with s.c. insulin was 14.0–19.8% (Leary et al., 2006).

Despite suffering from low bioavailabilities, several intranasal peptide formulations have been marketed without the use of permeation enhancers. Examples of marketed intranasal peptide products are salmon calcitonin (e.g., Miacalcin®), desmopressin (e.g., Stimate), nafarelin (Synarel), and buserelin (Suprefact®), which all show a bioavailability of approximately 2–3% (Costantino et al., 2007; Mathias and Hussain, 2010).

11.3.2 PULMONARY DELIVERY

Systemic delivery of proteins and peptides via the lung offers the potential for rapid absorption over a large absorptive surface area (approximately 100 m²) through the relatively thin epithelial barrier having moderate absorption capacity and moderate enzymatic activity. Different peptide- and protein-based drugs, such as leuprolide (9 amino acids), insulin (51 amino acids), and growth hormone (191 amino acids), have been shown to reach the systemic circulation after pulmonary administration (Patton and Platz, 1992).

Pulmonary drug delivery systems are drug–medical device combination products. Hence, an inhaler is needed for aerosolization of the drug product. The particle size distribution of the aerosol is a determinant factor for the deposition profile of the drug product in the lungs. The device should ideally generate an aerosol in the 0.5–5-μm mean aerodynamic diameter range (preferably approximately 1–3 μm) to facilitate deep lung deposition, which is considered optimal for systemic absorption of drugs. Furthermore, the aerosol exiting the device should be reproducible and have minimal influence on patient breathing pattern to achieve consistent dosing (Mathias and Hussain, 2010).

11.3.3 TRANSDERMAL DELIVERY

The outer layer of skin or stratum corneum is considered one of the least permeable barriers. Traditionally, low-dose drugs, which are relatively small (<500 g/mol) or moderately lipophilic (log P 1–3) (Benson and Namjoshi, 2008), such as estradiol, have been the ideal candidates for transdermal delivery providing convenient and steady delivery of drugs across intact skin over an extended period. A clear advantage of transdermal drug delivery is that dosing can be stopped simply by removing the delivery system from the skin, provided that no buildup of the drug in the skin has occurred.

Transdermal delivery of relatively large and hydrophilic molecules in therapeutically relevant amounts requires some degree of disruption of the stratum corneum. This can be done by electrical methods (iontophoresis and electroporation), mechanical methods (abrasion, thermal ablation, and perforation), and other energy-related techniques, such as ultrasound and needleless injection. One example of these techniques is thermal ablation, which selectively heats the skin surface to generate micron-scale perforations in the stratum corneum. The microporated area is covered with a patch from which drugs can bypass the stratum corneum and diffuse into the epidermis (Prausnitz and Langer, 2008).

11.3.4 ORAL DELIVERY

Owing to its convenience, the oral route is by far the most widely used route for drug administration and is in general very well accepted by patients, especially for chronic therapies. Unfortunately, oral delivery of macromolecular drugs is technically very complicated due to presystemic enzymatic degradation and poor permeability of the intestinal mucosa to macromolecules as well as first-pass metabolism in the liver. Pharmacokinetic variability with this administration route tends to be high due to gastrointestinal motility/transit time, food interaction, varying pH conditions, and variable amount of GI tract fluid. The low and variable oral bioavailability of macromolecular drugs, however, is more likely to be accepted than other noninvasive drug delivery technologies when taking into consideration the convenience of oral delivery and cost-effective manufacturing. The low permeability, in combination with a limited time for absorption, needs to be addressed by formulation development by including permeation enhancers as well as targeting to specific areas in the GI tract. Furthermore, protein engineering will have to be applied in order to design macromolecules more suitable for oral administration.

An important attribute in enhancing oral bioavailability in vivo is the simultaneous delivery of the drug and effective concentrations of the permeation enhancer to the preferred absorption site in a bolus manner according to Aungst (2000). This is why in vivo studies with local application of drug and enhancer often result in higher bioavailability than after oral dosing, because gastrointestinal motility and volume of fluid present in the GI tract will affect dilution and residence time at any particular site in the GI tract. Hence, the permeation enhancer should ideally remain in contact with the mucosa long enough to achieve maximal effect (Davis and Illum, 2003). Moreover, some targeting of the dosage form is often applied, such as an enteric coating, to prevent premature release in the hostile environment of the stomach. Some degree of controlled release is an advantage according to Bernkop-Schnürch (2010) in order to obtain a compromise between time for intimate contact and a high concentration gradient on the absorptive membrane. Furthermore, small-molecule-based permeation enhancers are quickly absorbed from the GI tract, leaving the protein/peptide behind (Bernkop-Schnürch, 2010).

Within the field of oral delivery, oral delivery of insulin is said to be the Holy Grail. For example, Biocon is currently in phase-3 clinical trials with chemically modified insulin (methoxy-PEG3-propionyl conjugated to the B chain of insulin at position LysB29) in a formulation containing a permeation enhancer (sodium caprate) (Hazra et al., 2010).

11.4 PHYSIOLOGICAL BARRIERS AND ABSORPTION PATHWAYS

A major obstacle to the nonparenteral administration of peptide and protein drugs is the physical barrier presented by the epithelium. Developed by the organism to protect it from invading microorganisms, toxins, and fluid loss, the epithelium restricts access to the systemic circulation. The epithelium consists of cells joined together in epithelial sheets and covers nearly all free surfaces of the human body. An epithelial sheet can be many cells thick as in the epithelium of the mouth and stomach or in the epidermal covering of the skin, or it can consist of a single layer of cells as in the lining of the bronchioles, the nasal cavity, and the intestinal tract. It can furthermore vary in its inherent tightness, ranging from the almost impassable blood–brain barrier to the more leaky epithelia found in the gut. Finally, epithelial cell sheets are polarized and contain an apical surface that is exposed to air, a luminal cavity, or a fluid and a basolateral surface that rests on vascularized connective tissue. The apical and the basolateral surfaces are usually chemically distinct from each other, and this polarization often plays an important physiological role for the epithelial function, which can vary from that of being a simple protective cell lining to having a plethora of biochemical functions, such as hormone secretion, nutrient absorption, and signal transduction (Alberts et al., 1998).

11.4.1 The Mucus Barrier

Most epithelia consist of a number of different cell types with distinct functions. Single-layered epithelia contain mucus-secreting cells such as the goblet cells found in the bronchi, nasal, and intestinal epithelia, whereas the multilayered epithelia of the mouth and stomach contain, or are adjacent to, specialized mucus-secreting glands like the salivary glands. Mucus-covered epithelia are correspondingly referred to as "mucosa." A mucus layer protects the apical cell surface against pathogens and foreign material; in the case of the intestinal mucosa, it also acts by lubricating the intestinal lumen to facilitate efficient passage of digested food. The primary component of mucus is a glycoprotein matrix made from hydrated glycosylated mucin subunits that entrap enzymes, electrolytes, and water. Since approximately 95% of the total mucus volume is water, the mucus layer is often described as an unstirred water layer, although this may be a misnomer due to the many dynamic processes that take place in the local mucus environment such as mucus flow during peristalsis, mechanical mixing of luminal fluids, and continuous mucus secretion by goblet cells as well as mucus sloughing off the epithelial surface.

Before reaching the underlying epithelial cells, a drug molecule must first diffuse across the mucus layer, which can vary in thickness from 10 to 150 μm for the nasal and intestinal mucosa to more than 500 μm for the stomach mucosa (MacAdam, 1993; Pappenheimer, 2001). The mucus layer acts as a filter for molecules with a molecular mass of 600–800 Da, allowing small hydrophilic molecules to rapidly diffuse across it, whereas larger or more hydrophobic molecules diffuse more slowly, if at all (Fricker and Drewe, 1996). The mucus layer may therefore be considered as a significant barrier to the absorption of protein and peptide drugs, since their high molecular weight and hydrophilic nature make them likely to interact with the mucus components that can potentially entrap them in the mucus and prevent them from reaching the underlying epithelia. A recent study analyzing the effects of the mucus layer on the intestinal absorption of a peptide has suggested, however, that the mucus layer mainly acts as an enzymatic barrier due to its high enzyme content and only secondarily as a diffusive barrier in the case of peptides (Aoki et al., 2005). Several research groups and pharmaceutical companies have attempted the use of both mucoadhesive and mucolytic agents to facilitate successful drug delivery across mucosa, and these approaches will be described in more detail below.

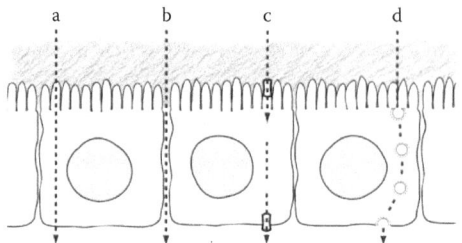

FIGURE 11.1 Absorption pathways through the epithelia. (a) Transcellular route of absorption through the epithelial cells. (b) Paracellular route of absorption between neighboring epithelial cells. (c) Carrier-mediated transport via membrane-associated transporters. (d) Transcytosis via pinocytosis or receptor-mediated endocytosis.

After successful penetration through the mucus layer, a drug molecule will have to cross the epithelial cell layer to gain access to the underlying vasculature in order to ensure systemic circulation. As illustrated in Figure 11.1, this can occur via four pathways: (a) passive transcellular diffusion, (b) passive paracellular diffusion, (c) active carrier-mediated transport, or (d) transcytosis, all of which are described in detail below.

11.4.2 THE TRANSCELLULAR PATHWAY

The phospholipid bilayer of the epithelial cell membrane is remarkably impermeable to most materials; consequently, passive transcellular diffusion normally occurs only for lipid-soluble small molecules. A drug will therefore need the appropriate physicochemical properties in terms of size, charge, hydrogen-bonding potential, and lipophilicity in order to partition through the cell membrane. The macromolecular size, charge, and hydrophilic nature of most peptide and protein drugs prevent them from permeating through the plasma membrane, although some notable exceptions such as cyclosporin A and desmopressin are known to occur (Hamman et al., 2005). One commonly employed strategy to facilitate transcellular absorption is the use of detergent-like absorption enhancers, which, through their solubilization of the cell membrane, increase the transcellular transport of peptides. This will be described in detail below.

11.4.3 THE PARACELLULAR PATHWAY

The hydrophilic nature of peptides and proteins makes them much more likely to be passively absorbed through the aqueous pores between the epithelial cells. Their macromolecular nature, however, will often prevent them from entering the intercellular space due to the presence of tight junctions. Tight junctions keep neighboring epithelial cells tightly bound to each other and act as a primary barrier to the diffusion of solutes through the intercellular space. Furthermore, tight junctions serve an important fence-like function in maintaining the polarity of the epithelial cells by restricting the free diffusion of receptors and transporters located on either the apical or the basolateral membranes (Tsukita et al., 2001). Several groups of trans-membrane proteins whose cytoplasmic terminals are anchored to the cytoskeleton are the structural components of tight junctions. The extracellular regions of these transmembrane proteins can interact with similar proteins on the adjacent cell, resulting in junctional complexes that seal the two cells together. The resulting tight junctions contain water-filled pores with estimated dimensions of between 0.3 and 1 nm, thus allowing water and small hydrophilic molecules to permeate through them, but precluding molecules with hydrodynamic radii of greater than 1 nm from being passively absorbed through the paracellular route (Hamman et al., 2005). Originally believed to be very static structures and "the cement that glues epithelial sheets together," tight junctions are now well recognized as very dynamic in nature and tight junction barrier integrity is regulated by a wide variety of both intracellular and extracellular factors (Deli, 2008; Johnson et al., 2008). Recent advances in tight junction barrier research have led to the discovery of several modulators of tight junction function with absorption-enhancing properties, and some examples of these are presented in Table 11.1.

TABLE 11.1
Classes of Permeation Enhancers

Salicylates

For example, sodium salicylate (Kajii et al., 1986; Nishihata et al., 1984)

N-acetylated α-amino acids

For example, N-[8-(2-hydroxybenzoyl)amino]caprylate (SNAC) (Leone-Bay et al., 1998)

Fatty acids and fatty acid esters

Medium chain fatty acids (e.g., sodium caprate) (Lindmark et al., 1998; Tomita et al., 1995)

Medium-chain mono- and diglycerides (Shima et al., 1998)

Acylcarnitines (LeCluyse et al., 1991; Requero et al., 1995)

Steroidal surfactants

Bile salts and bile salt derivates (e.g., sodium taurocholate, sodium taurodeoxycholate, and sodium taurodihydrofusidate)

Saponins

Synthetic surfactants

For example, sodium dodecyl sulfate and tetradecylmaltoside (Anderberg et al., 1992; Swenson et al., 1994)

Chelating agents

ethylenediaminetetraacetic acid (EDTA) and ethylene glycol tetraacetic acid (EGTA) (Raiman et al., 2003)

Peptide-based tight junction modulators

For example, zonula occludens toxin, AT-1002, Ac-ADTPPV-NH2 (6-mer peptide corresponding to the bulge in E-cadherin EC-1 domain), and C-terminal peptide of *Clostridium perfringens* enterotoxin (Daugherty and Mrsny, 1999; Deli, 2008; Fasano, 1998)

Cell-penetrating peptides

Short peptides with a high relative abundance of positively charged amino acids (e.g., penetratin, oligo-arginine, and transportan) (Lundberg et al., 2003)

Cationic polymers

Chitosan salts and derivates (e.g., N-trimethyl chitosan) (Kotze et al., 1999)

11.4.4   Carrier-Mediated Transport

In addition to passive diffusion, several active transport systems exist in epithelial cells. These active carriers ensure that some materials of benefit to the organism can be selectively transported across the epithelial barrier, whereas others are kept out. Two of the carriers that are of particular relevance for the delivery of peptide and protein drugs are the well-characterized peptide transporters PepT1 and PepT2, which are highly expressed on the apical membrane of the small intestine and are also found in high numbers in the lung and kidney epithelium (Rubio-Aliaga and Daniel, 2002). Their natural substrates are di- and tripeptides; however, they have also been found to be involved in the uptake of peptidomimetics (Walter et al., 1996). Several dipeptide prodrug derivatization strategies (i.e., acyclovir) have resulted in increased bioavailability. PepT1 and PepT2 are not likely to enable the membrane transport of larger peptides or proteins, however, due to the size limit on the pore that can be opened by the conformational change of the transporters (Washington et al., 2001). This size restriction is a general limitation of many epithelial carriers whose primary role is to transport small or premetabolized substrates across the cell membranes.

11.4.5   Transcytosis

One strategy that takes advantage of the presence of the solute carriers present on the epithelial membrane is the coupling of a carrier substrate to the drug macromolecule of interest. Although this will not necessarily cause carrier-mediated uptake due to the size limitation described above, the interaction between substrate and carrier will ensure that the coupled macromolecule is present in the immediate vicinity of the epithelial cell and can therefore potentially be taken up by pinocytosis. Pinocytosis is an unspecific form of endocytosis, in which the epithelial cell engulfs a part of its own plasma membrane along with a small amount of extracellular fluid, creating a vesicle that is trafficked into the interior of the cell along with its contents (Conner and Schmid, 2003). Another internalization process in epithelial cells is the process of receptor-mediated endocytosis, where the binding of a ligand to its receptor causes the ligand-bound receptor to be internalized via a membrane invagination, which is subsequently pinched off to form an intracellular endosomal vesicle. The rate of uptake through this process can be many orders of magnitude higher than would be expected from bulk phase uptake or pinocytosis. Endocytosis and pinocytosis processes are able to internalize large peptides and proteins; however, after uptake the vesicles undergo a complex process of endosomal sorting, which decides the fate of the endosomal contents, most of which ends up being degraded in the lysosomes (Conner and Schmid, 2003). If the vesicle avoids lysosomal degradation or intracellular sequestration, it can fuse with the basolateral membrane of the epithelial cell and release its contents for systemic distribution. Despite the limited control over intracellular vesicle trafficking, it has been shown that it is possible to utilize the process of transcytosis to increase the uptake of peptide and protein drugs, and examples of this will be presented below.

11.4.6 M CELLS

Some subtypes of mucosa (i.e., nasal and intestinal) contain lymphoid-associated tissue that is involved in the sampling of particulates originating from outside of the organism in order to present them to underlying lymphocytes, macrophages, and dendritic cells. In the intestinal epithelia, a subpopulation of cells known as M cells (also known as microfold or membraneous cells) is associated with the lymphoid tissue of the Peyer's patches, which are found in both the small and large intestine. The M cells have proved instrumental in initiating mucosal immunity against pathogens and viruses invading across epithelial barriers by facilitating the uptake and transport of particulate luminal antigens to underlying immune response-stimulating cells (Foster and Hirst, 2005). M cells are not covered by mucus and can therefore directly sample antigens from the intestinal lumen. Early research suggested that the Peyer's patches could be a potential route of entry for oral peptides and proteins in microparticles; however, it was subsequently found that their low uptake capacity prevented therapeutic levels of the drugs from being obtained (Brayden et al., 2005). Despite this shortcoming, it is currently believed that the Peyer's patches present a prime target for the delivery of oral vaccines in the form of either antigens or gene therapeutics since only a small amount of uptake of such an agent might be required to induce a memory-driven immune response (Brayden et al., 2005).

11.4.7 EFFLUX PUMPS

Besides the four main barriers described above, several epithelia express efflux pumps that can have a major influence on the absorption of drug molecules. Located primarily on the apical plasma membrane, the efflux pumps actively extrude endogenous substrates, toxins, and small molecules from the interior of the cell, but also macromolecular drugs can be exported. The most well-characterized efflux pump is the P-glycoprotein (P-gp), a member of the multidrug resistance protein family (MRP) that consists of ATP-dependent transporters that excrete organic molecules from cells. P-gp is highly expressed along the small and large intestinal epithelium, where it restricts the bioavailability of many orally administered drugs. The substrates of most efflux pumps like P-gp are generally uncharged and hydrophobic, and range in size from 200 to 1900 Da, which excludes the vast majority of macromolecular peptide and protein drugs; however, cyclosporine is a well-known exception that has been shown to be a substrate for P-gp (Schinkel and Jonker, 2003; Werle, 2008).

11.5 CHEMICAL BARRIERS

Protein/peptide drugs need to overcome chemical barriers in order to be effectively absorbed into the circulation. The two main barriers, the pH and the enzymatic machinery present in the GI tract, are described in more detail below. Although the enzymatic barrier of the GI tract is most active, proteolytic enzymes will also be encountered by the drugs delivered by other nonparenteral routes of administration.

11.5.1 pH

Drugs are exposed to a gradient of pH from neutral in the mouth to acidic in the stomach and then again to neutral pH in the intestine. The exact pH values for each compartment are not constant, but are influenced by species, time of day, metabolic state, disease state, and so on (Dressman et al., 2007; Kararli, 1995; McConnell et al., 2008). As pH is one of the crucial parameters defining solubility of both the active compounds and the excipients used in the final dosage form, it is important to deliver the drug to the compartment of choice in such a form that it can be solubilized and/or remain in solution. Furthermore, exposure to low pH conditions potentially leads to chemical modifications of proteins/peptides by, for example, deamidation of aparaginyl and glutaminyl residues, modification of tryptophanyl residues, or hydrolysis of peptide bonds. Aspartyl–prolyl peptide bonds are most prone to such hydrolysis (Inglis, 1983).

11.5.2 THE ENZYMATIC MACHINERY

The GI tract has evolved over the ages to break down as many proteins, carbohydrates, nucleic acids, and lipids as possible to derive the largest possible nutritional value from ingested food. As a result, powerful enzymatic machinery has developed to tackle these tasks. The digestive capacity of the GI tract differs from species to species. It is influenced by many factors, such as time of day, metabolic state, and disease state, even for a given species. The remainder of this chapter will focus on enzymes involved in protein/peptide degradation, since these are most relevant for degradation of protein/peptide drugs. However, enzymes responsible for lipid, nucleic acid, and carbohydrate metabolism can be involved in the degradation of inactive ingredients of the final dosage forms.

The stomach and intestine have by far the highest proteolytic enzyme activity. The stomach contains predominantly pepsin and gastricsin isoforms that have an acidic pH optimum (pH ~3) for protein digestion (Barrett et al., 2004). In contrast, the intestine possesses a variety of enzymes from several sources: pancreatic enzymes, enzymes associated with the epithelial cell layer (brush border enzymes), cellular enzymes released from the shed enterocytes or encountered during transcellular passage, and enzymes originating from bacteria living symbiotically in the intestine.

Enzymes secreted by the pancreas are produced by the exocrine pancreas as inactive zymogens and released into duodenum, where they are activated by the action of enteropeptidase and trypsin. Pancreatic secretions include mainly trypsin, chymotrypsin, elastase, and carboxypeptidases A and B (Table 11.2), and are regulated by gut-derived hormones, mainly cholecystokinin, in response to distension and/or the presence of food (Lewis and Williams, 1990). Consequently, amounts of pancreatic enzymes in the small intestine are several-fold higher in the fed state than in the fasting state. It was previously estimated that the pancreas secretes up to 50 g of proteins per day (Kukral et al., 1965).

Brush border enzymes are associated with the luminal side of the apical membrane of intestinal epithelial cells. These include both endopeptidases and exopeptidases (Table 11.2). In contrast to pancreatic enzymes, which act mainly on intact proteins and/or larger protein fragments, brush border enzymes act almost exclusively on smaller peptides. Furthermore, the amounts of these enzymes present in intestine are more or less constant, as they are not regulated by food intake to the same degree as the pancreatic enzymes.

TABLE 11.2
Substrate Specificity of Proteolytic Enzymes Commonly Found in Intestine

Enzyme	Site of Cleavage	Specificity (X =)
Luminally secreted protease		
Trypsin	H₂N-O— O— X↓— O— O-COOH	Arg, Lys
Chymotrypsin	H₂N-O— O— X↓— O— O-COOH	Phe, Tyr
Elastase	H₂N-O— O— X↓— O— O-COOH	Ala, Gly, Ile, Leu, Val
Carboxypeptidase A	H₂N-O— O— O— O— ↓X-COOH	Tyr, Phe, Ile, Thr, Glu, His, Ala
Carboxypeptidase B	H₂N-O— O— O— O— ↓X-COOH	Lys, Arg
Membrane-bound protease		
Aminopeptidase A	H₂N-X↓— O— O— O— O-COOH	Asp, Glu
Aminopeptidase N	H₂N-X↓— O— O— O— O-COOH	Many, but especially Ala, Leu
Aminopeptidase P	H₂N-X↓— O— O— O— O-COOH	Pro
Aminopeptidase W	H₂N-X↓— O— O— O— O-COOH	Trp, Tyr, Phe
γ-Glutamyl transpeptidase	H₂N-X↓— O— O— O— O-COOH	γ-Glutamic acid
Dipeptidyl peptidase IV	H₂N-O— X↓— O— O— O-COOH	Pro, Ala
Peptidylpeptidase A	H₂N-O— O— O— ↓X— X-COOH	His-Leu
Carboxypeptidase M	H₂N-O— O— O— O— ↓X-COOH	Lys, Arg
Carboxypeptidase P	H₂N-O— O— O— O— ↓X-COOH	Pro, Gly, Ala
γ-Glutamyl carboxypeptidase	H₂N-O— O— O— O— ↓X-COOH	γ-Glutamic acid
Endopeptidase-24.11	H₂N-O— O— ↓X— O— O-COOH	Hydrophobic amino acids
Endopeptidase-24.18	H₂N-O— ↓X↓— O— O-COOH	Aromatic amino acids
Enteropeptidase	H₂N-O— O— X↓— O— O-COOH	(Asp)-Lys

Source: Woodley, J.F., *Crit. Rev. Ther. Drug Carrier Syst.* 11, 61, 1994.

The spatial distribution of the enzymes in the intestine is not even. The highest concentration of both pancreatic enzymes and brush border enzymes is generally found in the duodenum and proximal jejunum, and is lower in the ileum. The colon exhibits only low levels of enzymatic activity by brush border enzymes and/or pancreatic enzymes. On the other hand, the concentration of bacteria and consequently also of enzymes released from these bacteria is the highest in colon.

For transcellular transport of proteins/peptides, enzymes from different cellular compartments might be involved in degradation. Lysosomal cathepsins are the most studied enzymes from this group.

11.6 STRATEGIES TO OVERCOME THE BARRIERS

In general terms, successful protein/peptide drug formulation ensures that sufficient amounts of the active compound survive the harsh environment of the GI tract and at the same time enhance absorption across the epithelial layer into circulation. This can be achieved by modifications of the protein/peptide itself and/or by delivering the protein/peptide in a formulation that protects the active compound from degradation in the GI tract.

11.6.1 Stabilization of Active Compounds

Stabilization of the active compound toward proteolytic degradation in the GI tract is commonly used to enhance oral bioavailability of protein/peptide drugs. This can be accomplished by substituting weak parts of the protein structure, such as amino acids that are substrates of endopeptidases (i.e., the cleavage hot spots), and/or stabilizing the N- and C-termini toward degradation by exopeptidases. Protein structure plays a very important role in determining which of the theoretical hot spots are actually cleaved by the enzymes. For example, cysteine-knot protein containing eight theoretical sites for trypsin cleavages was found relatively stable toward trypsin degradation, while a single theoretical cleavage site of chymotrypsin was sufficient to allow degradation by this enzyme (Werle et al., 2008). It is therefore beneficial to identify the degradation hot spots before introducing substitutions into the proteins. Hot spots can be identified by incubation of the active compound with purified enzymes (such as those described in Table 11.2) or various tissue extracts. Pancreatin, pancreatic extract containing all the enzymes secreted by pancreas, is commercially available and also described by the U.S. Pharmacopeia as an ingredient of simulated intestinal fluid. However, the pancreatic enzymes must be activated by enteropeptidase and/or trypsin before they reach full activity. Animal intestinal extracts are perhaps better suited for identification of hot spots, as these also contain brush border enzymes and bacterial enzymes. Although good qualitative results can be obtained by the methods described above, it is difficult to get the correct concentrations for all relevant enzymes to simulate the real in vivo situation. The following examples are not meant to give a comprehensive review of the field, but rather illustrate the different ways of protein/peptide stabilization.

Oral bioavailability of methionine enkephaline (YGGFM) is negligible. It can, however, be improved by N-methylation and/or by sulfoxide modification of the C-terminal methionine residue (Roemer et al., 1977; Su et al., 1985). Bioavailability of vasopressin (CYFQNCPRG, containing a disulfide bond between the two cysteine residues) was improved approximately 150-fold by removing the N-terminal amino group and replacing the arginine residue with D-arginine (Vilhardt and Lundin, 1986). Caco-2 cell transport of desmopressin esterified by pivalate was significantly enhanced relative to desmopressin alone; this prodrug of desmopressin was then quantitatively converted to desmopressin in human plasma (Kahns et al., 1993). Protection of the N- and C-termini of proteins can also be achieved by cyclization as demonstrated, for example, by Andersen et al. for insulin. Sensitivity of insulin toward carboxypeptidase degradation was completely abrogated by cyclization (Andersen et al., 2010). Cyclization of protein/peptides influences not only enzymatic stability but also absorption as shown, for example, by Pauletti et al. for a model hexapeptide; bioreversible cyclization of this hexapeptide based on ester prodrug improved Caco-2 cell penetration by a factor of 70 (Pauletti et al., 1997).

Another strategy for protein stabilization toward proteolytic degradation is to modify proteins by fatty acids and/or polyethyleneglycol (PEG) chains. For example, pegylation of interferon-α2b with PEG 40k greatly improves its stability toward trypsin (Ramon et al., 2005). Although the proteolytic stability can be significantly improved by introduction of the large PEG chains, their utility for oral administration is questionable due to their large hydrodynamic radii. Hexyl insulin monoconjugates, containing a small PEG chain (7 PEG units), were developed by Nobex Corporation to improve enzymatic degradation stability and to facilitate absorption by direct modification of the ε-amino group of the B29 lysine residue (Myers et al., 1997). The usefulness of this insulin analog for oral absorption has been tested in clinical trials (Clement et al., 2004; Kipnes et al., 2003). Similar results have also been reported for pegylated salmon calcitonin (Mansoor et al., 2005). Similarly, modification of insulin by 1,3-dipalmitoylglycerol results in improved stability toward degradation and also higher absorption from the large intestine (Hashimoto et al., 1989).

Many of these strategies are likely limited to synthetic peptides where chemose-lective reactions can be performed. Furthermore, although chemical modifications provide a powerful tool for improvement of enzymatic stability and/or absorption across membranes, the resulting compounds will very likely have modified biologi-cal properties such as potency, half-lives in circulation, clearance, immunogenicity, and toxicity, and the production costs will be significantly increased.

11.6.2 Use of Permeation Enhancers

As previously described, permeation enhancers may act through a transcellular and/ or a paracellular transport-enhancing mechanism. The transcellular pathway may be achieved by perturbation of the cell membrane either by complete solubiliza-tion of the cell membrane (irreversible) or by fluidization (reversible) of the lipid cell membrane. Solubilization is achieved above the critical micelle concentration of surfactants. During solubilization, the surfactants will partition into the membrane. The phospholipid membrane can accommodate surfactants up to a saturation point, after which the membrane will start to solubilize at an increasing surtactant/lipid ratio. This process was studied by simple light scattering studies after tritration of phospholipid vesicles with surfactants (Requero et al., 1995).

The fluidization phenomenon can be studied with model lipid membranes composed of pure phospholipids. Permeation enhancers, which stabilize the fluid phase, will reduce the transition temperature (melting point) of the phospholipids (see Figure 11.2), which can be measured by differential scanning calorimetry. One example of this class of permeation enhancers is the Emisphere proprietary com-pound N-[8-(2-hydroxybenzoyl)amino]caprylate (SNAC), which fluidizes rather than solubilizes cell membranes (Alani and Robinson, 2008; Leone-Bay et al., 1998).

The paracellular pathway can be achieved by modulating the tight junctions between the cells. However, many permeation enhancers may act on both pathways depending on the concentration of permeation enhancer. One example is sodium cap-rate, which dilates intestinal epithelial tight junctions at a concentration of 13–16 mM via activation of phospholipase C, leading to an increased intracellular calcium con-centration and contraction of the perijunctional actomyosin ring (Lindmark et al., 1998; Tomita et al., 1995). However, when caprate was incorporated at higher con-centrations into a rectal formulation containing ampicillin, there was some evidence that the paracellular route enhancement might not be the dominant mechanism at concentrations used in vivo (Leonard et al., 2006). Hence, it is more meaningful to classify permeation enhancers through their structure similarity rather than by

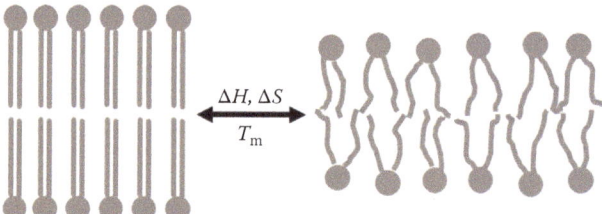

$\Delta H, \Delta S$
T_m

FIGURE 11.2   Phase behavior of phospholipid membranes above (liquid crystalline/fluid state) and below (gel state) the main transition temperature.

action. Various useful reviews of permeation enhancers can be found in the literature (Aungst, 2000; Bernkop-Schnürch, 2010; Hamman et al., 2005; Swenson and Curatolo, 1992; Whitehead and Mitragotri, 2008; Whitehead et al., 2008).

11.6.3   MUCOADHESION AND MUCOLYTICS

The mucus can potentially prevent peptide and protein drugs from being presented to the underlying epithelial membrane and thus prevent absorption. To overcome the mucus barrier, several strategies have been utilized.

The uptake of peptide and protein drugs is often limited by the short contact time between the drugs and the biological membrane across which they need to be absorbed. One method developed to retain the dosage form at the site of action is the use of mucoadhesion, which is based on the presence of mucus on several types of epithelia. In addition to decreasing the transit time of a drug delivery system, a further advantage of mucoadhesion is to ensure that the drug is released in proximity to the mucus-covered epithelia across which it needs to be absorbed. As previously described, the mucus layer consists mainly of water and mucins, which are hydrated macromolecular glycoproteins containing substructures linked by disulphide bonds. In addition, sialic acids and sulfates are often present in the mucus, resulting in an overall negative charge of the mucus layer (Edsman and Hagerstrom, 2005). Consequently, a well-established group of mucoadhesive agents are positively charged polymers like the cationic polysaccharide chitosan, which is able to bind the mucus layer via ionic interactions (Schipper et al., 1999). Another mucoadhesion strategy is the use of readily water-swellable poly(acrylates) known as carbomers, which adhere to mucosal surfaces by attracting water from the mucus layer and via hydrogen bonding at the molecular level (Lehr et al., 1992). A novel strategy has been the inclusion of thiol groups in the backbone of mucoadhesive polymers that facilitate interactions with the cysteine-rich subdomains of mucus. By forming disulfide bonds with the structural components of mucus, such thiomers are believed to be superior to existing mucoadhesive polymers whose function is based on ionic interactions or hydrogen bonding alone, since the thiomer adhesion due to covalent bonds is much stronger (Bernkop-Schnurch, 2005). However, the improved retention time of such polymers will, nevertheless, be limited by the natural mucus turnover.

A strategy diametrically opposed to the use of mucoadhesion is the use of mucolytic agents to remove the diffusion barrier presented by the mucus layer and thereby facilitate easy access of the peptide drugs to the underlying epithelia. Several different types of mucolytic agents have been tested, including mucolytic detergents like Triton X-100 and Tween 20 and more efficient mucolytic proteases like pronase and papain (Bernkop-Schnurch et al., 1999). The proteases were unfortunately also found to result in a rapid degradation of the coadministered peptide drug, suggesting that stabilization of the protein or peptide drug is a necessary requirement for the use of this type of mucolytics. N-acetyl cysteine, which is widely used for clinical medication and is found in several types of cough syrup due to its mucus-liquefying action, has recently been used to enhance the absorption of poorly absorbed hydrophilic compounds (Takatsuka et al., 2006). It acts by breaking the disulfide bonds that keep the mucus together, however, and this could consequently complicate its use with peptide or protein drugs containing one or more disulfides.

11.6.4 ACTIVE TRANSPORTERS

Several biotech companies utilize the natural transport systems present on epithelial surfaces to enhance the transepithelial transport of peptide and protein drugs. Due to the macromolecular size requirement, the strategy has often been to harness receptor-mediated endocytosis to facilitate transcytosis of the protein drug of choice. An example of such a receptor target is the neonatal Fc receptor (FcRn), which, in the placenta, is responsible for the transport of immunoglobulin G (IgG) from mother to fetus (Rubio-Aliaga and Daniel, 2002). FcRn binds the Fc fragment of IgG with high affinity at a slightly acidic pH but not at a physiological pH, which facilitates unidirectional transport of IgG. At the apical surface of an epithelial cell, FcRn binds IgG and transcytoses bound IgG to the basolateral membrane, where it is released into the extracellular space that is at a physiological pH (Roopenian and Akilesh, 2007). The epithelia of human lung and intestine have also been found to express FcRn; in a phase 1 clinical trial evaluating pulmonary delivery of an Fc-erythropoietin fusion protein, it was demonstrated that macromolecular therapeutic peptides and proteins can be delivered to humans via the lung, while retaining biological activity (Dumont et al., 2005). Furthermore, a proof-of-concept study in rats has demonstrated successful oral delivery of an Fc-follicle-stimulating hormone fusion protein (Low et al., 2005); however, it remains to be seen if this transcytosis pathway can be utilized in the human intestine.

Another pathway that has been targeted for drug delivery is the vitamin B_{12} (B_{12}) uptake pathway (Russell-Jones, 2001). B_{12} is synthesized by bacteria and is an essential nutrient for all animals. In humans, B_{12} is released from food by proteases in the stomach and subsequently bound and protected from degradation by a glycoprotein residing in the stomach. When the complex reaches the upper part of the small intestine, this glycoprotein is degraded by intestinal proteases and the released B_{12} is bound by a second glycoprotein and transported to the distal part of the small intestine. Here, the complex binds to the B_{12} receptor (cubulin) and B_{12} is transcytosed across the epithelia (Russell-Jones, 2001). The use of this pathway to deliver pharmaceuticals has received attention since the 1970s; however, earlier attempts were largely unsuccessful (Petrus et al., 2009). Not surprisingly, it was found that the conjugation site between B_{12} and the drug molecule is critically important to ensure that neither molecule obstructs the other (Petrus et al., 2009). Recent studies using

B_{12}-conjugated drugs have shown that it is in fact possible to utilize this pathway for the delivery of peptides and protein drugs; however, the lesser amount of B_{12} receptors sets the higher limit on the amount of drug molecules that can be absorbed via this route (Petrus et al., 2009). One strategy that has been successfully used to overcome this limitation is the use of B_{12}-coated nanoparticles, which when loaded with insulin caused a 70–75% reduction in plasma glucose (Chalasani et al., 2007).

11.6.5 NANOCARRIERS

Nanocarriers, or structures with a size range of 10–1000 nm, are increasingly being investigated for both pulmonary and oral drug delivery of proteins and peptides (des Rieux et al., 2006). The nanocarriers may partially protect the encapsulated drug from degradation by enzymes and improve cellular uptake via endocytosis. These nanocarriers include diverse systems like polymeric or lipid-based carriers such as liposomes, micelles, dendrimers, and nanoparticles, including nanospheres, nano-tubes, and nanocapsules. A wide variety of polymers, such as chitosan, albumin, alginate, poly-(lactic-co-glycolic acid), and poly(alkyl methacrylate), have been used for preparation of nanoparticles.

Factors such as size, shape, surface charge, and hydrophobicity will affect the intestinal absorption characteristics. Hence, enterocytes are able to endocytose nanoparticles in the size range of 50–100 nm, whereas the more specialized, but quantitatively fewer, M cells in the Peyer's patches are able to endocytose particles up to 5 µm (des Rieux et al., 2006). However, as a rule of thumb, particle sizes of less than 500 nm are required for internalization into the epithelia of the GI tract (Jung et al., 2000). With respect to surface charge, it has been claimed that positively charged nanoparticles have a higher ability than neutral and hydrophobic particles to cross the mucus layer and interact favorably with the negatively charged surface of the epithelial cells (Roger et al., 2010). Often, nanoparticles are functionalized by cell surface receptor ligands (e.g., vitamin B_{12}, Fc fragments, lectins, and ganglio-sides) in order to increase efficacy of uptake.

As previously described, different modes of cellular internalization of nanopar-ticles exist, the most predominant of which is the process of receptor-mediated endo-cytosis. After being endocytosed, nanocarriers have to be transcytosed or escape the endosomes to be able to enter the lymphatic or blood capillaries. This process might be directed with proper understanding of cellular trafficking of endocytosed material. For example, it has been claimed that a certain type of endosomes known as caveolae might avoid intracellular degradation and deliver their contents partly on the basolateral side of the enterocytes (Richard et al., 2003). However, in general, the capacity of transcytosis is considered very limited with respect to systemic delivery of drugs (Jung et al., 2000).

11.6.6 CELL-PENETRATING PEPTIDES

Cell-penetrating peptides (CPPs) are short peptides (<30 amino acids) derived from both natural and unnatural proteins facilitating a transport of a cargo to a cell (Table 11.3). A considerable body of knowledge about CPPs has been generated since

TABLE 11.3
Examples of Cell-Penetrating Enzymes

Peptides	Origin	Sequences	Cargo Types	References
Peptides derived from protein transduction domains				
Tat	HIV-Tat protein	PGRKKRRQRRPPQ	Protein/peptide/siRNA	Snyder and Dowdy (2005)
			Liposome/nanoparticle	Schwarze et al. (1999)
Penetratin	Homeodomain	RQIKIWFQNRRMKWKK	Peptide/siRNA/liposome	Joliot and Prochiantz (2004)
Transportan	Galanin-mastoparan	GWTLNSAGYLLGKINLKALAALAKKIL	Protein/PNA/siRNA	Pooga et al. (1998)
VP-22	HSV-1 structural protein	DAATATRGRSAASRPTERPRAPAR SASRPRRPVD	Protein	Elliott and O'Hare (1997)
Amphipathic peptides				
MPG	HIV Gp41-SV40 NLS	GALFLGFLGAAGSTMGAWSQPKKKRKV	siRNA/ODN/plasmid	Morris et al. (2008)
Pep-1	Trp-rich motif-SV40	NLS KETWWETWWTEWSQPKKKRKV	Protein/peptide	Gros et al. (2006)
MAP	Chimeric	KALAKALAKALA	Small molecule/plasmid	
SAP	Proline-rich motif	VRLPPPVRLPPPVRLPPP	Protein/peptide	Pujals et al. (2006)
PPTG1	Chimeric	GLFRALLRLLRSLWRLLLRA	Plasmid	Rittner et al. (2002)
Other cell-penetrating peptides: cationic peptides				
Oligoarginine	Chimeric	Arg8 or Arg9	Protein/peptide/siRNA/ ODN	Futaki et al. (2001), Wender et al. (2000)
hCT (9–32)	Human calcitonin	LGTYTQDFNKTFPQTAIGVGAP	Protein/plasmid DNA	Schmidt et al. (1998)
SynB	Protegrin	RGGRLSYSRRRFSTSTGR	Doxorubicin	Rousselle et al. (2001)
Pvec	Murine VE-cadherin	LLIILRRRIRKQAHAHSK	Protein/peptide	Elmquist et al. (2001)

Note: CPP, cell-penetrating peptide; NLS, nuclear localization sequence; PNA, peptide-nucleic acid; Tat, transcription-transactivating.

a surprisingly high uptake of HIV's *trans*-activating transcriptional activator (Tat) by culture cells was described in 1988 (Frankel and Pabo, 1988; Green and Loewenstein, 1988). There are two basic types of CPPs: the first type requires a covalent attachment of the cargo, while the second type is represented by short amphipathic peptides forming noncovalent complexes with its cargo during transmembrane transport (Morris et al., 2008). Endocytic pathways seem to be involved in the cellular uptake of CPPs, but many controversial reports have been published (Richard et al., 2003). Interested readers are referred to recent comprehensive reviews of the CPPs for further details (Edenhofer, 2008; Heitz et al, 2009).

While CPPs hold promise in many therapeutic areas for their efficient transfer into cells and lack of toxicity, their application for systemic administration also depends on the release from the cell to systemic circulation. An improvement (six- to eightfold) in absorption across Caco-2 monolayer for insulin-TAT conjugate has been reported (Liang and Yang, 2005), and similar results have been published after coadministration of insulin with oligoarginine in rats (Morishita et al., 2007).

11.7   CASE EXAMPLE OF INHALED INSULIN

The example of inhaled insulin development highlights the difficulties noninvasive delivery systems are faced with. The first inhaled insulin product to reach the market was Exubera, which was available in some European countries and on the U.S. market from September 2006 to October 2007. Exubera was manufactured as a spray-dried powder containing 60% insulin in a amorphous glassy matrix composed of mannitol, glycine, and citrate, resulting in room temperature stability of insulin (Weers et al., 2007; White et al., 2005). No permeation enhancers were used in the formulation. The insulin powder was packaged into single-dose alu blisters containing 1 or 3 mg (approximately 28 and 84 units), which was equivalent to approximately 3 and 8 units of s.c. fast-acting insulin, resulting in a relative bioavailability of approximately 10%. Dosing was based on titration in relation to glucose response and would require multiple inhalations if doses other than 3 and 8 units were required (Odegard and Capoccia, 2005)

Dosing was performed with a dry-powder inhaler, which dispersed the powder as an aerosol cloud into a 16-cm holding chamber (spacer) using compressed air (Figure 11.3). The patient then inhaled the aerosolized powder with a mass median aerodynamic diameter close to 3 μm by taking a slow deep breath (Becquemin and Chaumuzeau, 2010; Odegard and Capoccia, 2005). The development of the first-in-class Exubera platform presented significant challenges and consequently required a significant amount of resources within areas of formulation development, powder production, powder filling technology, and device development (Becquemin and Chaumuzeau, 2010).

The Exubera insulin was absorbed more quickly than subcutaneously administered regular human insulin in healthy subjects as shown by Rave et al. (2005). Hence, 6 mg of insulin (two 3-mg doses of Exubera) or s.c. injection of 18 units regular insulin resulted in a t_{max} of 55 and 148 minutes, respectively.

Although inhaled insulin provided a prandial metabolic control comparable to s.c. regular insulin (Heinemann, 2004), many diabetologists were skeptical about the

FIGURE 11.3   The Exubera™ inhaler.

need/advantages of inhaled insulin from the start (Heinemann et al., 2001). Moreover, some side effects have been associated with Exubera, such as a transient decrease in forced expiratory volume, a transient mild-to-moderate cough, and increased level of nonneutralizing insulin antibodies. The use of Exubera was also associated with lung cancer risk (Odegard and Capoccia, 2005; Wong, 2010).

One goal of treating patients with inhaled insulin was to make the insulin administration more convenient and to reduce the barrier to insulin treatment, allowing earlier intervention of type 2 diabetes with insulin therapy. The eventual termination of Exubera manufacture has been ascribed primarily to its failure to gain acceptance among patients and physicians. The failure of Exubera caused a cessation of nearly all other attempts to develop inhaled insulin formulations. Currently, there is only one company (MannKind) that continues with their Technosphere® insulin (Afrezza™). This technology is based on a carrier molecule (diketopiperazine) that gives an exceptional fast absorption of insulin from the lungs. Peak insulin plasma concentrations are observed 12–14 minutes after inhalation. According to Heinemann (2004), it is possible that postprandial glycaemic excursions are considerably better controlled with Technosphere insulin.

11.8   CONCLUDING REMARKS

Overcoming the multiple barriers presented by the human organism to nonparenteral delivery of peptide and protein drugs has so far proven very difficult. From the many unsuccessful attempts by both university researchers and large pharmaceutical companies, it has become clear that an interdisciplinary approach will be essential for the development of nonparenteral delivery systems. Only by combining an in-depth understanding of biological processes, formulation design, and protein engineering will the future delivery systems be able to meet the requirements for nonparenteral administration of peptide and protein drugs. Recent advances in the above-mentioned areas should, within a couple of years, enable the use of these types of delivery systems for improved patient compliance and, importantly, ensure a higher therapeutic efficacy of the administered drugs.

REFERENCES

Adjei, A., Sundberg, D., Miller, J., and Chun, A., 1992. Bioavailability of leuprolide acetate following nasal and inhalation delivery to rats and healthy humans. *Pharm. Res.* 9:244–249.

Alani, A.W.G., and Robinson, J.R., 2008. Mechanistic understanding of oral drug absorption enhancement of cromolyn sodium by an amino acid derivative. *Pharm. Res.* 25:48–54.

Alberts, B., Bray, D., Johnson, A., Lewis, J., Raff, M., Roberts, K., and Walter, P., 1998. *Essential Cell Biology*. New York & London: Garland Publishing, Inc.

Anderberg, E.K., Nystrom, C., and Artursson, P., 1992. Epithelial transport of drugs in cell culture. VII: Effects of pharmaceutical surfactant excipients and bile acids on transepithelial permeability in monolayers of human intestinal epithelial (Caco-2) cells. *J. Pharm. Sci.* 81:879–887.

Andersen, A.S., Palmqvist, E., Bang, S., Shaw, A.C., Hubalek, F., Ribel, U., and Hoeg-Jensen, T., 2010. Backbone cyclic insulin. *J. Peptide Sci.* 16:473–479.

Aoki, Y., Morishita, M., and Takayama, K., 2005. Role of the mucous/glycocalyx layers in insulin permeation across the rat ileal membrane. *Int. J. Pharm.* 297:98–109.

Arbit, E., and Kidron, M., 2009. Oral insulin: The rationale for this approach and current developments. *J. Diabetes Sci. Technol.* 3:562–567.

Aungst, B.J., 2000. Intestinal permeation enhancers. *J. Pharm. Sci.* 89:429–442.

Barrett, A.J., Rawlings, N.D., and Woessner, J.F., 2004. *Handbook of Proteolytic Enzymes*, 2nd ed. London: Elsevier.

Becquemin, M.H., and Chaumuzeau, J.P., 2010. Inhaled insulin: A model for pulmonary systemic absorption? *Rev. Malad. Respir.* 27:e54–e65.

Behl, C.R., Pimplaskar, H.K., Sileno, A.P., deMeireles, J., and Romeo, V.D., 1998. Effects of physicochemical properties and other factors on systemic nasal drug delivery. *Adv. Drug Deliv. Rev.* 29:89–116.

Benson, H.A.E., and Namjoshi, S., 2008. Proteins and peptides: Strategies for delivery to and across the skin. *J. Pharm. Sci.* 97:3591–3610.

Bernkop-Schnurch, A., 2005. Thiomers: A new generation of mucoadhesive polymers. *Adv. Drug Deliv. Rev.* 57:1569–1582.

Bernkop-Schnürch, A., 2010. Low molecular mass permeation enhancers in oral delivery of macromolecular drugs. In *Oral Delivery of Macromolecular Drugs*, Ed. A. Bernkop-Schnürch, 86–101. New York: Springer.

Bernkop-Schnurch, A., Valenta, C., and Daee, S.M., 1999. Peroral polypeptide delivery. A comparative in vitro study of mucolytic agents. *Arzneimittelforschung* 49:799–803.

Brayden, D.J., Jepson, M.A., and Baird, A.W., 2005. Keynote review: Intestinal Peyer's patch M cells and oral vaccine targeting. *Drug Discov. Today* 10:1145–1157.

Chalasani, K.B., Russell-Jones, G.J., Yandrapu, S.K., Diwan, P.V., and Jain, S.K., 2007. A novel vitamin B12-nanosphere conjugate carrier system for peroral delivery of insulin. *J. Control. Release* 117:421–429.

Clement, S., Dandona, P., Still, J.G., and Kosutic, G., 2004. Oral modified insulin (HIM2) in patients with type 1 diabetes mellitus: Results from a phase I/II clinical trial. *Metabolism* 53:54–58.

Conner, S.D., and Schmid, S.L., 2003. Regulated portals of entry into the cell. *Nature* 422:37–44.

Costantino, H.R., Illum, L., Brandt, G., Johnson, P.H., and Quay, S.C., 2007. Intranasal delivery: Physicochemical and therapeutic aspects. *Int. J. Pharm.* 337:1–24.

Daugherty, A.L., and Mrsny, R.J., 1999. Regulation of the intestinal epithelial paracellular barrier. *Pharm. Sci. Technol. Today* 2:281–287.

Davis, S.S., and Illum, L., 2003. Absorption enhancers for nasal drug delivery. *Clin. Pharmacokinet.* 42:1107–1128.

Deli, M.A., 2008. Potential use of tight junction modulators to reversibly open membranous barriers and improve drug delivery. *Biochim. Biophys. Acta* 1788:892–910.

des Rieux, A., Fievez, V., Garinot, M., Schneider, Y.J., and Préat, V., 2006. Nanoparticles as potential oral delivery systems of proteins and vaccines: A mechanistic approach. *J. Control. Release* 116:1–27.

Dressman, J.B., Vertzoni, M., Goumas, K., and Reppas, C., 2007. Estimating drug solubility in the gastrointestinal tract. *Adv. Drug Deliv. Rev.* 59:591–602.

Dumont, J.A., Bitonti, A.J., Clark, D., Evans, S., Pickford, M., and Newman, S.P., 2005. Delivery of an erythropoietin-Fc fusion protein by inhalation in humans through an immunoglobulin transport pathway. *J. Aerosol Med.* 18:294–303.

Edenhofer, F., 2008. Protein transduction revisited: Novel insights into the mechanism underlying intracellular delivery of proteins. *Curr. Pharm. Des.* 14:3628–3636.

Edsman, K., and Hagerstrom, H., 2005. Pharmaceutical applications of mucoadhesion for the non-oral routes. *J. Pharm. Pharmacol.* 57:3–22.

Elliott, G., and O'Hare, P., 1997. Intercellular trafficking and protein delivery by a herpesvirus structural protein. *Cell* 88:223–233.

Elmquist, A., Lindgren, M., Bartfai, T., and Langel, U., 2001. VE-cadherin-derived cell-penetrating peptide, pVEC, with carrier functions. *Exp. Cell Res.* 269:237–244.

Fasano, A., 1998. Novel approaches for oral delivery of macromolecules. *J. Pharm. Sci.* 87:1351–1356.

Forst, T., Hohberg, C., Schöndorf, T., Borchert, M., Forst, S., Roth, W., Dehos, B., and Pfützner, A., 2009. Time-action profile and patient assessment of inhaled insulin via the Exubera device in comparison with subcutaneously injected insulin aspart via the FlexPen device. *Diabetes Technol. Therap.* 11:87–92.

Foster, N., and Hirst, B.H., 2005. Exploiting receptor biology for oral vaccination with biodegradable particulates. *Adv. Drug Deliv. Rev.* 57:431–450.

Frankel, A.D., and Pabo, C.O., 1988. Cellular uptake of the Tat protein from human immunodeficiency virus. *Cell* 55:1189–1193.

Fricker, G., and Drewe, J., 1996. Current concepts in intestinal peptide absorption. *J. Pept. Sci.* 2:195–211.

Frokjaer, S., and Otzen, D.E., 2005. Protein drug stability: A formulation challenge. *Nat. Rev. Drug Discov.* 4:298–306.

Futaki, S., Suzuki, T., Ohashi, W., Yagami, T., Tanaka, S., Ueda, K., and Sugiura, Y., 2001. Arginine-rich peptides. An abundant source of membrane-permeable peptides having potential as carriers for intracellular protein delivery. *J. Biol. Chem.* 276:5836–5840.

Green, M., and Loewenstein, P.M., 1988. Autonomous functional domains of chemically synthesized human immunodeficiency virus tat trans-activator protein. *Cell* 55:1179–1188.

Gros, E., Deshayes, S., Morris, M.C., Aldrian-Herrada, G., Depollier, J., Heitz, F., and Divita, G., 2006. A non-covalent peptide-based strategy for protein and peptide nucleic acid transduction. *Biochim. Biophys. Acta* 1758:384–393.

Hamman, J.H., Enslin, G.M., and Kotze, A.F., 2005. Oral delivery of peptide drugs: Barriers and developments. *BioDrugs* 19:165–177.

Hashimoto, M., Takada, K., Kiso, Y., and Muranishi, S., 1989. Synthesis of palmitoyl derivatives of insulin and their biological activities. *Pharm. Res.* 6:171–176.

Hazra, P., Adhikary, L., Dave, N., Khedkar, A., Manjunath, H.S., Anantharaman, R., and Iyer, H., 2010. Development of a process to manufacture PEGylated orally bioavailable insulin. *Biotechnol. Prog.* 26:1695–1704.

Heinemann, L., 2002. Variability of insulin absorption and insulin action. *Diabetes Technol. Ther.* 4:673–682.

Heinemann, L., 2004. Current status of the development of inhaled insulin. *Br. J. Diabetes Vasc. Dis.* 4:295–301.

Heinemann, L., Pfützner, A., and Heise, T., 2001. Alternative routes of administration as an approach to improve insulin therapy: Update on dermal, oral, nasal and pulmonary insulin delivery. *Curr. Pharm. Des.* 7:1327–1351.

Heitz, F., Morris, M.C., and Divita, G., 2009. Twenty years of cell-penetrating peptides: From molecular mechanisms to therapeutics. *Br. J. Pharmacol.* 157:195–206.

Inglis, A.S., 1983. Cleavage at aspartic acid. *Methods Enzymol.* 91:324–332.

Johnson, P.H., Frank, D., and Costantino, H.R., 2008. Discovery of tight junction modulators: Significance for drug development and delivery. *Drug Discov. Today* 13:261–267.

Joliot, A., and Prochiantz, A., 2004. Transduction peptides: From technology to physiology. *Nat. Cell Biol.* 6:189–196.

Jung, T., Kamm, W., Breitenbach, A., Kaiserling, E., Xiao, J.X., and Kissel, T., 2000. Biodegradable nanoparticles for oral delivery of peptides: Is there a role for polymers to affect mucosal uptake? *Eur. J. Pharm. Biopharm.* 50:147–160.

Kahns, A.H., Buur, A., and Bundgaard, H., 1993. Prodrugs of peptides. 18. Synthesis and evaluation of various esters of desmopressin (dDAVP). *Pharm. Res.* 10:68–74.

Kajii, H., Horie, T., Hayashi, M., and Awazu, S., 1986. Effects of sodium salicylate and caprylate as adjuvants of drug absorption on isolated rat small intestinal epithelial cells. *Int. J. Pharm.* 33:253–255.

Kararli, T.T., 1995. Comparison of the gastrointestinal anatomy, physiology, and biochemistry of humans and commonly used laboratory-animals. *Biopharm. Drug Dispos.* 16:351–380.

Kipnes, M., Dandona, P., Tripathy, D., Still, J.G., and Kosutic, G., 2003. Control of postprandial plasma glucose by an oral insulin product (HIM2) in patients with type 2 diabetes. *Diabetes Care* 26:421–426.

Kotze, A.F., Thanou, M.M., Lueben, H.L., De Boer, A.G., Verhoef, J.C., and Junginger, H.E., 1999. Enhancement of paracellular drug transport with highly quaternized N-trimethyl chitosan chloride in neutral environments: In vitro evaluation in intestinal epithelial cells (Caco-2). *J. Pharm. Sci.* 88:253–257.

Kukral, J.C., Adams, A.P., and Preston, F.W., 1965. Protein producing capacity of human exocrine pancreas—Incorporation of S35 methionine in serum and pancreatic juice protein. *Ann. Surg.* 162:63–73.

Leary, A.C., Stote, R.M., Cussen, K., O'Brien, J., Leary, W.P., and Buckley, B., 2006. Pharmacokinetics and pharmacodynamics of intranasal insulin administered to patients with type 1 diabetes: A preliminary study. *Diabetes Technol. Therap.* 8:81–88.

LeCluyse, E.L., Appel, L.E., and Sutton, S.C., 1991. Relationship between drug absorption enhancing activity and membrane perturbing effects of acylcarnitines. *Pharm. Res.* 8:84–87.

Lehr, C.M., Bouwstra, J.A., Kok, W., De Boer, A.G., Tukker, J.J., Verhoef, J.C., Breimer, D.D., and Junginger, H.E., 1992. Effects of the mucoadhesive polymer polycarbophil on the intestinal absorption of a peptide drug in the rat. *J. Pharm. Pharmacol.* 44:402–407.

Leonard, T.W., Lynch, J., McKenna, M.J., and Brayden, D.J., 2006. Promoting absorption of drugs in humans using medium-chain fatty acid-based solid dosage forms: GIPET. *Expert. Opin. Drug Deliv.* 3.5:685–692.

Leone-Bay, A., Paton, D.R., Variano, B., Leipold, H., Rivera, T., Miura-Fraboni, J., Baughman, R.A., and Santiago, N., 1998. Acylated non-α-amino acids as novel agents for the oral delivery of heparin sodium, USP. *J. Control. Release* 50:41–49.

Lewis, L.D., and Williams, J.A., 1990. Cholecystokinin—A key integrator of nutrient assimilation. *News Physiol. Sci.* 5:163–167.

Liang, J.F., and Yang, V.C., 2005. Insulin-cell penetrating peptide hybrids with improved intestinal absorption efficiency. *Biochem. Biophys. Res. Commun.* 335:734–738.

Lindmark, T., Kimura, Y., and Artursson, P., 1998. Absorption enhancement through intracellular regulation of tight junction permeability by medium chain fatty acids in Caco-2 cells. *J. Pharmacol. Exp. Ther.* 284:362–369.

Low, S.C., Nunes, S.L., Bitonti, A.J., and Dumont, J.A., 2005. Oral and pulmonary delivery of FSH-Fc fusion proteins via neonatal Fc receptor-mediated transcytosis. *Hum. Reprod.* 20:1805–1813.

Lundberg, M., Wikstrøm, S., and Johansson, M., 2003. Cell surface adherence and endocytosis of protein transduction domains. *Mol. Ther.* 8:143–150.

MacAdam, A., 1993. The effect of gastro-intestinal mucus on drug absorption. *Adv. Drug Deliv. Rev.* 11:201–220.

Mansoor, S., Youn, Y.S., and Lee, K.C., 2005. Oral delivery of mono-PEGylated sCT (Lys18) in rats: Regional difference in stability and hypocalcemic effect. *Pharm. Dev. Technol.* 10:389–396.

Mathias, N.R., and Hussain, M.A., 2010. Non-invasive systemic drug delivery: Developability considerations for alternate routes of administration. *J. Pharm. Sci.* 99:1–20.

McConnell, E.L., Basit, A.W., and Murdan, S., 2008. Measurements of rat and mouse gastrointestinal pH, fluid and lymphoid tissue, and implications for in-vivo experiments. *J. Pharm. Pharmacol.* 60:63–70.

Morishita, M., Kamei, N., Ehara, J., Isowa, K., and Takayama, K., 2007. A novel approach using functional peptides for efficient intestinal absorption of insulin. *J. Control. Release* 118:177–184.

Morris, M.C., Deshayes, S., Heitz, F., and Divita, G., 2008. Cell-penetrating peptides: From molecular mechanisms to therapeutics. *Biol. Cell.* 100:201–217.

Myers, S.R., Yakubu-Madus, F.E., Johnson, W.T., Baker, J.E., Cusick, T.S., Williams, V.K., Tinsley, F.C., Kriauciunas, A., Manetta, J., and Chen, V.J., 1997. Acylation of human insulin with palmitic acid extends the time action of human insulin in diabetic dogs. *Diabetes* 46:637–642.

Nishihata, T., Higuchi, T., and Kamada, A., 1984. Salicylate-promoted permeation of cefoxitin, insulin and phenylalanine across red cell membrane. Possible mechanism. *Life Sci.* 34:437–445.

Odegard, P.S., and Capoccia, K.L., 2005. Inhaled insulin: Exubera. *Ann. Pharmacother.* 39:843–853.

Owens, D.R., 2002. New horizons—Alternative routes for insulin therapy. *Nat. Rev. Drug Discov.* 1:529–540.

Pappenheimer, J.R., 2001. Role of pre-epithelial "unstirred" layers in absorption of nutrients from the human jejunum. *J. Membr. Biol.* 179:185–204.

Patton, J.S., and Platz, R.M., 1992. Pulmonary delivery of peptides and proteins for systemic action. *Adv. Drug Deliv. Rev.* 8:179–196.

Pauletti, G.M., Gangwar, S., Siahaan, T.J., Aubé, J., and Borchardt, R.T., 1997. Improvement of oral peptide bioavailability: Peptidomimetics and prodrug strategies. *Adv. Drug Deliv. Rev.* 27:235–256.

Petrus, A.K., Fairchild, T.J., and Doyle, R.P., 2009. Traveling the vitamin B12 pathway: Oral delivery of protein and peptide drugs. *Angew. Chem. Int. Ed. Engl.* 48:1022–1028.

Pooga, M., Soomets, U., Hallbrink, M., Valkna, A., Saar, K., Rezaei, K., Kahl, U., Hao, J.X., Xu, X.J., Wiesenfeld-Hallin, Z., Hokfelt, T., Bartfai, T., and Langel, U., 1998. Cell penetrating PNA constructs regulate galanin receptor levels and modify pain transmission in vivo. *Nat. Biotechnol.* 16:857–861.

Prausnitz, M.R., and Langer, R. 2008. Transdermal drug delivery. *Nat. Biotechnol.* 26:1261–1268.

Pujals, S., Fernandez-Carneado, J., Lopez-Iglesias, C., Kogan, M.J., and Giralt, E., 2006. Mechanistic aspects of CPP-mediated intracellular drug delivery: Relevance of CPP self-assembly. *Biochim. Biophys. Acta* 1758:264–279.

Raiman, J., Törmälehto, S., Yritys, K., Junginger, H.E., and Mönkkönen, J., 2003. Effects of various absorption enhancers on transport of clodronate through Caco-2 cells. *Int. J. Pharm.* 261:129–136.

Ramon, J., Saez, V., Baez, R., Aldana, R., and Hardy, E., 2005. PEGylated interferon-alpha 2b: A branched 40K polyethylene glycol derivative. *Pharm. Res.* 22:1374–1386.

Rave, K., Bott, S., Heinemann, L., Sha, S., Becker, R.H.A., Willavize, S.A., and Heise, T., 2005. Time-action profile of inhaled insulin in comparison with subcutaneously injected insulin lispro and regular human insulin. *Diabetes Care* 28:1077–1082.

Requero, M.A., Goni, F.M., and Alonso, A., 1995. The membrane-perturbing properties of palmitoyl-coenzyme A and palmitoylcarnitine. A comparative study. *Biochemistry* 34:10400–10405.

Richard, J.P., Melikov, K., Vives, E., Ramos, C., Verbeure, B., Gait, M.J., Chernomordik, L.V., and Lebleu, B., 2003. Cell-penetrating peptides. A reevaluation of the mechanism of cellular uptake. *J. Biol. Chem.* 278:585–590.

Rittner, K., Benavente, A., Bompard-Sorlet, A., Heitz, F., Divita, G., Brasseur, R., and Jacobs, E., 2002. New basic membrane-destabilizing peptides for plasmid-based gene delivery in vitro and in vivo. *Mol. Ther.* 5:104–114.

Roemer, D., Buescher, H.H., Hill, R.C., Pless, J., Bauer, W., Cardinaux, F., Closse, A., Hauser, D., and Huguenin, R., 1977. A synthetic enkephalin analogue with prolonged parenteral and oral analgesic activity. *Nature* 268:547–549.

Roger, E., Lagarce, F., Garcion, E., and Benoit, J.P., 2010. Biopharmaceutical parameters to consider in order to alter the fate of nanocarriers after oral delivery. *Nanomedicine* 5:287–306.

Roopenian, D.C., and Akilesh, S., 2007. FcRn: The neonatal Fc receptor comes of age. *Nat. Rev. Immunol.* 7:715–725.

Rousselle, C., Smirnova, M., Clair, P., Lefauconnier, J.M., Chavanieu, A., Calas, B., Scherrmann, J.M., and Temsamani, J., 2001. Enhanced delivery of doxorubicin into the brain via a peptide-vector-mediated strategy: Saturation kinetics and specificity. *J. Pharmacol. Exp. Ther.* 296:124–131.

Rubio-Aliaga, I., and Daniel, H., 2002. Mammalian peptide transporters as targets for drug delivery. *Trends Pharmacol. Sci.* 23:434–440.

Russell-Jones, G.J., 2001. The potential use of receptor-mediated endocytosis for oral drug delivery. *Adv. Drug Deliv. Rev.* 46:59–73.

Schinkel, A.H., and Jonker, J.W., 2003. Mammalian drug efflux transporters of the ATP binding cassette (ABC) family: An overview. *Adv. Drug Deliv. Rev.* 55:3–29.

Schipper, N.G., Varum, K.M., Stenberg, P., Ocklind, G., Lennernas, H., and Artursson, P., 1999. Chitosans as absorption enhancers of poorly absorbable drugs. 3: Influence of mucus on absorption enhancement. *Eur. J. Pharm. Sci.* 8:335–343.

Schmidt, M.C., Rothen-Rutishauser, B., Rist, B., Beck-Sickinger, A., Wunderli-Allenspach, H., Rubas, W., Sadee, W., and Merkle, H.P., 1998. Translocation of human calcitonin in respiratory nasal epithelium is associated with self-assembly in lipid membrane. *Biochemistry* 37:16582–16590.

Schwarze, S.R., Ho, A., Vocero-Akbani, A., and Dowdy, S.F., 1999. In vivo protein transduction: Delivery of a biologically active protein into the mouse. *Science* 285:1569–1572.

Shima, M., Kimura, Y., Adachi, S., and Matsuno, R., 1998. The relationship between transport-enhancement effects and cell viability by capric acid sodium salt, monocaprin, and dicaproin. *Biosci. Biotechnol. Biochem.* 62:83–86.

Snyder, E.L., and Dowdy, S.F., 2005. Recent advances in the use of protein transduction domains for the delivery of peptides, proteins and nucleic acids in vivo. *Expert Opin. Drug Deliv.* 2:43–51.

Su, K.S., Campanale, K.M., Mendelsohn, L.G., Kerchner, G.A., and Gries, C.L., 1985. Nasal delivery of polypeptides I: Nasal absorption of enkephalins in rats. *J. Pharm. Sci.* 74:394–398.

Swenson, E.S., and Curatolo, W.J., 1992. Intestinal permeability enhancement for proteins, peptides and other polar drugs: Mechanisms and potential toxicity. *Adv. Drug Deliv. Rev.* 8:39–92.

Swenson, E.S., Milisen, W.B., and Curatolo, W., 1994. Intestinal permeability enhancement: Efficacy, acute local toxicity, and reversibility. *Pharm. Res.* 11:1132–1142.

Takatsuka, S., Kitazawa, T., Morita, T., Horikiri, Y., and Yoshino, H., 2006. Enhancement of intestinal absorption of poorly absorbed hydrophilic compounds by simultaneous use of mucolytic agent and non-ionic surfactant. *Eur. J. Pharm. Biopharm.* 62:52–58.

Tomita, M., Hayashi, M., and Awazu, S., 1995. Absorption-enhancing mechanism of sodium caprate and decanoylcarnitine in Caco-2 cells. *J. Pharmacol. Exp. Ther.* 272:739–743.

Tsukita, S., Furuse, M., and Itoh, M., 2001. Multifunctional strands in tight junctions. *Nat. Rev. Mol. Cell Biol.* 2:285–293.

Vilhardt, H., and Lundin, S., 1986. In vitro intestinal transport of vasopressin and its analogues. *Acta Physiol. Scand.* 126:601–607.

Walter, E., Kissel, T., and Amidon, G.L., 1996. The intestinal peptide carrier: A potential transport system for small peptide derived drugs. *Adv. Drug Deliv. Rev.* 20:33–58.

Washington, N., Washington, C., and Wilson, C.G., 2001. *Physiological Pharmaceutics*. Boca Raton, FL: CRC Press.

Weers, J.G., Tarara, T.E., and Clark, A.R., 2007. Design of fine particles for pulmonary drug delivery. *Expert Opin. Drug Deliv.* 4:297–313.

Wender, P.A., Mitchell, D.J., Pattabiraman, K., Pelkey, E.T., Steinman, L., and Rothbard, J.B., 2000. The design, synthesis, and evaluation of molecules that enable or enhance cellular uptake: Peptoid molecular transporters. *Proc. Natl. Acad. Sci. USA* 97:13003–13008.

Werle, M., 2008. Natural and synthetic polymers as inhibitors of drug efflux pumps. *Pharm. Res.* 25:500–511.

Werle, M., Kolmar, H., Albrecht, R., and Bernkop-Schnurch, A., 2008. Characterisation of the barrier caused by luminally secreted gastro-intestinal proteolytic enzymes for two novel cystine-knot microproteins. *Amino Acids* 35:195–200.

White, S., Bennett, D.B., Cheu, S., Conley, P.W., Guzek, D.B., Gray, S., Howard, J., Malcolmson, R., Parker, J.M., Roberts, P., Sadrzadeh, N., Schumacher, J.D., Seshadri, S., Sluggett, G.W., Stevenson, C.L., and Harper, N.J., 2005. EXUBERA: Pharmaceutical development of a novel product for pulmonary delivery of insulin. *Diabetes Technol. Therap.* 7:896–906.

Whitehead, K., Karr, N., and Mitragotri, S., 2008. Discovery of synergistic permeation enhancers for oral drug delivery. *J. Control. Release* 128:128–133.

Whitehead, K., and Mitragotri, S., 2008. Mechanistic analysis of chemical permeation enhancers for oral drug delivery. *Pharm. Res.* 25:1412–1419.

Wong, T.W., 2010. Design of oral insulin delivery systems. *J. Drug Target.* 18:79–92.

Woodley, J.F., 1994. Enzymatic barriers for GI peptide and protein delivery. *Crit. Rev. Ther. Drug Carrier Syst.* 11:61–95.

12 Immunogenicity of Therapeutic Proteins

Grzegorz Kijanka, Wim Jiskoot, Melody Sauerborn, Huub Schellekens, and Vera Brinks

CONTENTS

12.1 INTRODUCTION

The use of therapeutic proteins originates in the 19th century, when polyclonal horse antitetanus and antidiphtheria immunoglobulins were used for the first time. Then in the 1920s, bovine insulin and porcine insulin were introduced for the treatment of diabetes. Since then, insulin and many more protein drugs, including growth factor, erythropoietin, and a whole range of monoclonal antibodies, have become major therapeutic drugs. Currently, a substantial portion of newly approved drugs is constituted by therapeutic proteins. A common feature of all therapeutic proteins is their immunogenicity, the potential to evoke an adverse immune response. In the field of protein drugs, the presence of antidrug antibodies (ADAs) is often referred to as seroprevalence, whereas the term "immunogenicity" is referred to as a quantitative measurement of ADA titers (Sominada et al., 2007). However, this distinction is not a general rule. Therefore, in this chapter the term "immunogenicity" will be used in a general manner, describing the potential of the therapeutic protein to evoke an adverse immune response.

The immunogenicity of the first therapeutic proteins, such as insulin, has been explained by their nonhuman origin, resulting in a classical immune response (see Section 12.3.2). Protein drugs of the next generation, for example, plasma-derived clotting factors (Jacquemin and Saint-Remy, 1998) and growth hormone extracted from the pituitary glands of human cadavers (Milner, 1985), were deprived of this drawback, and therefore application of these human-derived proteins was believed to solve the problem of immunogenicity. Surprisingly, it did not. Moreover, their potential to spread infections like HIV and Creutzfeld–Jacob resulted in restriction of application of those drugs. The real breakthrough in protein drug development took place in the 1980s, when DNA recombination methods were introduced in the production of therapeutic proteins. This drastically opened the possibility of large-scale production of highly purified, therapeutic proteins homologous to endogenous proteins. Although those are assumed to be nonimmunogenic, we are now still far away from the elimination of unwanted immune responses and understanding the exact mechanisms underlying immunogenicity of therapeutic proteins that closely mimic the structure of endogenous proteins. Many hypotheses trying to explain immunogenicity are based on assumptions rather than facts, which additionally blurs our insight into immunogenicity.

In this chapter, the current knowledge in understanding immunogenicity of therapeutic proteins is summarized. Insight into potential risk factors, mechanisms underlying unwanted immune responses, as well as the available tools and assays to determine immunogenic potential of (new) therapeutic proteins are described.

12.2 IMMUNOGENICITY AS A CLINICAL PROBLEM

Almost all patients treated with a nonhuman therapeutic protein developed an antibody response against the therapeutic protein (Schellekens, 2003). The development of recombinant human proteins, together with the improvement of production processes, allowed a reduction of adverse reactions associated with protein drug therapy. However, immunogenicity is still an important issue that has to be included in the assessment of a protein drug before its introduction onto the market. Moreover, the development of new, more sensitive ADA-detecting assays revealed that protein drugs induce an antibody response in a much higher percentage of patients than was initially believed (van Schouwenburg et al., 2010).

12.2.1   Clinical Significance of Unwanted Immune Responses

Immunogenicity of therapeutic proteins can have a number of clinical consequences that differ in their impact on therapy outcome or patient health, ranging from no obvious clinical consequences to life-threatening conditions. In the development of an immune response, usually binding ADAs (BAbs) are observed first (Chamberlain, 2002). Such antibodies recognize drug epitopes that are not involved in receptor recognition or substrate processing, and often lead to rapid clearance of the therapeutic protein–antibody complexes from the blood stream. However, some drugs in complex with antibodies show a prolonged half-life (Moore and Leppert, 1980). BAbs do not always lead to lower efficiency of therapy. In contrast to BAbs, neutralizing ADAs (NAbs) recognize the active site of the therapeutic protein with high affinity and inhibit substrate processing or its receptor interaction (Chamberlain, 2002). Figure 12.1 shows the comparison between BAbs and NAbs. How NAbs affect treatment strongly depends on their titers. When NAbs are produced in low or moderate titers, their influence on therapy can be similar to that of BAbs, whereas in patients with high NAb titers, the effect of treatment can be completely inhibited (Sominada et al., 2007; van der Voort et al., 2010).

Since most of the protein drugs are recombinant human proteins, it is possible that ADAs interact with corresponding endogenous proteins. Such an interaction would correlate with a serious risk of autoimmune disease, for example, pure red-cell aplasia (PRCA) in the case of erythropoietin (Pollock et al., 2008), as described later (Section 12.2.2). Therefore, if there is suspicion of ADA production, for example, significant reduction of treatment efficiency or increased disease symptomatology,

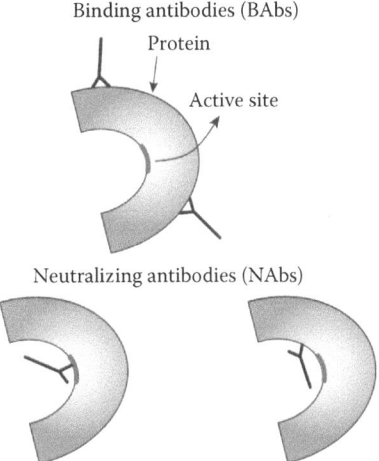

Binding antibodies (BAbs)

Protein

Active site

Neutralizing antibodies (NAbs)

FIGURE 12.1   Schematic representation of the difference in binding between binding antidrug antibodies (BAbs) and neutralizing antidrug antibodies (NAbs). BAbs recognize a part of the protein that is not involved in the enzymatic activity or receptor recognition (active site). Therefore, BAbs do not block the activity of the protein. In contrast, NAbs recognize the active site or an epitope nearby, directly or indirectly blocking the activity of the protein.

the levels of ADAs, including neutralizing activity, should be tested. If the risk of severe side effects is serious, the treatment has to be terminated and an alternative therapy applied. Physicians are recommended to consider switching the treatment of patients with multiple sclerosis (MS) from interferon beta (IFNβ) to alternatives (e.g., glatiramer acetate or natalizumab) as soon as NAbs are detected (Palman et al., 2010). However, before deciding to terminate treatment, the balance between therapy outcome and possible risk of antibody formation has to be considered. For example, in patients with MS anti-IFNβ NAb titers, especially low or moderate ones, have a tendency to decrease over time. Deciding too early that the treatment should be terminated can bereave the patient of the positive effect of IFNβ, and might even lead to a faster disease progression (Palman et al., 2010).

12.2.2   EXAMPLES OF IMMUNOGENICITY

Introducing protein drugs into the clinic has been a real breakthrough in the treatment of a number of diseases, such as rheumatoid arthritis (monoclonal antibody [Markham and Lamb, 2000]), anemia (erythropoietin [Ludwig et al., 1992]), diabetes (insulin [Macleod, 1922]), and hepatitis (interferon alpha [Smith et al., 1982]). However, the beneficial effect of therapeutic proteins is often significantly reduced by the unwanted immune response against the drug. In this section, four examples of protein drugs with clinical consequences of an unwanted immune response will be described.

The most well-known therapeutic protein with severe immunogenicity issues is erythropoietin. Immunogenicity of this therapeutic protein shook the belief that therapeutic proteins are very safe, and began the discussion about the significance of immunogenicity in the clinic. Antibody formation against erythropoietin can induce PRCA when antibodies against the drug cross-neutralize endogenous erythropoietin. Although PRCA incidence has been reported since erythropoietin therapy was introduced for patients suffering from renal failure, PRCA was extremely rare and a connection between the disease and treatment was not clear. In 1998, Johnson & Johnson introduced reformulated erythropoietin named Eprex® for the therapy of patients with chronic renal failure. Following European Medicines Agency (EMA) guidance, human serum albumin (HSA) was replaced by polysorbate 80 and glycine as stabilizers (outside the United States). Between 1998 and 2004, the occurrence of PRCA significantly increased and, in the majority of patients, was related to Eprex® treatment (Casadevall et al., 2002; Pollock et al., 2008). A number of studies correlated immunogenicity to the use of noncoated rubber stoppers of ready-to-use syringes with Eprex®. According to the manufacturer, leachates containing phenolic derivates, extracted by polysorbate 80 from uncoated rubber, would have acted like an adjuvant inducing an immune response (Sharma, 2007). Since thousands of patients undergo Eprex® therapy and PRCA developed only in a small percentage of them, many scientists question the importance of adjuvant properties of leachates as the main immunogenic factor (Pollock et al., 2008; Schellekens, 2007). More probably, the usage of polysorbate 80 led to lower stability of the protein, which could have induced immunogenic impurities, for example, as a result of improper handling. However, the exact mechanism of erythropoietin-associated PRCA remains unknown.

IFNβ is the major drug used as the first-line therapy of MS. Although it does not cure the disease, it reduces the number and strength of relapses of MS, significantly improving patients' quality of life. The first available formulation of IFNβ was Betaseron®, produced in bacteria (*Escherichia coli*) and introduced on the market in 1993. The first reports showing its immunogenicity were published already in the same year. Betaseron® and biosimilar Extavia® (which has been recently marketed) are formulations of the IFNβ-1b type, which differs from the endogenous protein in three points: (1) it lacks N-terminal methionine, (2) the cysteine from residue 17 is replaced by serine, and (3) it is not glycosylated. In contrast, IFNβ-1a, the active compound in Avonex® (Biogen) and Rebif® (Merck Serono), is produced in Chinese hamster ovary cells and does not differ in amino-acid sequence from endogenous IFNβ. Although Avonex® and Rebif® show lower seroprevalence (percent of ADA-positive patients) than Betaseron®, Rebif® has been shown to induce the highest titers of NAbs among those drugs (Bertolotto et al., 2004; Ross et al., 2000; Sominada et al., 2007). The difference in immunogenicity between the products is probably related not only to protein structure, but also to their activity, formulation, route, and frequency of administration (van Beers et al., 2010a). Betaseron® contains HSA as a stabilizer, whereas Avonex® and Rebif® are available in a formulation with and without HSA. Betaseron® and Avonex® with HSA are sold in the lyophilized form, which has to be redissolved directly before injection. HSA-free Avonex® and Rebif® are supplied as liquids in ready-to-use syringes. Betaseron® and Rebif® are administered subcutaneously (s.c.) every other day and three times a week, respectively, whereas Avonex® is injected intramuscularly (i.m.) every week. Overall, the immune response against IFNβ develops within 3–12 months of treatment. After 24 months, if treatment is not terminated, antibody titers have a tendency to decrease (Hartung et al., 2007; van der Voort et al., 2010). However, if high titers of NAbs are developed, the therapy outcome is significantly altered (van der Voort et al., 2010). Therefore, a major item in the society of MS experts is the question whether treatment should be continued or terminated when NAbs are detected (Hartung et al., 2007; Herndon et al., 2005; Palman et al., 2010).

Monoclonal antibodies are the fastest growing class of biopharmaceuticals. Due to their nature (possession of highly variable antigen-binding regions), they can be designed to selectively bind almost any target. The first monoclonal antibodies used in the clinic were of murine origin, very immunogenic, and often induced severe anaphylactic or anaphylactoid reactions (Dillman, 1999). Thanks to the recombinant technology developed in the 1980s, most monoclonal antibodies nowadays present in the clinic are humanized or fully human proteins (Kang and Saif, 2007). In general, the more human the monoclonal antibody is, the less immunogenicity it evokes. However, a human antibody can also show a significant immunogenicity. After a few months of treatment with infliximab, a chimeric antibody against tumor necrosis factor alpha (TNF-α), antibodies are found in 20% to 60% of the patients (depending on study) (Cassinotti and Travis, 2009). Adalimumab, a fully human antibody against the same therapeutic target, has been shown to evoke antibody production in a similar percentage of patients (van Schouwenburg et al., 2010; West et al., 2008).

The last group of therapeutic proteins described in this section are bone morphogenic proteins (BMPs) belonging to the family of growth factors. There are two types of BMPs available on the market, BMP-2 and BMP-7. They are used to induce spinal fusion or are given to treat long nonunion fractures. In contrast to the majority of other protein drugs, which are used mainly in the therapy of chronic diseases, BMPs are used as single treatment, by surgical implantation. Depending on the study, BMPs were shown to evoke unwanted immune responses in 40–60% percent of the patients. Although severe side effects are very rare, development of anti-BMP antibodies significantly reduces the success of treatment (Hwang et al., 2009).

12.3   IMMUNOLOGICAL MECHANISM UNDERLYING IMMUNOGENICITY OF THERAPEUTIC PROTEINS

The immune system is complex and evolved to protect organisms from all kinds of pathogens, such as viruses and bacteria. The wide range of infectious agents forced the development of a very sophisticated defense system that includes the cooperation of many proteins and cell types. Thanks to many years of intensive studies, our knowledge about the immune responses against many antigens has drastically improved, although new findings still bring more insight into the complexity of the immune system.

12.3.1   INNATE AND ADAPTIVE IMMUNE SYSTEM

The first line of defense against pathogens is the innate immune system composed of physical barriers and phagocytic cells, which are responsible for preventing invasion of pathogens and for the removal of pathogens from the organism, respectively. A typical feature of the cellular component of the innate immune system is its rapid, but rather unspecific manner of antigen removal. The efficiency of the innate system strongly depends on the identification of invaders, which is based on the recognition of foreign compounds on the surface of pathogens by mannose receptors, scavenger receptors, and Toll-like receptors (TLRs) (Seong and Matzinger, 2004). When the innate system cannot sufficiently fight the infection, the adaptive system comes into action. In contrast to the innate system, the adaptive system is capable of mounting highly specific responses. The main players of adaptive immune response are T and B lymphocytes.

A very important feature of the adaptive immune system is tolerance for self-antigens. During development, immature lymphocytes undergo selection based on their affinity to self-epitopes in the thymus (T lymphocytes) and in the bone marrow (B lymphocytes) (Li and Boussiotis, 2006; Melchers and Rolink, 2006). This mechanism of T and B cell selection is known as central tolerance. However, since it is virtually impossible to present all possible self-epitopes in the thymus or bone marrow, some autoreactive cells may escape the elimination. Those self-reactive T and B lymphocytes are controlled by a mechanism known as peripheral tolerance. Peripheral tolerance is maintained mainly due to the lack of the danger signal, usually derived from other immune cells, needed for T or B cell activation (Li and Boussiotis, 2006). In addition, specialized regulatory cells (both B and T type) are capable of eliminating self-reacting immune cells from periphery, creating an additional safety mechanism.

12.3.2 IMMUNE REACTION AGAINST THERAPEUTIC PROTEINS

There is no universal immune mechanism underlying the immunogenicity of therapeutic proteins. Mostly, the immune response against protein drugs is explained as a classical immune response against foreign epitopes or as a breakage of tolerance. A classical immune response is usually triggered by external antigens and involves the T cell-dependent (TD) activation of B cells (Figure 12.2). This type of immune

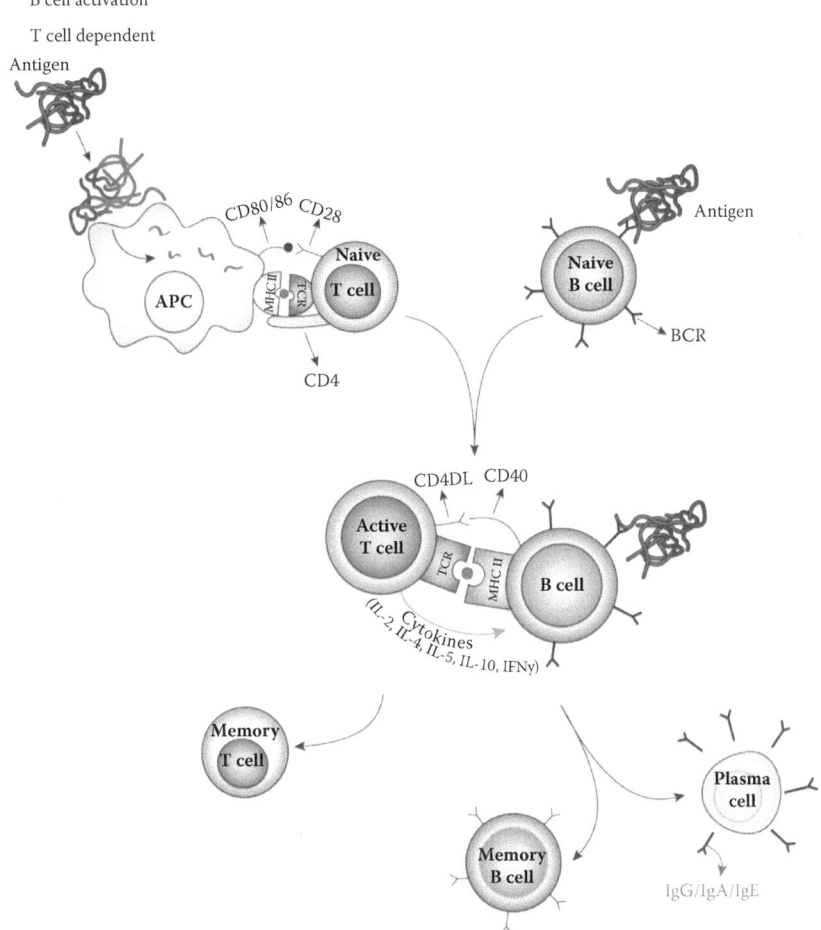

FIGURE 12.2 Schematic representation of the B cell activation pathway in a T cell-dependent manner. Antigen is caught and digested by an antigen-presenting cell (APC). The APC then presents short peptides, created during digestion, to T cells via major histocompatibility complex class II(MHC II) molecules. The interaction between MHC II and T-cell receptor (TCR) is the first activation signal for T cells. Upon additional stimulation (i.e., costimulation via CD4 and CD80/86 interaction with CD28), T cells become activated. Simultaneously, B cells interact with an antigen via their B cell receptors (BCRs). However, without additional stimulation from direct contact with T cells and T cell-derived cytokines, a B cell remains in a state of anergy. Activated B cells undergo a maturation process and differentiate into antibody-secreting plasma cells or memory cells.

mechanism is found when treating patients with therapeutic proteins having a non-human origin. However, most of the recombinant proteins are human analogs. They likely lead only to a classical immune response in patients who have a genetic disorder and thus lack tolerance for a certain protein (e.g., hemophiliacs with a factor VIII gene defect [Jacquemin and Saint-Remy, 1998]).

In general, the classical immune response involves a fast and robust immune reaction leading to the production of NAbs as well as memory formation. For example, Muromonab-CD3, a murine antibody that was the first monoclonal antibody approved for usage in humans, induces production of NAbs in all patients within two weeks after administration (Singh, 2011).

While the classical immune response is quite well understood, the breaking of tolerance by human derived or recombinant proteins remains a mystery. Tolerance for self-derived epitopes has to be at least temporally disturbed to develop a reaction against those types of protein drugs, for example, due to the presence of aggregates that might activate B cells via a T cell-independent (TI) pathway. Development of an immune response against recombinant or human proteins usually takes a few months and leads to the production of NAbs only in a part of all ADA-positive patients. Clinical and experimental data show also that during the proposed breaking of tolerance IgG is produced, while memory B cells are not formed (van Beers et al., 2010b). Moreover, the immune response against the majority of protein drugs does not lead to severe side effects.

12.3.3 T Cell-Independent Immune Response

The TI response is triggered by molecules or structures that are typical for lower organisms like bacteria or viruses. The T cell-independent activation of B cells can be divided into two types: polyclonal, induced by bacterial molecules like lipopolysaccharide (LPS)—Type I, and not polyclonal, induced by repetitive structures—Type II (Figure 12.3). As in the TD immune response, a strong danger signal has to be provided to activate B cells and therefore induce antibody production. In a TI type I response, binding of antigens like LPS or bacterial DNA to TLRs directly leads to B cell activation. A TI type II response is elicited by an antigen-induced change of B cell receptor (BCR) reorganization on the naive B lymphocyte, leading to crosslinking or dissociation of BCRs (Goldsby et al., 2006). To induce such a reorganization of BCRs, a special type of antigen is required. The typical TI type II antigen is a polymer or an aggregate with a molecular weight above 100 kDa, possessing on its surface repetitive epitopes, which form a two-dimensional network with a spacing of about 50–100 Å (5–10 nm) between epitopes (Vos et al., 2000a). The TI pathway leads to direct B lymphocyte activation and antibody production. However, in contrast with the TD pathway, the B lymphocytes do not differentiate into memory B cells and maturation of the antibody response does not occur or is very limited. Therefore, a TI immune response is characterized by the formation of IgM as the major antibody class (Goldsby et al., 2006).

It is important to note that the TD and TI response are not mutually exclusive. In fact, most immune responses are a combination of a TD and a TI mechanism. It is known that the TI immune response can be modulated by cytokines produced by

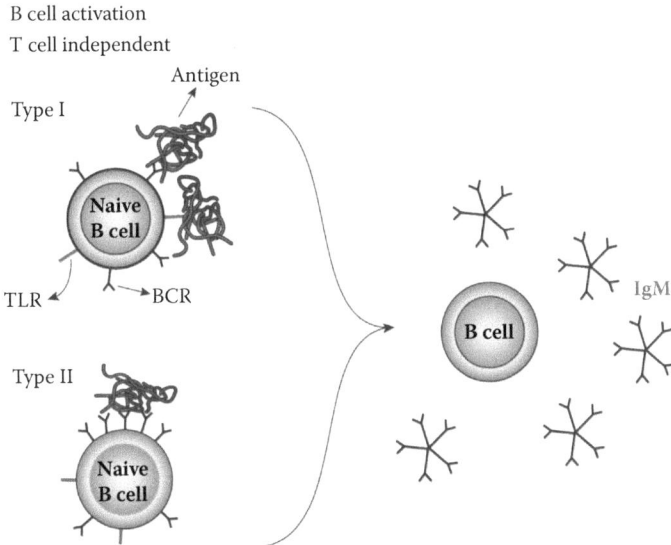

FIGURE 12.3 Schematic representation of the B cell activation pathway in a T cell-independent manner. The first step in the TI activation of a B cell is the interaction of an antigen with the BCR. To fully activate a B cell, an additional danger signal is needed. TI type I antigens bind to a receptor designed to recognize bacterial structures, for example, toll-like receptor (TLR). In contrast, TI type II antigens bind to several BCRs changing their distribution on the membrane and leading to BCR cross-reaction. To bind more than one BCR antigen, repetitive epitopes have to be present. These repetitive epitopes can be part of the polymeric structure of an antigen (e.g., sugar moieties of polysaccharides) or can be displayed by protein aggregates.

T cells (Vos et al., 2000b). Reorganization of BCRs, induced by TI type II antigens, also leads to a much faster development of a TD immune response (Snapper et al., 1995). Moreover, a few years ago the production of memory B cells via a TI response was observed (Obukhanych and Nussenzweig, 2006). The insight to date on therapeutic protein immunogenicity will be discussed in Section 12.4.

12.4 WHY ARE THERAPEUTIC PROTEINS IMMUNOGENIC?

Explanations why therapeutic proteins evoke an antibody response have been changing over time. For the first therapeutic proteins, their foreign origin was believed to underlie the immunogenicity, for example, a classical immune response against bovine insulin (Schellekens, 2003). As recombinant human proteins and human antibodies also turned out to be immunogenic, we now realize that the mechanisms of antibody formation are much more complex than a distinction between "self" and "nonself." Since our knowledge on the mechanisms underlying immunogenicity of current therapeutic proteins is limited, factors potentially affecting immunogenicity should be studied for each (class of) therapeutic protein. In general, there are three categories of factors that can affect immunogenicity: patient, product, and treatment related (Table 12.1).

TABLE 12.1

Factors Influencing the Immunogenicity of Therapeutic Proteins

Factors Influencing Immunogenicity		
Product Related	**Therapy Related**	**Patient Related**
1. Source of protein	1. Dose	1. Genetic background
2. Post-translational modifications	2. Dosing schedule	(e.g., nonsense mutations and
3. Excipients	3. Duration of therapy	HLA haplotype)
4. Impurities (including degradation	4. Comedication	2. Age
products such as aggregates)		3. Type of disease
5. Pegylation		

12.4.1 PATIENT-RELATED FACTORS

The significance of patient-related factors influencing immunogenicity of a protein drug is often underestimated during drug development. However, clinical data show that most therapeutic proteins induce an unwanted immune response only in some patients, suggesting that patient characteristics may be crucial for immunogenicity. For example, the genetic background of patients can have a significant impact on the treatment outcome.

Some diseases, for example, hemophilia A, are correlated with various mutations of the gene encoding the endogenous protein (in case of hemophilia A, factor VIII) (Fakharzadeh and Kazazian, 2000). Approximately 20% of patients with severe hemophilia A develop antibodies during treatment with factor VIII. Those patients present the most severe form of hemophilia A, which is a direct result of large deletions or nonsense mutations in the factor VIII gene. As a consequence, patients are not tolerant to functional factor VIII. In contrast, patients with mild or moderate hemophilia A, who do produce functional factor VIII, are much less prone to the development of antibodies against factor VIII (Fakharzadeh and Kazazian, 2000). Moreover, a significant part of protein therapies have been designed to replace or to supplement endogenous protein deficiency, mainly enzymes. Such therapies will always be associated with a higher risk of immunogenicity compared to therapies for patients that do not have any (genetic) abnormalities in endogenous protein expression (Lidove et al., 2010; Richards, 2002).

Not only lack of functional protein expression but also patients' human leukocyte antigen (HLA) haplotype can influence the immunogenicity of therapeutic proteins. Diabetic patients with the human leukocyte antigen D-related 7 (HLA-DR7) have been reported to produce higher titers of insulin antibodies than other diabetic patients, whereas patients who are HLA-DR3 homozygous produce lower titers (Reeves et al., 1984). Also in MS patients treated with IFNβ, an association between MHC class II haplotype and the level of anti-IFNβ antibody production was found. In vitro studies showed that the IFNβ molecule contains T cell epitopes that can be recognized and presented on B cells expressing DRB1*0701 MHC II allele (Barbosa et al., 2006). Clinical data confirm that patients whose antigen-presenting cells (APCs) express DRB1*0701

MHC II allele are more prone to develop an unwanted immune response than patients who lack this MHC II allele (Goldsby et al., 2006).

ADA production can also be influenced by the type of disease. Patients who suffer from chronic infectious diseases and whose immune systems are in an activated state are more likely to produce antibodies against interferon alpha (IFNα) than patients suffering from cancer (Antonelli and Dianzani, 1999). In contrast, patients with impaired immune system may be less prone to the development of ADAs than patients with an intact immune system (Schellekens, 2002).

12.4.2 Product-Related Factors

The production of a protein is a complicated, multistep process. During product development, many factors that may influence the immunogenicity of the final product have to be taken into consideration. In this section, the most important product-related factors will be described.

One of the most important product-related factors that can enhance the probability of an immune response against the protein drug is its origin (Schellekens, 2003). The immune mechanisms underlying immunogenicity of nonhuman proteins are usually explained on the basis of the sequence differences between the product and the endogenous protein. However, other factors, for example, protein aggregation, can be equally important for triggering the immune system. Nonhuman proteins are recognized as foreign antigens and evoke a response via the TD pathway. This type of immune response is responsible for the immunogenicity associated with the first animal-derived therapeutic proteins as well as with the immune reaction against plant or bacterial proteins.

Although sequence variation between endogenous and exogenous proteins seems to be an important risk factor for immunogenicity, 100% sequence homology does not lead to absence of immunogenicity. The use of human tissue-derived proteins like growth hormone (Moore and Leppert, 1980) or human IFNβ (Fierlbeck et al., 1994) is accompanied by clinically significant immune responses.

All recombinant therapeutic proteins, despite the similarity of their sequences to natural proteins, evoke an immune response. In fact, some of the recombinant human proteins have been found to be more immunogenic than their human tissue-derived counterparts; for example, whereas natural IL-2 shows only minimal immunogenicity, the recombinant IL-2 produced in *E. coli* cultures induces an immune response in over 50% of patients (Prümmer, 1997). This phenomenon can be explained by differences in post-translational modifications between bacteria and mammals, such as glycosylation (not present in bacteria), methylation, or acetylation of protein. Differences in post-translational modifications can result in the exposure of epitopes on endogenous proteins, which are normally covered by polysaccharides. These different modifications themselves may also be recognized as foreign and thereby trigger an immune response. Lack of glycosylation is one of the possible reasons explaining higher immunogenicity of IFNβ-1b compared to IFNβ-1a formulations.

One of the crucial steps in product development is formulation development. In some formulations HSA is used. HSA is generally considered as a nonimmunogenic excipient with very good stabilizing characteristics. However, the addition of HSA was correlated with formation of aggregates and increased immunogenicity of IFN-α-2a formulations (Palleroni et al., 1997). When HSA and IFN-α-2a are mixed, aggregation between the therapeutic protein and HSA may occur. Moreover, addition of bovine serum albumin (BSA) to IFN-α-2b in molar ratio equal or higher than 5:1 (BSA:IFN-α-2b) significantly disrupted the secondary structure of IFN-α-2b(Johnston et al., 2010). HSA may have a similar impact on protein in formulations. Despite the possible immunological consequences of adding HSA as an excipient, its removal has to be carefully considered as it may lead to a less robust formulation (cf. PRCA case with Eprex®; see Section 12.2.2).

During recent years, several articles have been published that show the potential immunogenicity of protein aggregates. Formation of aggregates can occur due to many physical or chemical processes, such as oxidation, protein cleavage, or conformational changes (Mahler et al., 2009; Philo and Arakawa, 2009). Oxidation can change the epitope structure and induce faster degradation of the protein. Misfolding, degradation, or aggregation may be also induced by leachates from containers, silica droplets, and contact with glass (Sharma, 2007). Deglycosylation of the protein backbone can unmask hydrophobic epitopes that are hidden under the polysaccharide residues (Seong and Matzinger, 2004).

A high content of aggregates has been correlated with increased immunogenicity of different IFNβ formulations: Betaferon® (high aggregate content) is more immunogenic than Avonex® (low aggregate content) (Bertolotto et al., 2004; van Beers et al., 2010a). The presence of aggregates has been shown to correlate with increased immunogenic potential of human-derived growth factor. Patients treated with highly aggregated hGH (50–70% of aggregates) developed persistent antibody production, which remained after switching to nonaggregated hGH (Moore and Leppert, 1980). On the other hand, patients treated with a formulation of hGH containing low amounts of aggregates (<5%) stopped producing antibodies when they received a nonaggregated product (Moore and Leppert, 1980).

Chemical conjugation of proteins with polyethylene glycol (PEG) might affect their immunogenicity. Pegylation of the protein chain usually does not lead to conformational changes, while it does prolong the protein's circulation time. In addition, it may shield immunogenic sites, like glycosylation, and therefore reduce the risk of an immune reaction. For example, pegylation successfully decreased immunogenicity of therapeutic enzymes such as asparaginase (Dipak et al., 2010) or arginase (Chirino et al., 2004), without affecting their enzymatic properties. However, studies with pegylated rhIFNα2a have shown that although its immunogenicity was decreased and serum half-life increased, the activity of the drug in vitro was lowered by 93% (Chirino et al., 2004). Moreover, it has been claimed that pegylation might lead to the production of anti-PEG antibodies, but the influence of PEG on immunogenicity is not proven.

12.4.3   ADA Detection: Assay-Related Factors

The identification of ADA-responsive patients and thus the observed immunogenicity of the therapeutic proteins is significantly influenced by the assays used to detect ADAs. Lack of standardized assays and protocols results in high variability in ADA detection between studies. The reported percentage of patients developing ADAs against Betaseron® ranged from 30% up to 97% (Bertolotto et al., 2004).

The seroprevalence can be underestimated when ADAs are measured in serum in the presence of drug, especially when ADA titers are relatively low and drug circulation time is long. Recently, a report showing the impact of assay improvement on seroprevalence of patients treated with Adalimumab® has been published. Acidic treatment used to dissociate ADAs from the drug resulted in an increase of ADA-positive samples from 5% to almost 90% (van Schouwenburg et al., 2010). This example reveals the problem of lack of ADA assay standardization. Therefore, both FDA and EMA prepared guidelines in which recommendations for sample collection, preparation, and ADA assay formats are enclosed (European Medicines Agency, 2007; Food and Drug Administration, 2009).

12.4.4 THERAPY-RELATED FACTORS

In addition to patient and product characteristics, design of therapy can significantly influence the immunogenicity of protein drugs. By changing the dose, frequency, and route of administration, it is possible to efficiently decrease the drug's immunogenicity. Moreover, additional drugs often used to increase the treatment efficiency may have impact on the immunogenicity of therapeutic protein. For instance, co-medication with immunosuppressant drugs can result in lower or less frequent immune responses against therapeutic proteins (Schellekens, 2002).

Proteins can be administrated by many different routes that may enhance or decrease their immunogenic potential. In general, the probability of unwanted immune response is the highest when protein is administered s.c., followed by the i.m. and intravenous route (Ross et al., 2000; Schellekens, 2003; Singh, 2011). The immune response against both IFNβ-1a (Rebif®) and IFNβ-1b (Betaseron®) was higher after s.c. injection than after i.m. administration (Perini et al., 2001; Ross et al., 2000). Clinical data suggest a correlation between increased immunogenicity and increased frequency of drug administration or its dose (Singh, 2011). However, clinical data are not always consistent for all therapeutic drugs. Whereas a higher dose was correlated with higher immunogenicity in patients treated s.c. with rhIFNβ (Ross et al., 2000), the administration of a higher dose of Adalimumab led to a decreased percentage of patients with measurable ADAs (Miyasaka et al., 2008).

Duration of treatment is also correlated with immunogenicity of protein drugs. Long-lasting, chronic therapy is usually accompanied with a higher risk of an unwanted immune reaction than acute treatment (Schellekens, 2003; Singh, 2011). However, some drugs, like murine monoclonal antibody Muromonab-CD3 or human recombinant BMPs, can induce a significant unwanted immune response after only single administration (Hwang et al., 2009; Singh, 2011).

12.5 PRECLINICAL ASSESSMENT OF THE IMMUNOGENICITY OF THERAPEUTIC PROTEINS

Protein drugs are the leading group of medicines with respect to their increasing market value. More and more companies are increasing their efforts in developing new protein drugs as well as generating biosimilars. Simultaneously, both the regulatory bodies and the industry put a higher emphasis on the understanding and prediction of

TABLE 12.2
Comparison of the Different Approaches for Immunogenicity Prediction

Tools	Assays	Used for	Features
In silico	T cell epitope prediction B cell epitope prediction Prediction of aggregate-prone sequences	Primary screening in drug discovery phase	Time efficient Relatively cheap Outcome depends on input data that are at best partly known
In vitro	MHC II binding assays T cell activation assays B cell activation assays	Validation of epitopes found in silico Screening of formulated products	Relatively easy Time consuming Performed in artificial environment
In vivo	Nontransgenic animals Nonhuman primates Transgenic animals	Evaluation of in silico and in vitro screening approaches Evaluation of formulated product	Expensive Time consuming Encompasses the entire immune system Translation to immunogenicity in patients questionable

their immunogenicity. In general, three types of approaches are being used to predict potential immunogenicity of protein drugs: in silico, in vitro, and in vivo methods (Table 12.2). All methods suffer from lack of data about the exact mechanism underlying immunogenicity, and their presumed predictive value is largely based on assumptions derived from the mechanisms described by classical immunology. For example, almost all in silico algorithms consider the unwanted immune response against therapeutic proteins as vaccine-like. As such, in silico prediction of immunogenic T cell epitopes is based on the TD pathway. Whereas this may be useful in antigen selection for vaccine development, in case of therapeutic proteins it can lead to mis- or over-interpretation of experimental data. Moreover, in silico methods are purely based on the structure (mainly primary) of the protein and do not take into account how product quality (e.g., presence of impurities such as aggregates and host cell proteins) and treatment regime (route of administration, concomitant medication, etc.) affect the risk of an immune response. The aim of this section is to present existing tools for immunogenicity prediction and their usefulness in the process of protein drug development.

12.5.1 In Silico Tools

The first in silico algorithm for the determination of a protein's immunogenic properties was presented in the 1980s (De Groot et al., 2008). This algorithm calculates the immunogenic properties of proteins on the basis of their relative hydrophobicity, as determined by their amino acid structure. Since then, new tools for immunogenicity prediction in silico have been developed and improved. Nowadays a number of algorithms designed for immunogenicity prediction are available. The majority of algorithms were developed for the prediction of T cell rather than B cell epitopes.

This is explained by the fact that T cells recognize short, linear sequences of amino acid chains presented on MHC II molecules, whereas most epitopes recognized by B cells have three-dimensional structures and/or are discontinuous in nature. Moreover, new structural B cell epitopes can be created if the structure of a protein is disrupted, for example, due to oxidation or aggregation.

Years of intensive studies have revealed the mechanism of peptide:MHC II interaction. The most important amino acids in the peptide sequence, which determine binding to the MHC II pocket, are called anchoring residues. The remaining amino acids are recognized by the TCR and therefore create T cell epitopes. At the moment there are three types of computational models that are used for the prediction of peptide:MHC II complex formation. Most of the in silico statistical methods calculate the probability of peptide:MHC II complex formation on the basis of known interactions between different amino acid sequences and HLA allotypes. Less sophisticated methods like the Syfpeithi database (Rammensee et al., 1999) rely only on sequence similarities between known binders and potentially immunogenic, studied antigens. Other developed tools create a scoring matrix that not only is based on the similarities between binding peptides, but also scores each amino acid depending on its position and character. Such an approach is applied in tools like Tepitope (Bian and Hammer, 2004) or Rankprep (Reche et al., 2004). More recent algorithms like ANNprep and Comprep (Lata et al., 2007), which are based on artificial neural networks (ANNs), have been developed. The superiority of ANN-based tools over score matrices is due to their ability to adjust the identification of immunogenic epitopes under impact of newly acquired data on immunogenic epitopes. Therefore, ANNs can be used not only for epitope prediction, but also to search for unknown binding patterns. The last and the most recent in silico tools are structure-based methods that aim to predict the possibility of peptide binding on the basis of the three-dimensional structure of the binding grooves of MHC II complexes derived from crystallographic studies. Therefore, such tools, for example Epibased (Desmet et al., 2005), are less dependent on the experimental data input than other in silico tools.

Whereas T cells recognize only linear, outstretched sequences, 90% of the epitopes recognized by B cells are three-dimensional structures (Ansari and Raghava, 2010). Although most of the in silico tools focus on the prediction of T cell epitopes, there are some models attempting to predict B cell epitopes, for example, Preditop (Pellequer and Westhof, 1993) and Bepitop (Odorico and Pellequer, 2003). They use propensity scales, corresponding to hydrophilicity, accessibility, flexibility, or secondary structure propensities, to assess which amino acids on the surface of an antigen are most likely involved in the epitope. In general, prediction of B cell epitopes lacks power since even a slight change in the tertiary or quaternary structure of the protein can lead to formation of neoepitopes.

In general, in silico methods for T and B cell epitope prediction are a promising field that will definitely allow to significantly reduce the number of potential drugs for further in vitro and in vivo validation in the future. Nevertheless, even the most sophisticated in silico tools will not eliminate the necessity of in vitro and in vivo methods for immunogenicity screening of therapeutic proteins. So far, in silico methods have a tendency to overestimate the number of potential immunogenic peptides. For example, they do not take into account that some epitopes can be hidden inside the core of the protein or covered by polysaccharides. Such an epitope can give a high immunogenic score in silico, but this will not be confirmed in vivo. Moreover, the influence of other cells, for example, regulatory cells, is omitted.

Although the prediction of T cell or B cell epitopes are the tools most often referred to for immunogenicity prediction and reduction, the increasing understanding of the immunogenic potency of aggregates has led to the incorporation of algorithms designed to predict aggregate prone sequences of proteins into the panel of immunogenicity prediction tools. Two main approaches are used to predict the ability of the protein to form aggregates. The first approach, for example, Aggrescan (Conchillo-Solé et al., 2007), TANGO (Fernandez-Escamilla et al., 2004), and PAGE (Tartaglia et al., 2005), bases the prediction only on the amino acid sequence of the protein. Less sophisticated algorithms (e.g., Aggrescan), base their prediction on sequences that are known to form β-cross aggregates; that is, they score the susceptibility to form β-cross aggregates by assessing the possibility of forming cross-reactive β-sheets. More sophisticated tools belonging to the first approach also include the character of every amino acid residue (e.g., possession of aromatic ring, charge, solubility, and hydrophobicity) that builds the linear peptide of a minimum length of 5–9 amino acids. The second approach creates algorithms based on molecular simulation techniques. They are designed to identify the hydrophobic patches on the surface of the protein, for example, SAP (Chennamsetty et al., 2009) and a method described by Pechmann (Pechmann and Vendruscolo, 2010). In contrast to prediction tools from the first group (e.g., Aggrescan or TANGO), SAP and Pechmann's algorithm can predict potential hydrophobic patches also in the form of the three-dimensional structures (e.g., on the surface of protein molecule) and not only as the linear sequence.

12.5.2   In Vitro Tools

The majority of in vitro assays, similar to the computational methods, are based on assumptions that immunogenicity against therapeutic proteins is T cell dependent and, hence, that protein drugs have to possess T cell epitopes. A number of different approaches have been designed to study T cell activation or antigen binding to MHC class II in vitro.

To validate in silico predictions, two types of MHC II-based in vitro assays can be used. The first is to assess the possibility of binding of the potentially immunogenic peptides to MHC class II molecules, which can be performed by incubation of peptides with a lymphoblastoid cell line of B cells (De Groot and Moise, 2007). Since those cells express different MHC class II alleles, it is possible to predict which sequence fragment of the protein is most immunogenic as well as which haplotype is correlated with increased risk of an unwanted immune response. In the second in vitro assay, which is a competition epitope binding assay, recombinant MHC class II is immobilized on a plate and then incubated with class II MHC epitope complexes (De Groot and Moise, 2007; McMurry et al., 2007). If the affinity of an epitope is higher for the immobilized allele than for the soluble one, the peptide will be detected on the plate after washing. That will provide information on which MHC class II haplotype might be correlated with a higher risk of developing an immune response against the epitope. Although the information derived from both assays could be very valuable in predicting and validating T cell epitopes, both approaches have important drawbacks. They are time and material consuming; covering of all MHC class II haplotypes seems to be unlikely; and, especially in the case of the competition assays, they are performed in nonphysiological conditions that can influence the outcome. Moreover, like in silico assays, these tests focus only on one aspect of product-related factors contributing to immunogenicity.

In contrast to the assays described in the previous paragraph, T cell-based in vitro assays are less complex and can give wider prediction potential, since not only separate peptides but also intact proteins, including formulated ones, can be used.

In T cell-based in vitro assays, T cells are isolated from the blood of a naive or drug-exposed subject and are used alone or mixed with other peripheral blood mononuclear cells (De Groot and Moise, 2007; Hobeika et al., 2005). While cells from a naive subject can give information about immunogenic properties of a protein, cells from a previously exposed patient can be used to detect and expand memory T cells. Then, depending on the approach, protein- or peptide-induced proliferation of T cells can be measured, for example, by measuring the uptake of labeled substrates (e.g., radio-labeled thymidine) incorporated into newly formed cells. In addition, protein- or peptide-induced production of activation markers such as interleukin 2 or IFNγ can be measured. While the proliferation assay gives only quantitative information on the proliferation of T cells and thus which proteins or peptides can activate T cells, measuring which activation markers (e.g., cytokines) are produced gives additional data about the nature of the immune response. Irrespective of the method used, the T cell response to the protein of interest has to be compared to the basic activity of nonstimulated cells obtained from the same subject (since T cell activation state can differ between individuals) (De Groot and Moise, 2007; Hobeika et al., 2005).

Despite many advantages of the present T cell assays, some drawbacks have to be pointed out. Similarly to the MHC class II assays, full coverage of MHC II haplotypes is difficult to obtain. Also, since, in general, T cells are less susceptible for development of response in vitro than in vivo, depletion of suppressing T cells is often necessary to unmask the response of other kinds of T cells.

A very interesting strategy for the discovery of T cell epitopes, MHC-Associated Peptide Proteomics (MAPPs), was established by Röhn et al. (2005). In this technique, class II MHC–epitope complexes are isolated from DCs that were previously stimulated with a protein in vitro or in vivo. The peptides are then dissociated from the MHC II molecule and analyzed in a system composed of a two-dimensional capillary liquid chromatography and tandem mass spectrometry system (LC-MS/MS). The most powerful features of this technique are as follows: (1) it provides the researcher with the immunodominant epitopes, (2) it takes into account the posttranslational modifications of MHC II molecules, which may affect the peptide recognition and/or selection, (3) it includes those MHC II alleles that (still) are not covered by in silico tools, and (4) it can be performed on fully formulated protein (Kropshofer and Singer, 2006; Röhn et al., 2005). However, MAPPs is a very specialized technique that requires trained personnel as well as complex calibration and validation of the analytical system. Like other techniques, MAPPs also generate a few percent of false positives and negatives due to the presentation of peptides that do not act as T cell epitopes (false positives) and due to mass spectrometry detection system limitations (false negatives); for example, 2–3% of peptides cannot be detected by ion trap mass spectrometry (Kropshofer and Singer, 2006).

In the previous paragraphs, the in vitro methods for T cell epitope prediction are described. The identification of B cell epitopes in cell-based approaches is also possible, but it is burdened with many disadvantages. The most important one, which practically eliminates B cell assays from the set of in vitro prediction tools, is the nature of B cell activation and the fact that in vitro responses of B cells are much weaker than in vivo responses. Only potent TI antigens such as LPS or TNP-Ficoll are potent enough to measurably activate naive B cells in vitro (Mond et al., 1995). In the case of weaker TI antigens, a costimulation from another agent/cell type is needed in order to obtain measurable activation of B cells. It was shown that the presence of T cells increases the speed of TI B cell activation; however, addition of T cells into a B cell assay makes it impossible to distinguish between direct activation of B cells by the protein (aggregates) or activation by MHC II-derived or structural epitopes. Therefore, B cell-based assays are not commonly used as prediction tools.

12.5.3 IN VIVO TOOLS

Immunogenicity prediction might also be performed in vivo. In general, animal models, especially nonhuman primates and transgenic rodents, allow more accurate prediction of immunogenic properties of therapeutic drugs than in silico and in vitro tools do. In silico and in vitro models simplify the immune system and therefore may lead to a wrong estimation of immunogenic potential (Hobeika et al., 2005). Moreover, the physicochemical characteristics of a drug can change after administration. This is not detected in in silico and in vitro models. The number of unknown variables related to drug administration makes it impossible to mimic this process in silico or in vitro with 100% accuracy. For example, the microenvironment of the injection spot can have a tremendous impact on immunogenicity; different routes of administration can be compared in vivo but cannot be recreated in vitro (Ross et al., 2000). Therefore, animal models still remain an important part of immunogenicity prediction.

Mouse and rat models have been widely used to assess immunogenic properties of protein drugs. However, the prediction power of those models is rather weak. Since human proteins are exogenous antigens, animals will generally develop a classical TD immune response. Therefore, direct extrapolation of immunogenicity of human-derived or recombinant proteins from those models to humans is difficult and unreliable. In contrast, nonhuman primates were found to be useful in the prediction of the relative immunogenicity of some human proteins such as growth hormone or thrombopoietin (Wierda et al., 2001). However, the reliability of nonhuman primates is limited to proteins with a conserved structure.

Better models to study immunogenicity are transgenic animals. Two different transgenic models are nowadays commonly used in preclinical studies: human HLA transgenic mouse models and animals transgenic for a particular protein of interest. The HLA transgenic models are animals with a deficient murine MHC class II that do express human HLA alleles. These mice are routinely used for the evaluation of an immune response against vaccines (Depil et al., 2006; Man et al., 1995). They show a clear correlation to effects in humans in terms of a vaccine-induced T-dependent immune response. This correlation suggests that these animals might also have predictive value when assessing immunogenicity of therapeutic drugs of both human and nonhuman origin. The existence of several transgenic strains expressing the most common HLA alleles allows us to correlate HLA haplotype with a potential risk of treatment. However, all human therapeutic proteins are external antigens for HLA transgenic animals; therefore, direct extrapolation of the immunogenicity from those animals to patients cannot be done.

The other group of transgenic animal models in immunogenicity research includes mice designed to be immune tolerant to a native human protein of interest. These mice have the human gene of the protein of interest encoded in their genome and therefore express this human protein. However, by treating these transgenic mice with the protein drug of interest, their tolerance against the protein can be broken and antibody production is initiated. Importantly, the mechanism of the antibody response in transgenic mice is likely similar to that in humans.

Several immune-tolerant transgenic animal models have been used to predict immunogenicity. In addition, they are used to evaluate critical product-related parameters leading to the immune response against protein drugs. The first animal models implemented in immunogenicity research were transgenic mice producing human plasminogen activator (tPA) (Stewart et al., 1989). Although these animals were tolerant to the natural tPA, they were capable of producing antibodies against a modified form of tPA (single amino acid substitution). Since then several other transgenic mouse models were developed and validated. For example, studies using transgenic animals tolerant for human IFNα (Hermeling et al., 2006), IFNβ (van Beers et al., 2010c), and insulin (Ottesen et al., 1994) have been and are being carried out. Several of those studies revealed that presence of aggregates is an important risk factor for unwanted immune responses. Both IFNα and IFNβ, when present in the monomer form, are not able to induce an immune response in transgenic mice. In contrast, some, but not all, formulations containing aggregated forms of those proteins elicit antibodies.

Although transgenic animals expressing human proteins of interest are valuable models for immunogenicity studies, they still have some important disadvantages. In contrast to HLA transgenic animals, they express murine MHC II alleles; therefore, even though they are tolerant for a human protein, during the immune response against aggregated or altered protein a different pattern of epitopes can be recognized compared to patients. Moreover, many available therapeutic drugs contain HSA as stabilizer. During the assessment of immunogenic properties of HSA-containing products, caution is needed as the impact of HSA (a foreign protein for the transgenics, but obviously not for human patients) on the immune response against the protein drug is not known.

12.6 HOW TO DECREASE/ELIMINATE THE IMMUNOGENICITY OF THERAPEUTIC PROTEINS

The existence of even the most sophisticated and accurate tools to predict immunogenicity would be worthless without the availability of techniques that allow elimination of immunogenic sites. Before the age of recombinant technology, the immunogenic potential of protein drugs could be reduced only by optimizing the production process, including formulation, and assuring proper storage and administration conditions. In fact, immunogenicity of the first protein drugs was significantly reduced after improving their purification and storage conditions.

The development of recombinant technology opened a new possibility of rational protein modification in order to achieve higher stability and reduced immunogenicity. In combination with solving the sequence of proteins (at both DNA and amino acid levels), recombinant technology allows direct elimination of the most immunogenic parts of the protein sequence that are thought to be responsible for triggering the immune response.

12.6.1 ELIMINATION OF T CELL EPITOPES

The elimination of T cell epitopes is one of the most potent ways of limiting immunogenicity of therapeutic proteins; for example, it is used for reducing immunogenicity of factor VIII (Jones et al., 2009; Song et al., 2010). Although this method has been mainly applied for nonhuman protein drugs, in principle it can be used for all types of protein drugs. While nonhuman proteins contain a number of potential MHC II-binding peptide sequences, usually only a few of them are dominant. A small change in these potential epitopes, such as replacing one amino acid residue, can significantly reduce its recognition by immune cells. For example, a single amino acid substitution (glutamic acid to serine or alanine) in the T cell epitope of staphylokinase lowered the titers of produced antibodies by 35% (He et al., 2010). Also, immunogenicity of the human-derived or the recombinant human proteins can be lowered by applying directed mutagenesis to remove potentially immunogenic epitopes. A good example is human-derived IFNβ, in which the possibility of T cell activation in vitro was inhibited by a point mutation (Yeung et al., 2004). Moreover, the immune response of BALB/cByJ mice against mutated human IFNβ was significantly lower than against the nonmutated form (Stewart et al., 1989). However, lower immunogenicity of this mutated protein has not been confirmed in patients. During T cell epitope removal, caution should be applied since every mutation in the protein might change the conformation or the activity, or increase susceptibility to aggregation.

In the case of monoclonal antibodies, another approach has been undertaken to lower immunogenicity. Because many monoclonal antibodies were (partially) foreign, these proteins are subjected to humanization, a process during which almost the whole murine sequence, except complementarity-determining regions (CDRs), is replaced with its human counterpart (Jones et al., 2009; Roque-Navarro et al., 2003). Monoclonal antibodies are especially suitable for that modification due to their characteristic conserved structure including the possession of Fab and Fc regions. Since the specificity and the affinity of an antibody are dependent only on the sequence and conformation of the Fab region, the rest of the residues can be modified in order to reduce its immunogenicity or increase its stability (Harding et al., 2010). Modification of amino acid residues not involved in epitope binding is a common approach to increasing stabilization and safety of monoclonal antibodies. Humanization of murine antibodies by complete exchange of the murine Fc region by its human counterpart was the first successful attempt in reducing immunogenicity of the first therapeutic monoclonal antibodies. The humanization of these chimeric monoclonal antibodies was pushed even further by the development of humanized antibodies, with only the CDRs being nonhuman, and fully human antibodies. However, even a fully human antibody contains foreign CDRs, which cannot be removed without loss of the antibody's activity (Harding et al., 2010).

12.6.2 ELIMINATION OF AGGREGATION-PRONE SEQUENCES

The basic approach underlying elimination of aggregation-prone sequences is similar to T cell epitope removal. Directed mutations allow changing the nature of aggregate-prone sequences and thereby limit the possibility to form aggregates. Elimination of aggregate-prone sequences suffers from the same drawbacks as elimination of T cell epitopes. Any modification of the protein structure can influence the specificity, activity, and conformation of the protein. A change in protein conformation can lead to the formation of neoepitopes, which can be more efficiently recognized by immune cells. Elimination of aggregate-prone sites is especially complicated in monoclonal antibodies as they are often localized in the CDRs (Wang et al., 2010).

12.7 CONCLUSIONS

In the near future, therapeutic proteins will dominate the drug market. However, to fully benefit from their therapeutic potential, deeper knowledge about all aspects of immunogenicity is needed. During the past few years, great progress in understanding the importance of the immunogenicity and accompanied risk factors has been made. The development of new, more sensitive assays for immunogenicity development revealed the commonness of unwanted immune responses even for drugs that until recently were considered hardly immunogenic. Despite this progress, the exact immunological mechanism of immunogenicity of especially recombinant proteins still remains poorly understood and further intensive studies are needed in order to understand the mechanism of immunogenicity and to rationally design less immunogenic protein therapeutics.

REFERENCES

Ansari, H., Raghava, G. 2010. Identification of conformational B-cell epitopes in an antigen from its primary sequence. *Immunome Res* 6:6.

Antonelli, G., Dianzani, F. 1999. Development of antibodies to interferon beta in patients: Technical and biological aspects. *Eur Cytokine Netw* 10:413–422.

Barbosa, M., Vielmetter, J., Chu, S., Smith, D., Jacinto, J. 2006. Clinical link between MHC class II haplotype and interferon-beta (IFN-B) immunogenicity. *Clin Immunol* 118:42–50.

Bertolotto, A., Deisenhammer, F., Gallo, P., Sölberg Sørensen, P. 2004. Immunogenicity of interferon beta: Differences among products. *J Neurol* 251:II/15–II/24.

Bian, H., Hammer, J. 2004. Discovery of promiscuous HLA-II-restricted T cell epitopes with TEPITOPE. *Methods* 34:468–475.

Casadevall, N., Nataf, J., Viron, B., et al. 2002. Pure red-cell aplasia and antierythropoietin antibodies in patients treated with recombinant erythropoietin. *N Engl J Med* 346:469–475.

Cassinotti, A., Travis, S. 2009. Incidence and clinical significance of immunogenicity to infliximab in Crohn's disease: A critical systematic review. *Inflamm Bowel Dis* 15:1264–1275.

Chamberlain, P. 2002. Immunogenicity of therapeutic proteins. Part 1: Causes and clinical manifestations of immunogenicity. *Regul Rev* 5:4–9.

Chennamsetty, N., Voynov, V., Kayser, V., Helk, B., Trout, B. 2009. Design of therapeutic proteins with enhanced stability. *Proc Natl Acad Sci USA* 106:11937–11942.

Chirino, A., Ary, M., Marshall, S. 2004. Minimizing the immunogenicity of protein therapeutics. *Drug Discov Today* 9:82–90.

Conchillo-Solé, O., DeGroot, N., Avilés, F., Vendrell, J., Daura, X., Ventura, S. 2007. AGGRESCAN: A server for the prediction and evaluation of "hot spots" of aggregation in polypeptides. *BMC Bioinform* 8:65.

De Groot, A., McMurry, J., Moise, L. 2008. Prediction of immunogenicity: In silico paradigms, ex vivo and in vivo correlates. *Curr Opin Pharmacol* 8:620–626.

De Groot, A., Moise, L. 2007. Prediction of immunogenicity for therapeutic proteins: State of the art. *Curr Opin Drug Discov Devel* 10:332–340.

Depil, S., Angyalosi, G., Moralès, O., et al. 2006. Peptide-binding assays and HLA II transgenic Abeta degrees mice are consistent and complementary tools for identifying HLA II-restricted peptides. *Vaccine* 24:2225–2229.

Desmet, J., Meersseman, G., Boutonnet, N., Pletinckx, J., De Clercq, K., Debulpaep, M., Braeckman, T., Lasters, I. 2005. Anchor profiles of HLA-specific peptides: Analysis by a novel affinity scoring method and experimental validation. *Proteins* 58:53–69.

Dillman, R. 1999. Infusion reactions associated with the therapeutic use of monoclonal antibodies in the treatment of malignancy. *Cancer Metastasis Rev* 18:465–471.

Dipali, S., Bicol, M., Sathy, V. 2010. Delivery of therapeutic proteins. *J Pharm Sci* 99:2557–2575.

European Medicines Agency. 2007. Guideline on immunogenicity assessment of biotechnology derived therapeutic proteins. Doc. Ref. EMEA/CHMP/BMWP/14327/2006. http://www.ema.europa.eu/docs/en_GB/document_library/Scientific_guideline/2009/09/WC500003946.pdf (accessed December 1, 2011).

Fakharzadeh, S., Kazazian, H. 2000. Correlation between factor VIII genotype and inhibitor development in hemophilia A. *Semin Thromb Hemost* 26:167–171.

Fernandez-Escamilla, A., Rousseau, F., Schymkowitz, J., Serrano, L. 2004. Prediction of sequence-dependent and mutational effects on the aggregation of peptides and proteins. *Nat Biotechnol* 22:1302–1306.

Fierlbeck, G., Schreiner, T., Schaber, B., Walser, A., Rassner, G. 1994. Neutralizing interferon beta antibodies in melanoma patients treated with recombinant and natural interferon beta. *Cancer Immunol Immunother* 39:263–269.

Food and Drug Administration. 2009. Guidance for industry assay development for immunogenicity testing of therapeutic proteins. http://www.fda.gov/downloads/Drugs/GuidanceComplianceRegulatoryInformation/Guidances/UCM192750.pdf (accessed December 1, 2011).

Goldsby, R., Kindt, T., Osborne, R., Kuby, J. 2006. *Immunology*, 6th ed. W.H. Freeman & Co, New York.

Harding, F., Stickler, M., Razo, J., Dubridge, R. 2010. The immunogenicity of humanized and fully human antibodies: Residual immunogenicity resides in the CDR regions. *MAbs* 2:256–265.

Hartung, H., Polman, C., Bertolotto, A., et al. 2007. Neutralising antibodies to interferon beta in multiple sclerosis: Expert panel report. *J Neurol* 254:827–837.

He, J., Xu, R., Chen, X., Jia, K., Zhou, X., Zhu, K. 2010. Simultaneous elimination of T-cell and B-cell epitope by structure-based mutagenesis of single Glu80 residue within recombinant staphylokinase. *Acta Biochim Biophys Sin* 42:209–215.

Hermeling, S., Schellekens, H., Maas, C., Gebbink, M., Crommelin, D., Jiskoot, W. 2006. Antibody response to aggregated human interferon alpha2b in wild-type and transgenic immune tolerant mice depends on type and level of aggregation. *J Pharm Sci* 95:1084–1096.

Herndon, R., Rudick, R., Munschauer, F., et al. 2005. Eight-year immunogenicity and safety of interferon beta-1a-Avonex treatment in patients with multiple sclerosis. *Mult Scler* 11:409–419.

Hobeika, A., Morse, M., Osada, T., Ghanayem, M., Niedzwiecki, D., Barrier, R., Lyerly, H., Clay, T. 2005. Enumerating antigen-specific T-cell responses in peripheral blood: A comparison of peptide MHC Tetramer, ELISpot, and intracellular cytokine analysis. *J Immunother* 28:63–72.

Hwang, C., Vaccaro, A., Lawrence, J., Hong, J., Schellekens, H., Alaoui-Ismaili, M., Falb, D. 2009. Immunogenicity of bone morphogenetic proteins. *J Neurosurg Spine* 10:443–451.

Jacquemin, M., Saint-Remy, J. 1998. Factor VIII immunogenicity. *Haemophilia* 4:552–557.

Johnston, M., Nemr, K., Hefford, M.A. 2010. Influence of bovine serum albumin on the secondary structure of interferon alpha 2b as determined by far UV circular dichroism spectropolarimetry. *Biologicals* 38:314–320.

Jones, T., Crompton, L., Carr, J., Baker, M. 2009. Deimmunization of monoclonal antibodies. *Methods Mol Biol* 525:405–423.

Kang, S., Saif, M. 2007. Infusion-related and hypersensitivity reactions of monoclonal antibodies used to treat colorectal cancer—identification, prevention, and management. *J Support Oncol* 5:451–457.

Kropshofer, H., Singer, T. 2006. Overview of cell-based tools for pre-clinical assessment of immunogenicity of biotherapeutics. *J Immunotoxicol* 3:131–136.

Lata, S., Bhasin, M., Raghava, G. 2007. Application of machine learning techniques in predicting MHC binders. *Methods Mol Biol* 409:201–251.

Li, L., Boussiotis, V. 2006. Physiologic regulation of central and peripheral T cell tolerance: Lessons for therapeutic applications. *J Mol Med* 84:887–899.

Lidove, O., West, M., Pintos-Morell, G., et al. 2010. Effects of enzyme replacement therapy in Fabry disease—A comprehensive review of the medical literature. *Genet Med* 12:668–679.

Ludwig, H., Fritz, E., Leitgeb, C., Krainer, M., Kührer, I., Sagaster, P., Umek, H. 1992. Erythropoietin treatment for chronic anemia of selected hematological malignancies and solid tumors. *Ann Oncol* 4:161–167.

Macleod, J. 1922. Insulin in diabetes: A general statement of the physiological and therapeutic effects of insulin. *BMJ* 2:833–835.

Mahler, H., Friess, W., Grauschopf, U., Kiese, S. 2009. Protein aggregation: Pathways, induction factors and analysis. *J Pharm Sci* 98:2909–2934.

Man, S., Newberg, M., Crotzer, V., Luckey, C., Williams, N., Chen, Y., Huczko, E., Ridge, J., Engelhard, V. 1995. Definition of a human T cell epitope from influenza A non-structural protein 1 using HLA-A2.1 transgenic mice. *Int Immunol* 7:597–605.

Markham, A., Lamb, H. 2000. Infliximab: A review of its use in the management of rheumatoid arthritis. *Drugs* 59:1341–1359.

McMurry, J., Gregory, S., Moise, L., Rivera, D., Buus, S., De Groot, A. 2007. Diversity of *Francisella tularensis* Schu4 antigens recognized by T lymphocytes after natural infections in humans: Identification of candidate epitopes for inclusion in a rationally designed tularemia vaccine. *Vaccine* 25:3179–3191.

Melchers, F., Rolink, A. 2006. B cell tolerance—How to make it and how to break it. *Curr Top Microbiol Immunol* 305:1–23.

Milner, R. 1985. Growth hormone. *BMJ* 291:1593–1594.

Miyasaka, N., The CHANGE Study Investigators. 2008. Clinical investigation in highly disease-affected rheumatoid arthritis patients in Japan with adalimumab applying standard and general evaluation: The CHANGE study. *Mod Rheumatol* 18:252–262.

Mond, J., Vos, Q., Lees, A., Snapper, C. 1995. T-cell independent antigens. *Curr Opin Immunol* 7:349–354.

Moore, W., Leppert, P. 1980. Role of aggregated human growth hormone (hGH) in development of antibodies to hGH. *J Clin Endocrinol Metab* 51:691–697.

Obukhanych, T., Nussenzweig, M. 2006. T-independent type II immune responses generate memory B cells. *J Exp Med* 203:305–310.

Odorico, M., Pellequer, J. 2003. BEPITOPE: Predicting the location of continuous epitopes and patterns in proteins. *J Mol Recognit* 16:20–22.

Ottesen, L., Nillson, P., Jami, J., Weilguny, D., Dührkop, M., Bucchini, D., Havelund, S., Fogh, J. 1994. The potential immunogenicity of human insulin and insulin analogues evaluated in a transgenic mouse model. *Diabetologia* 12:1178–1185.

Palleroni, A., Aglione, A., Labow, M., Brunda, M., Pestka, S., Sinigaglia, F., Garotta, G., Alsenz, J., Braun, A. 1997. Interferon immunogenicity: Preclinical evaluation of interferon-alpha 2a. *J Interferon Cytokine Res* 17:s23–s27.

Palman, C., Bertolotto, A., Deisenhammer, F., et al. 2010. Recommendations for clinical use of data on neutralising antibodies to interferon-beta therapy in multiple sclerosis. *Lancet Neurol* 9:740–750.

Pechmann, S., Vendruscolo, M. 2010. Derivation of a solubility condition for proteins from an analysis of the competition between folding and aggregation. *Mol Biosyst* 6:2490–2497.

Pellequer, J., Westhof, E. 1993. PREDITOP: A program for antigenicity prediction. *J Mol Graph* 11:204–210.

Perini, P., Facchinetti, A., Bulian, P., Massaro, A., Pascalis, D., Bertolotto, A., Biasi, G., Gallo, P. 2001. Interferon-beta (INF-beta) antibodies in interferon-beta1a- and interferon-beta1b-treated multiple sclerosis patients. Prevalence, kinetics, cross-reactivity, and factors enhancing interferon-beta immunogenicity in vivo. *Eur Cytokine Netw* 12:56–61.

Philo, J., Arakawa, T. 2009. Mechanisms of protein aggregation. *Curr Pharm Biotechnol* 10:348–351.

Pollock, C., Johnson, D., Hörl, W., et al. 2008. Pure red cell aplasia induced by erythropoiesis-stimulating agents. *Clin J Am Soc Nephrol* 3:193–199.

Prümmer, O. 1997. Treatment-induced antibodies to interleukin-2. *Biotherapy* 10:15–24.

Rammensee, H., Bachmann, J., Emmerich, N., Bachor, O., Stevanović, S. 1999. SYFPEITHI: Database for MHC ligands and peptide motifs. *Immunogenetics* 50:213–219.

Reche, P., Glutting, J., Zhang, H., Reinherz, E. 2004. Enhancement to the RANKPEP resource for the prediction of peptide binding to MHC molecules using profiles. *Immunogenetics* 56:405–419.

Reeves, W., Gelsthorpe, K., Van der Minne, P., Torensma, R., Tattersall, R. 1984. HLA phenotype and insulin antibody production. *Clin Exp Immunol* 57:443–448.

Richards, S. 2002. Immunologic considerations for enzyme replacement therapy in the treatment of lysosomal storage disorders. *Clin Appl Immunol Rev* 2:241–253.

Röhn, T., Reitz, A., Paschen, A., Nguyen, X., Schadendorf, D., Vogt, A., Kropshofer, H. 2005. A novel strategy for the discovery of MHC class II-restricted tumor antigens: Identification of a melanotransferrin helper T-cell epitope. *Cancer Res* 65:10068–10078.

Roque-Navarro, L., Mateo, C., Lombardero, J., Mustelier, G., Fernández, A., Sosa, K., Morrison, S., Pérez, R. 2003. Humanization of predicted T-cell epitopes reduces the immunogenicity of chimeric antibodies: New evidence supporting a simple method. *Hybrid Hybridomics* 22:245–257.

Ross, C., Clemensen, K., Svenson, M., Sorenses, P., Koch-Henriksen, N., Skovgaard, G., Bentzen, K. 2000. Immunogenicity of interferon-beta in multiple sclerosis patients: Influence of preparation, dosage, dose frequency, and route of administration. Danish Multiple Sclerosis Study Group. *Ann Neurol* 48:706–712.

Schellekens, H. 2002. Immunogenicity of therapeutic proteins: Clinical implications and future prospects. *Clin Ther* 24:1720–1740.

Schellekens, H. 2003. Immunogenicity of therapeutic proteins. *Nephrol Dial Transplant* 18:1257–1259.

Schellekens, H. 2007. Lesson learned from Eprex-associated pure red cell aplasia. *Kidney Blood Press Res* 30:9–12.

Seong, S., Matzinger, P. 2004. Hydrophobicity: An ancient damage-associated molecular pattern that initiates innate immune responses. *Nat Rev Immunol* 4:469–478.

Sharma, B. 2007. Immunogenicity of therapeutic proteins. Part 2: Impact of container closures. *Biotechnol Adv* 25:318–324.

Singh, S. 2011. Impact of product-related factors on the immunogenicity of biotherapeutics. *J Pharm Sci* 100:354–387.

Smith, C., Kitchen, L., Scullard, G., Robinson, W., Gregory, P., Merigan, T. 1982. Vidarabine monophosphate and human leukocyte interferon in chronic hepatitis B infection. *JAMA* 247:2261–2265.

Snapper, C., Kehry, M., Castle, B., Mond, J. 1995. Multivalent, but not divalent, antigen receptor cross-linkers synergize with CD40 ligand for induction of Ig synthesis and class switching in normal murine B cells. A redefinition of the TI-2 vs T cell-dependent antigen dichotomy. *J Immunol* 154:1177–1187.

Sominada, A., Rot, U., Suoniemi, M., Deisenhammer, F., Hillert, J., Fogdell-Hahn, A. 2007. Interferon beta preparations for the treatment of multiple sclerosis patients differ in neutralizing antibody seroprevalence and immunogenicity. *Mult Scler* 13:208–214.

Song, C., Martin, W., DeGroot, A., Scott, D. 2010. Deimmunization of HLA-DR epitopes in the C2 domain of human FVIII. *J Immunol* 184:144.29.

Stewart, T., Hollingshead, P., Pitts, S., Chang, R., Martin, L., Oakley, H. 1989. Transgenic mice as a model to test the immunogenicity of proteins altered by site-specific mutagenesis. *Mol Biol Med* 6:275–281.

Tartaglia, G., Cavalli, A., Pellarin, R., Caflisch, A. 2005. Prediction of aggregation rate and aggregation-prone segments in polypeptide sequences. *Protein Sci* 14:2723–2734.

van Beers, M., Jiskoot, W., Schellekens, H. 2010a. On the role of aggregates in the immunogenicity of recombinant human interferon beta in patients with multiple sclerosis. *J Interferon Cytokine Res* 30:767–775.

van Beers, M., Sauerborn, M., Gilli, F., Brinks, V., Schellekens, H., Jiskoot, W. 2010b. Aggregated recombinant human interferon beta induces antibodies but no memory in immune-tolerant transgenic mice. *Pharm Res* 27:1812–1824.

van Beers, M., Sauerborn, M., Gilli, F., Hermeling, S., Brinks, V., Schellekens, H., Jiskoot, W. 2010c. Hybrid transgenic immune tolerant mouse model for assessing the breaking of B cell tolerance by human interferon beta. *J Immunol Methods* 352:32–37.

van der Voort, L., Gilli, F., Bertolotto, A., Knol, D., Uitdehaag, B., Polman, C., Killestein, J. 2010. Clinical effect of neutralizing antibodies to interferon beta that persist long after cessation of therapy for multiple sclerosis. *Arch Neurol* 67:402–407.

van Schouwenburg, P., Bartelds, G., Hart, M., Aarden, L., Wolbink, G., Wouters, D. 2010. A novel method for the detection of antibodies to adalimumab in the presence of drug reveals "hidden" immunogenicity in rheumatoid arthritis patients. *J Immunol Methods* 362:82–88.

Vos, Q., Lees, A., Wu, Z., Snapper, C., Mond, J. 2000a. B-cell activation by T-cell-independent type 2 antigens as an integral part of the humoral immune response to pathogenic microorganisms. *Immunol Rev* 176:154–170.

Vos, Q., Snapper, C., Mond, J. 2000b. Th1 versus Th2 cytokine profile determines the modulation of in vitro T cell-independent type 2 responses by IL-4. *Int Immunol* 12:1337–1345.

Wang, X., Singh, S., Kumar, S. 2010. Potential aggregation-prone regions in complementarity-determining regions of antibodies and their contribution towards antigen recognition: A computational analysis. *Pharm Res* 27:1512–1529.

West, R., Zelinkowa, Z., Wolbink, G., Kuipers, E., Stokkers, P., van der Woude, C. 2008. Immunogenicity negatively influences the outcome of adalimumab treatment in Crohn's disease. *Aliment Pharmacol Ther* 28:1122–1126.

Wierda, D., Smith, H., Zwickl, C. 2001. Immunogenicity of biopharmaceuticals in laboratory animals. *Toxicology* 158:71–74.

Yeung, P., Chang, J., Miller, J., Barnett, C., Stickler, M., Harding, F. 2004. Elimination of an immunodominant CD4+ T cell epitope in human IFN-B does not result in an in vivo response directed at the subdominant epitope. *J Immunol* 172:6658–6665.

13 Biosimulation of Peptides and Proteins

Tue Søeborg, Christian Hove Rasmussen, Erik Mosekilde, and Morten Colding-Jørgensen

CONTENTS

13.1 BIOSIMULATION: AN INTRODUCTION

The use of biological simulation models is likely to play a rapidly increasing role in the coming years as part of the industry's efforts to continuously improve the pharmaceutical formulation of peptides, proteins, and other drugs and to optimize clinical effect, bioavailability, shelf life, and so on. Models of a similar type are used practically in all other industries and the results are significant reductions in developing time and costs.

Biosimulation is *the behavior of a biological system translated into mathematics and investigated by means of computer simulation.* The translation into mathematics often takes the form of differential equations, and computer simulation allows for the investigation of the dynamics of these equations for different parameter combinations. It is important, however, that the translation into mathematical equations represents the biological mechanisms of the investigated phenomena as directly as possible, considering both the timescale over which the phenomena unfold and the physiological level at which they take place.

Biosimulation represents a process- and mechanism-oriented approach rather than a data-driven methodology commonly used in the pharmaceutical industry. This implies that parameters and interactions are defined in such a way that they have a direct physiological meaning and can be measured independently of the model (or determined from other sources of information). The idea is that all relevant information about the drug and its biological effects can be incorporated into a consistent, quantitative description and that this description can be used to predict kinetics as well as clinical effects under conditions that have not yet been examined. The biosimulation approach allows new information to be interpreted directly in the context of already established knowledge. This reduces the required amount of experimental work significantly and increases the efficiency of the development process.

The traditional methods of pharmacokinetics and pharmacodynamics have been developed with a primary view on exploiting a given set of data as well as possible in the accurate determination of specific parameters. This occurs through fitting of parameterized curves to the data. In this connection, the advantage of using physiologically defined parameters is that it allows the models to be continuously improved and expanded as additional information becomes available or new problems become of interest. One of the purposes of this chapter is to precisely demonstrate how an insulin absorption model can be established step by step and how it can be extended subsequently in order to address different problems.

A biosimulation model is both more and less than a specific set of data it is meant to describe. It is more than the data, because it builds on a consistent hypothesis about the processes that have generated the data, established, in general, on the basis of significant biological experience. However, the model is also less than a particular set of data, because all such sets will display variations and deviations from the model predictions. In every particular case, the model constructor must, therefore, decide whether the deviations are associated with processes that can be neglected in the present context or whether they require modifications of the model.

In connection with the discussion of parameter accuracy, it is important to realize that biological systems generally involve a significant number of negative feedback regulations. Each such negative feedback tends to stabilize the system and reduces its sensitivity to parameter variations, thus allowing different individuals to function more or less in the same way in spite of significant parameter differences. If the model represents the biological feedback mechanisms correctly, it will display a similar reduced sensitivity to precise parameter values. If the model does not represent the feedback regulations, it is not a good model.

In this chapter, we shall demonstrate the use of biosimulation to predict insulin absorption kinetics after subcutaneous administration. Following a brief discussion of subcutaneous injection of soluble and suspended substances, we first present a simple analysis of the chemical equilibrium conditions between soluble insulin monomers, dimers, and hexamers. This leads us to establish a model that describes the dynamics of the chemical reaction processes when absorption, diffusion, and degradation of insulin species occur simultaneously. Simulations are performed to show the effects on the absorption kinetics of injecting insulin in different doses and concentrations.

   To further broaden the scope of the analysis, the dynamics associated with dissolution and recrystallization of insulin is introduced in the insulin absorption model, and simulations are performed to illustrate how heaps of crystalline insulin are formed and dissolved in the subcutaneous tissue. These simulations provide a partial explanation to the higher absorption rate variability and lower bioavailability of crystalline insulin as compared to regular, human insulin. Finally, the model is used to quantitatively describe how temporal variations in blood perfusion at the injection site influences bioavailability and absorption rates of insulin mixtures.

13.2   SUBCUTANEOUS INJECTIONS

Before describing the absorption kinetics of subcutaneously administered peptides and proteins, it is useful to consider what actually happens when a substance is injected into the subcutaneous tissue. Hewitt (1954) investigated the mechanisms of subcutaneous injection by administering 0.2 mL of a soluble or suspended dye to mice. The results showed that the volume of the colored region of the soluble dye was approximately 2.3 mL, or 11 times greater than the injected volume. Following injection of suspended dye, the colored space was approximately 0.02 mL or 10 times less than the injected volume, but with a weak trace of solute reaching the same space as the soluble dye.

   These observations can be explained by considering the following: In subcutaneous tissue, fat cells and other constituents take up much of the space. The remaining extracellular space, available for diffusion, accounts for approximately 10% as reported by Linde and Chisholm (1975) and as reviewed by Crandall et al. (1997). Thus, the injected volume will spread over a larger space than can be accounted for by the actual injected volume, consistent with the findings of Hewitt (1954). Injected solute or soluble substances will diffuse around the cells in the irregular and winded extracellular space. This restricted diffusion can be described by the so-called tortuosity factor, λ. A simple way of introducing λ is to reduce the diffusion constants (see Section 13.3.2), typically by a factor of λ around 1.5–1.7 (Nicholson and Sykova, 1998; Sharkawy et al., 1997; Shen and Chen, 2007).

   Subcutaneous injection of suspended particles will, dependent on particle size, result in a different scenario. Particles above a certain size will be filtered by the tissue and retained near the injection site, while the solute will reach more or less the same spreading as pure solute as observed by Hewitt (1954). In a formulation of U100* crystalline neutral protamine Hagedorn (NPH) insulin, the crystals are estimated to account for about 3.6% of the total volume. With complete separation, the crystals are retained to the innermost 3.6% of the volume (33% of the radius), while the solute will spread and reach the same space as when injecting pure solute, if the tissue is intact. With partial separation, the crystals will reach between 33% and 100% of the radius, typically in an irregular fashion.

   The particles in a pharmaceutical formulation intended for subcutaneous administration should normally not be larger than approximately 40 μm (DeFelippis and Akers, 2000). With larger particles, high doses, and/or high injection pressures, the tissue may burst due to clogging of the mesh in the tissue. Consequently, the central

---

* U100 is defined as 100 U ml⁻¹. U is insulin units with 1 U = 6.00 nmol (Vølund, 1993).

part of the depot will mainly consist of unaltered injection fluid without fat cells, and the total volume becomes smaller. After the injection, a large part of the fluid will sieve outwards, so the crystals will end up with a distribution resembling that in the previous case with no bursting.

The practical implications of the subcutaneous tissue acting like a sieve that retains injected particles will be described in more detail in Section 13.4.

13.3 ABSORPTION KINETICS AND SELF-ASSOCIATION

The complex physicochemical properties of insulin provide an obvious field for the use of biosimulation. In the following, the dynamics of soluble insulin absorption kinetics is split into subsystems, which are described individually. Subsequently, the subsystems are combined into an overall system for simulation purposes.

The first biosimulation model of insulin absorption kinetics was published by Mosekilde et al. (1989) and several others have built on the same principles, see, for example, Trajanoski et al. (1993), Wach et al. (1995), Ploughman et al. (2003), Tarín et al. (2005), Li and Kuang (2009), and Søeborg et al. (2009, 2012). Apart from these models, several nonphysiological descriptions of insulin kinetics have been published.

13.3.1 Oligomer Equilibria

Regular, human insulin may be regarded as a chemical equilibrium between insulin monomers, dimers, and hexamers. One hexamer consists of three dimers, which may break into a total of six monomers (Pekar and Frank, 1972). This may be written as

$$6\,\text{M} \overset{K_{MD}}{\leftrightarrow} 3\,\text{D} \overset{K_{DH}}{\leftrightarrow} \text{H} \tag{13.1}$$

where M, D, and H represent the insulin monomers, dimers, and hexamers, respectively, and K_{MD} and K_{DH} are the equilibrium constants between monomers and dimers, and between dimers and hexamers, respectively. Assuming that all species (monomers, dimers, and hexamers) are in mutual equilibrium at a given time, the total concentration of insulin becomes

$$C_T = C_M + C_D + C_H = C_M + K_{MD}C_M^2 + K_{MD}K_{MD}^3 C_M^6 \tag{13.2}$$

where C_T, C_M, C_D, and C_H are the concentrations of total insulin, insulin monomers, dimers, and hexamers, respectively, all measured in monomer concentration units. As evident from Equation 13.2, an insulin solution of high concentration contains relatively more insulin hexamers than a low concentrated solution of insulin.

To increase the shelf life of pharmaceutical formulations of insulin, auxiliary substances such as zinc and phenol and/or phenol-like substances are present in the vial. These substances shift the chemical equilibrium toward insulin hexamers, which are more chemically and physically stable than the insulin dimers and monomers

(Brange, 1987). The presence of the auxiliary substances leads to an increase in the value of K_{DH} relative to the situation where no auxiliary substances are present. With zinc and phenol and/or phenol-like substances, insulin is predominantly found in the R_6 hexamer state. Without phenol and/or phenol-like substances, the conformation of the hexamer shifts into the less stable T_6 hexamer that may further dissociate into dimers (Derewenda et al., 1989; Hassiepen et al., 1999; Rahuel-Clermont et al., 1997).

After subcutaneous administration, the auxiliary substances are rapidly eliminated. This lowers the value of K_{DH} and shifts the equilibrium between the insulin species toward monomers and dimers (Hvidt, 1991). To describe and quantify the absorption kinetics, K_{DH} and K_{MD} now have to be known at every point in time. Only the value of K_{DH} is assumed to change dependent on the presence or absence of auxiliary substances, because the auxiliary substances predominantly affect the state of the hexamers. The values of K_{DH} and K_{MD} are available from the literature, although they are associated with significant uncertainty related to differences in study design, insulin type, method of analysis, and so on, as reviewed in Søeborg et al. (2009).

The relative distributions of insulin species as functions of the total insulin concentration according to Equation 13.2 in the vial (with auxiliary substances) and in the subcutaneous depot (without auxiliary substances) are given in Figure 13.1a and b.

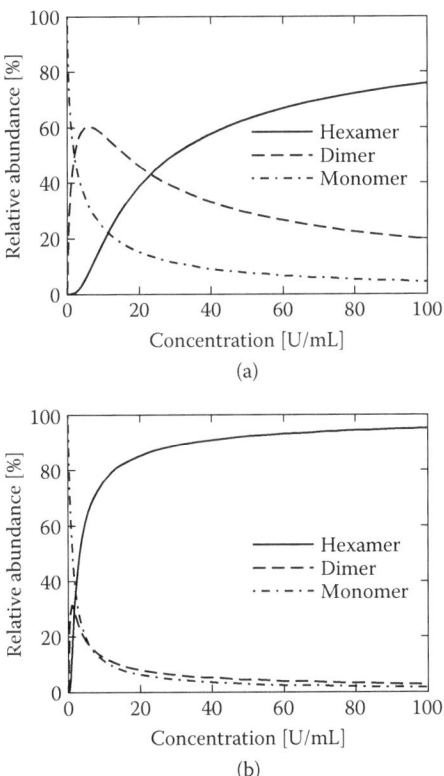

FIGURE 13.1   Distribution of insulin species as a function of total concentration in the vial (a) and in the subcutaneous depot (b).

13.3.2 Absorption, Diffusion, and Degradation

Owing to the difference in size, subcutaneously administered insulin monomers are absorbed into the bloodstream at a faster rate than insulin dimers, which again are absorbed at a faster rate than insulin hexamers (Brange et al., 1990). Limited absorption of insulin hexamers has been found to take place (Kurtzhals and Ribel, 1995), but it may be through the lymphatic system as suggested by Charman et al. (2001).

When insulin monomers and dimers are absorbed, the chemical equilibrium described in Equations 13.1 and 13.2 is shifted leading to dissociation of insulin hexamers into dimers and monomers. Furthermore, the different insulin species diffuse into the subcutaneous tissue at different rates due to the difference in size (monomers diffuse at a faster rate than dimers and hexamers, respectively) (Bocian et al., 2008; Lin and Larive, 1995; Oliva et al., 2000).

The subcutaneous bioavailability of soluble insulin is around 65% (Søeborg et al., 2012). This means that some of the subcutaneously administered insulin is degraded before being absorbed. The mechanisms determining the bioavailability of soluble insulin are not known. Coadministration of protease inhibitors has been shown to increase the subcutaneous bioavailability of soluble insulin (Takeyama et al., 1991), implying that enzymatic degradation of insulin takes place in the depot. On the other hand, subcutaneous administration of the protease inhibitor aprotinin was found to increase the subcutaneous blood flow (SBF) (Williams et al., 1983). The enhanced bioavailability of subcutaneously administered insulin following treatment with protease inhibitors could, therefore, be due to reduced enzymatic degradation in the depot, or it could be due to faster absorption from the depot leaving less time for the insulin to undergo some unspecific, time-dependent degradation.

In the following, and as suggested by Hori et al. (1983), it is assumed that the subcutaneous degradation of soluble insulin follows a first-order reaction. Finally, it is assumed that the degradation is nonsaturable under physiological conditions. Taken together, this means that the degradation, like absorption, is proportional to the concentrations of the insulin species.

13.3.3 Simulating the Kinetics

The above discussion represents an example of how a complex system can be divided into smaller and less complex subsystems as described in Section 13.1. The combined system describing the dynamics of regular, human insulin becomes

$$\frac{dC_H}{dt} = P_{DH}\left(K_{DH}C_D^3 - C_H\right) + B_H C_H + D_H \nabla^2 C_H - A_H C_H \tag{13.3}$$

$$\frac{dC_D}{dt} = -P_{DH}\left(K_{DH}C_D^3 - C_H\right) + B_D C_D + D_D \nabla^2 C_D - A_D C_D \tag{13.4}$$

where P_{DH} is a rate constant, B_H and B_D are the absorption constants of hexamers and dimers, respectively, D_H and D_D are the diffusion constants of hexamers and dimers, respectively, ∇^2 is the Laplacian operator (related to the spatial diffusion), and A_H and A_D are the degradation constants of hexamers and dimers, respectively. Equation 13.4, describing the dynamics of insulin dimers, includes also the dynamics of insulin monomers, so the term "dimer" is used as a common descriptor of both dimers and monomers. This approach is a way to simplify the system and it has previously been shown that it can be used without affecting the overall behavior of the system (Mosekilde et al., 1989; Søeborg et al., 2009).

Combining subsystems, as in Equations 13.3 and 13.4, enables simulations of insulin absorption kinetics to be performed. Such simulations reveal that injecting high doses or high concentrations of insulin results in slower absorption kinetics. A high dose implies that a large volume is injected. A large volume increases relatively slower by diffusion than a small volume. This makes the dilution of the insulin slower, so more insulin is kept on the slowly absorbed hexameric form. In the same way, more concentrated insulin will, according to Equation 13.2 and Figure 13.1, contain relatively more hexamers than less concentrated insulin. The effect of different doses is shown in Figure 13.2, where simulations of the disappearance of 1 and 20 U of U100 insulin are given. The effect of concentration is shown in Figure 13.3, where simulations of the disappearance of 20 U of U40 and U100 insulin are given.

Figures 13.2 and 13.3 show the remaining nondegraded insulin in the subcutaneous depot, meaning that the removal of insulin is due to both absorption and degradation. Experimental disappearance curves based on radiolabelled insulin do not show directly the removal of insulin but rather the removal of the radiolabel from the depot. This removal will probably also include labels attached to insulin degradation products, and therefore the method would also detect some of the degradation products, giving a seemingly slower removal of insulin. A detailed description of the above system including simulations of insulin in plasma is given in Søeborg et al. (2009).

The same system may also be used for optimizing a pharmaceutical formulation of insulin. With detailed knowledge of the effect(s) of the auxiliary substances on

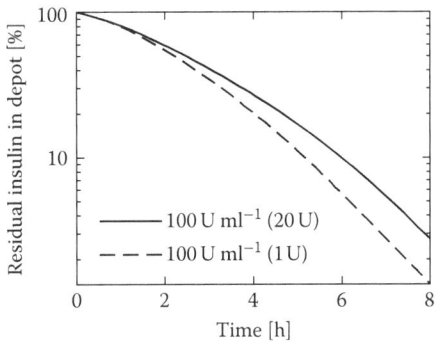

FIGURE 13.2 Disappearance of 20 and 1 U of U100 regular, human insulin from the subcutaneous depot following injection.

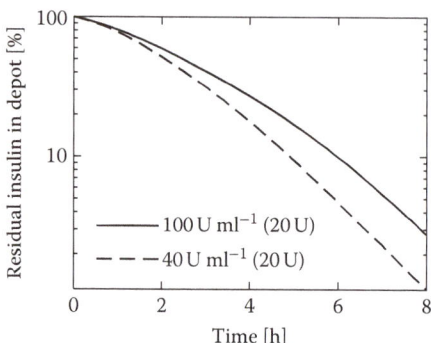

FIGURE 13.3 Disappearance of 20 U of U40 and U100 regular, human insulin from the subcutaneous depot following injection.

K_{DH} over time, the amount and composition of auxiliary substances in the formulation can be optimized with regards to shelf life as well as to pharmacokinetics.

13.4 HEAP FORMATION OF SUSPENSIONS

Subcutaneous administration of insulin is associated with great variability, which is caused by variations in local blood flow, temperature, and so on (Heinemann, 2002; Hildebrandt et al., 1985). Furthermore, the absorption of intermediate- and long-acting insulin suspensions is more variable than the corresponding absorption of fast-acting, soluble insulins (Heinemann, 2002). The reason for this difference in variability is not well understood. In the following, it is demonstrated how the use of a biosimulation model can describe and explain at least part of the mechanisms behind the greater variability associated with insulin suspensions.

As an example of a widely used insulin suspension, NPH insulin is considered. This insulin is comprised of water (50%), insulin hexamers, and the basic peptide, protamine (Krayenbühl and Rosenberg, 1946; Norrman et al., 2007). Its dynamics is described by

$$\frac{\mathrm{d}C_{\mathrm{NPH}}}{\mathrm{d}t} = -\beta C_{\mathrm{NPH}}\left(1 - \alpha C_{\mathrm{H}}C_{\mathrm{P}}\right) - A_{\mathrm{NPH}}C_{\mathrm{NPH}} \tag{13.5}$$

$$\frac{\mathrm{d}C_{\mathrm{P}}}{\mathrm{d}t} = \beta C_{\mathrm{NPH}}\left(1 - \alpha C_{\mathrm{H}}C_{\mathrm{P}}\right) + B_{\mathrm{P}}C_{\mathrm{P}} + D_{\mathrm{P}}\nabla^{2}C_{\mathrm{P}} - A_{\mathrm{P}}C_{\mathrm{P}} \tag{13.6}$$

where C_{NPH} and C_{P} are the concentrations of NPH insulin and protamine, respectively. β is a constant determining the rate of dissolution of the NPH crystals into insulin hexamers, and α is the NPH inhibition constant that slows down the breakdown of NPH crystals if the concentration of hexamers and/or protamine is high (as it is in the vial). The first part of Equation 13.5 resembles the Noyes–Whitney equation, described, for example, in Florence and Attwood (1998). A_{NPH} is a constant describing the breakdown of NPH crystals in the subcutaneous tissue due to macrophage activity (Søeborg et al., 2012). Equations 13.5 and 13.6 may be combined with Equations 13.3 and 13.4 to describe the dynamics of a system of NPH insulin (which dissociates into hexamers) or a biphasic mixture of NPH insulin and soluble insulin. Such mixtures are much used in the treatment of insulin-dependent diabetic patients and have been commercially available since the 1960s (Schlichtkrull et al., 1965).

13.4.1 FILTRATION IN SUBCUTIS

After the injection of suspended NPH insulin into the subcutaneous depot, the tissue will act as a sieve and allow the injected solute to quickly spread out in the tissue while the insulin crystals are left behind as concentrated heaps near the site of injection (see Figure 13.4; Markussen et al. [1996]).

13.4.2 SURFACE-DEPENDENT DISSOLUTION

The heaps are assumed to dissolve, thereby releasing insulin hexamers from the surface. The hexamers further dissociate into dimers and monomers, and are absorbed. Furthermore, macrophages appear at the injection site approximately 3 hours after the administration (Hasselager, personal communication). The macrophages degrade the NPH heaps, also from the surface (Markussen et al., 1996). Thus, both dissolution and degradation take place from the same surface as shown in Figure 13.5.

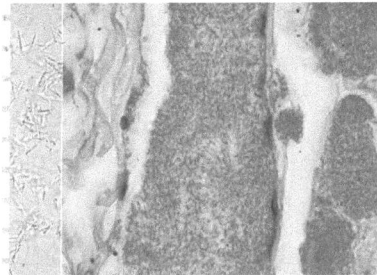

FIGURE 13.4 **(See color insert.)** A formulation of U200 NPH insulin is shown in the left panel. The same formulation 1 hour following administration into the subcutaneous tissue in a pig is shown in the right panel (same scale).

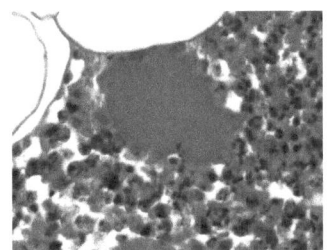

FIGURE 13.5 **(See color insert.)** The NPH insulin heap (solid red) is surrounded by macrophages (purple) and adipose tissue (white).

The shape of the heaps, which presumably is determined by the local environment in the subcutaneous depot, affects the insulin absorption kinetics. Assuming that the degradation and dissolution take place at a certain rate on the entire surface as suggested by Brooke (1973), the smallest dimension determines the time for a complete disappearance of a heap, for example, radius rather than length for long, thin cylinders.

Structures with a high volume-to-surface ratio will therefore survive longer than structures with a small value of this ratio. This is illustrated in Figure 13.6, where the "moose" and the "pine" represent the fate of two NPH heaps over time. Both contain the same amount of insulin (the same number of pixels), but the "pine" has a much larger "surface area" (boundary toward white pixels) than the "moose." As a result, the "pine" is completely dissolved 14 hours before the "moose" (degradation not included) and gives a completely different insulin plasma profile (Figure 13.7) and hence a different clinical effect.

The two structures could represent two injections of insulin in the same patient.

FIGURE 13.6 (See color insert.) The *moose* and the *pine* represent two identical "doses" (number of pixels) of NPH insulin in the subcutaneous depot. The two heaps have different "surface areas" (boundary toward white pixels) resulting in different kinetics. The figure shows the dissolution of the two NPH insulin heaps over time. The corresponding plasma insulin curves are given in Figure 13.7.

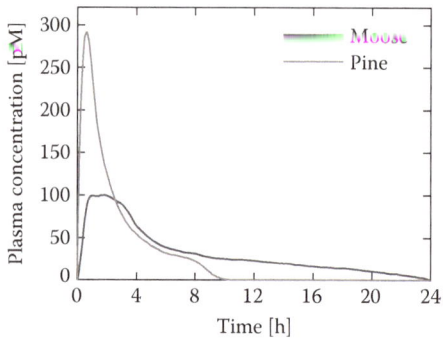

FIGURE 13.7 (See color insert.) Plasma insulin profiles corresponding to the *moose* and the *pine* shown in Figure 13.6.

The variability associated with insulin, and insulin suspensions in particular, may therefore be hard to control. This can result in serious and even life-threatening conditions for insulin-dependent diabetic patients. One way to minimize the variability is to eliminate the blood flow–dependent absorption rate by binding of the insulin molecule to albumin as described in Section 13.5.

13.4.3 BIOAVAILABILITY

Degradation by macrophages is assumed to take place at a certain rate (A_{NPH} in Equation 13.5), whereas dissolution is a balanced process, where the surrounding concentrations of hexamers and protamine determine the net rate of dissolution. This means that if the surrounding concentrations of hexamers and/or protamine are high, the dissolution of the heaps will be slow. Since the degradation rate of the macrophages is not affected, the bioavailability will be lower (Søeborg et al., 2012). This has been shown clinically in type 1 diabetic patients and in healthy volunteers with increasing NPH/NPL* to insulin aspart and insulin lispro ratios resulting in decreasing bioavailability of the mixtures (Chen et al., 2004; Heise et al., 1998, 2008; Thorisdottir et al., 2009).

13.5 EFFECTS OF LOCAL PERFUSION

After subcutaneous administration, insulin will diffuse to the capillary walls and permeate into the blood stream. The SBF directly affects the absorption rate of regular, human insulin as described, for example, by Hildebrandt et al. (1985). Furthermore, SBF varies a great deal due to exercise, temperature, massage, smoking, and so on, see, for example, Koivisto (1980) and Hildebrandt (1991). In the following, a quantitative description of the SBF-dependent absorption of regular, human insulin is presented.

13.5.1 PERMEATION AND FLOW

Based on a study by Hildebrandt and Birch (1988), Claessen and Mortensen (2009) determined how the insulin absorption constant, B, depends on the local SBF. This dependence is expressed by

$$B = \frac{\alpha_{MM}SBF}{K_M + SBF} \tag{13.7}$$

where α_{MM} and K_M are the Michaelis–Menten constants. Claessen and Mortensen, hereafter, incorporated Equation 13.7 in the system given in Equations 13.3 and 13.4 to be able to simulate the effects of variations in SBF on absorption of subcutaneously administered insulin.

---

* NPL is the neutral protamine Hagedorn formulation of the rapid-acting insulin analogue insulin lispro.

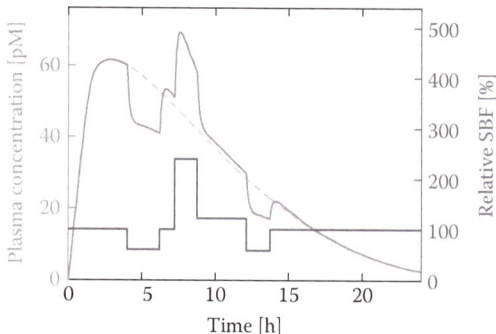

FIGURE 13.8 (**See color insert.**) Simulated insulin plasma profile and the corresponding relative subcutaneous blood flow (SBF) during five phases of sleep following a subcutaneous injection of NPH insulin 4 hours prior to the first phase of sleep. (Modified from Claessen, S., and Mortensen, T., Local subcutaneous blood flow and the effect on insulin absorption. Technical University of Denmark, Copenhagen, 2009.)

Sindrup et al. (1992) studied the SBF during five phases of sleep in healthy volunteers and found it to increase 140% for approximately 100 minutes during the so-called hyperemic phase relative to the SBF just before going to bed. Data from this study was used to simulate the corresponding effects on the insulin concentration in plasma following a subcutaneous injection of NPH insulin 4 hours prior to the first phase of sleep. The simulation is shown in Figure 13.8.

Subcutaneous administration of the insulin analogue, insulin detemir, has been shown to be associated with reduced pharmacokinetic and pharmacodynamic variations compared to conventional medium- and long-acting insulins (Hermansen et al., 2001; Vague et al., 2003). This is because insulin detemir binds reversibly to albumin in plasma and in the subcutaneous tissue. In this way, albumin keeps the concentration of free insulin in the blood very low. The absorption rate is, thus, mainly determined by diffusion across the capillary wall and is much less dependent on changes in the local blood as compared to the conventional types of insulin (Kurtzhals and Colding-Jørgensen, 2004).

13.5.2 Variability in Absorption Rate and Bioavailability

Changes in the SBF affect insulin bioavailability—with the exception of insulin detemir and similar insulins as described earlier. With lower blood flow (corresponding to reduced absorption constants B_H and B_D in Equations 13.3 and 13.4), more insulin is degraded in the subcutaneous tissue before it is absorbed into the blood stream. This is illustrated in Figure 13.9 for regular, human insulin and NPH insulin. From the figure, it is noted that the bioavailability of NPH insulin is relatively more sensitive to changes in blood flow compared to regular, human insulin. In the case of increased blood flow, it is because the faster absorption of soluble species has the additional effect of increasing the dissolution of NPH heaps (due to lower hexamer presence), thereby decreasing the time available for macrophage degradation. This effect is reversed for decreased blood flow. A detailed description and discussion of the bioavailability of soluble and suspended insulin can be found in Søeborg et al. (2012).

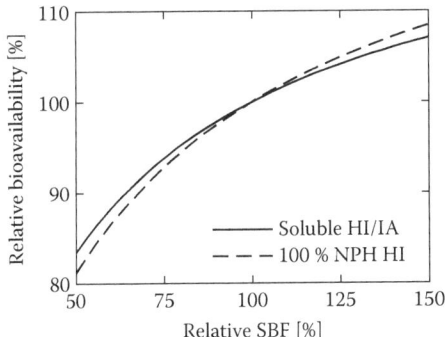

FIGURE 13.9   The relative bioavailability of regular, human insulin and NPH insulin as a function of relative subcutaneous blood flow.

Figure 13.9 shows how, in particular, low SBF results in a substantial change in insulin bioavailability compared to normal flow.

13.6   CONCLUSIONS

The use of biosimulation as a tool in pharmaceutical formulation of peptides and proteins has been demonstrated using regular, human insulin and insulin suspensions as examples.

Furthermore, as shown in Section 13.3, it has been demonstrated how the complex absorption kinetics of regular, human insulin can be broken down into subsystems, which are relatively simple to describe. These subsystems or modules are subsequently connected and the entire system is simulated. As more data becomes available or new pharmaceutical formulations are developed, new modules can be constructed and combined with the existing ones. As an example, the module describing diffusion in the subcutaneous tissue will apply to all subcutaneously administered soluble substances with only the value of the diffusion constant changed. The collection of subsystems can thus be combined in various ways into more or less complex systems intended to answer specific questions.

The use of biosimulation for the investigation of insulin variability and bioavailability has also been demonstrated. A description of heap formation of insulin suspensions in the subcutaneous tissue was formulated. It shows how the shape of these heaps affect the variability of insulin suspensions and also explains at least part of the greater variability and lower bioavailability associated with insulin suspensions compared to regular, human insulin. Finally, how changes in the local SBF affects insulin bioavailability was simulated.

It should be emphasized that the simulations give a simplified version of the biological phenomena taking place. The systems described in this chapter do not, for example, take into account all of the factors that may change the SBF and thus affect the insulin absorption kinetics. Instead, biosimulation should be used to increase the understanding of the mechanisms responsible for the dynamics in question. Scenarios may be simulated and the outcome may be used for planning or understanding experiments or clinical trials or for optimizing pharmaceutical formulations with regards to, for example, shelf life, pharmacokinetics, and pharmacodynamics.

ACKNOWLEDGMENTS

Erik Hasselager is highly acknowledged for providing the histological samples shown in this chapter. This work was supported by the European Union through the Network of Excellence BioSim, Contract No. LSBH-CT-2004-005137.

REFERENCES

Bocian, W., Sitkowski, J., Tarnowska, A., Bednarek, E., Kawecki, R., Kozminski, W., and Kozerski, L. 2008. Direct insight into insulin aggregation by 2D NMR complemented by PFGSE NMR. *Proteins* 71:1057–1065.

Brange, J. 1987. *Galenics of Insulin*, 1st ed. Springer-Verlag, Berlin Heidelberg.

Brange, J., Owens, D. R., Kang, S., and Volund, A. 1990. Monomeric insulins and their experimental and clinical implications. *Diabetes Care* 13:923–954.

Brooke, D. 1973. Dissolution profile of log-normal powders: Exact expression. *J. Pharm. Sci.* 62:795–798.

Charman, S., McLennan, D., Edwards, G., and Porter, C. 2001. Lymphatic absorption is a significant contributor to the subcutaneous bioavailability of insulin in a sheep model. *Pharm. Res.* 18:1620–1626.

Chen, J. W., Lauritzen, T., Christiansen, J. J., Jensen, L. H., Clausen, W. H. O., and Christiansen, J. S. 2004. Pharmacokinetic profiles of biphasic insulin aspart 30/70 and 70/30 in patients with type 1 diabetes: A randomized double-blinded crossover study. *Diabet. Med.* 22:273–277.

Claessen, S., and Mortensen, T. 2009. Den lokale blodgennemstrømning i subcutis' betydning for insulinabsorption (Local subcutaneous blood flow and the effect on insulin absorption). Technical University of Denmark, Copenhagen (in Danish).

Crandall, D., Hausman, G., and Kral, J. 1997. A review of the microcirculation of adipose tissue: Anatomic, metabolic and anigiogenic perspectives. *Microcirculation* 4:211–232.

DeFelippis, M. R., and Akers, M. J. 2000. Peptides and proteins as parental suspensions: An overview of design, development, and manufacturing considerations. In *Pharmaceutical Formulation Development of Peptides and Proteins*, eds. S. Frøkjær and L. Hovgaard, 113–144. Taylor & Francis, London.

Derewenda, U., Derewenda, Z., Dodson, E. J., Dodson, G. G., Reynolds, C. D., Smith, G. D., Sparks, C., and Swenson, D. 1989. Phenol stabilizes more helix in a new symmetrical zinc insulin hexamer. *Nature* 338:594–596.

Florence, A., and Attwood, D. 1998. *Physicochemical Principles of Pharmacy*, 3rd ed. Macmillan Press, Ebbw Vale.

Hassiepen, U., Federwisch, M., Mulders, T., and Wollmer, A. 1999. The lifetime of insulin hexamers. *Biophys. J.* 77:1638–1654.

Heinemann, L. 2002. Variability of insulin absorption and insulin action. *Diabet. Techol. Ther.* 4:673–682.

Heise, T., Eckers, U., Kanc, K., Nielsen, J. N., and Nosek, L. 2008. The pharmacokinetic and pharmacodynamic properties of different formulations of biphasic insulin aspart: A randomized, glucose clamp, crossover study. *Diabetes Technol. Ther.* 10:479–485.

Heise, T., Weyer, C., Serwas, A., Heinrichs, S., Osinga, J., Roach, P., Woodworth, J., Gudat, U., and Heinemann, L. 1998. Time-action profiles of novel premixed preparations of insulin lispro and NPL insulin. *Diabetes Care* 21:800–803.

Hermansen, K., Madsbad, S., and Perrild, H. 2001. Comparison of the soluble basal insulin analog insulin detemir with NPH insulin. *Diabetes Care* 24:296–301.

Hewitt, H. 1954. The mechanics of subcutaneous injection. *Br. J. Exp. Path.* 35:35–40.

Hildebrandt, P. 1991. Subcutaneous absorption of insulin in insulin-dependent diabetic patients. Influence of species, physico-chemical properties of insulin and physiological factors. *Dan. Med. Bull.* 38:337–346.

Hildebrandt, P., and Birch, K. 1988. Basal rate subcutaneous insulin infusion: Absorption kinetics and relation to local blood flow. *Diabet. Med.* 5:434–440.

Hildebrandt, P., Birch, K., Sestoft, L., and Nielsen, S. L. 1985. Orthostatic changes in subcutaneous blood flow and insulin absorption. *Diabetes Res.* 2:187–190.

Hori, R., Komada, F., and Okumura, K. 1983. Pharmaceutical approach to subcutaneous dosage forms of insulin. *J. Pharm. Sci.* 72:435–439.

Hvidt, S. 1991. Insulin association in neutral solutions studied by light scattering. *Biophys. Chem.* 39:205–213.

Koivisto, V. 1980. Sauna-induced acceleration in insulin absorption from subcutaneous injection site. *Br. Med. J.* 280:1411–1413.

Krayenbühl, C., and Rosenberg, T. 1946. Crystalline protamine insulin. *Reports of the Steno Memorial Hospital and the Nordisk Insulinlaboratorium* 1:61–73.

Kurtzhals, P., and Colding-Jørgensen, M. 2004. Albumin binding of insulin detemir reduces the risk for hypoglycaemic events. *Diabetes* 53:A477.

Kurtzhals, P., and Ribel, U. 1995. Action profile of cobalt(III)-insulin. A novel principle of protraction of potential use for basal insulin delivery. *Diabetes* 44:1381–1385.

Li, J., and Kuang, Y. 2009. Systematically modeling the dynamics of plasma insulin in subcutaneous injection of insulin analogues for type 1 diabetes. *Math. Biosci. Eng.* 6:41–58.

Lin, M., and Larive, C. K. 1995. Detection of insulin aggregates with pulsed-field gradient nuclear magnetic resonance spectroscopy. *Anal. Biochem.* 229:214–220.

Linde, B., and Chisholm, G. 1975. The interstitial space of adipose tissue as determined by single injection and equilibration techniques. *Acta. Physiol. Scand.* 95:383–390.

Markussen, J., Havelund, S., Kurtzhals, P., Andersen, A. S., Halstrøm, J., Hasselager, E., Larsen, U. D., et al. 1996. Soluble, fatty acid acylated insulins bind to albumin and show protracted action in pigs. *Diabetologia* 39:281–288.

Mosekilde, E., Jensen, K., Binder, C., Pramming, S., and Thorsteinsson, B. 1989. Modeling absorption kinetics of subcutaneously injected soluble insulin. *J. Pharmacokin. Biopharm.* 17:67–87.

Nicholson, C., and Sykova, E. 1998. Extracellular space structure revealed by diffusion analysis. *Trends Neurosci.* 21:207–215.

Norrman, M., Hubálek, F., and Schluckebier, G. 2007. Structural characterization of insulin NPH formulations. *Eur. J. Pharm. Sci.* 30:414–423.

Oliva, A., Farina, J., and Llabres, M. 2000. Development of two high-performance liquid chromatographic methods for the analysis and characterization of insulin and its degradation products in pharmaceutical preparations. *J. Chrom. B Biomed. Sci. Appl.* 749:25–34.

Pekar, A. H., and Frank, B. H. 1972. Conformation of proinsulin. Comparison of insulin and proinsulin self-association at neutral pH. *Biochem.* 11:4013–4016.

Ploughmann, S., Hejlesen, O. K., and Cavan, D. A. 2003. Implementation of a new insulin model covering both injection and pump delivery in DiasNet. In *Proceedings of the 25th Annual International Conference of the IEEE EMBS*, 1264–1267. Cancun, Mexico.

Rahuel-Clermont, S., French, C., Kaarsholm, N., and Dunn, M. F. 1997. Mechanisms of stabilization of the insulin hexamer through allosteric ligand interactions. *Biochemistry* 36:5837–5845.

Schlichtkrull, J., Munck, O., and Jersild, M. 1965. Insulin rapitard and insulin actrapid. *Acta Med. Scand.* 177:103–113.

Sharkawy, A., Klitzman, B., Truskey, G., and Reichert, W. 1997. Engineering the tissue which encapsulates subcutaneous implants. I. Diffusion properties. *J. Biomed. Mater. Res.* 37:401–412.

Shen, L., and Chen, Z. 2007. Critical review of the impact of tortuosity on diffusion. *Chem. Eng. Sci.* 62:3748–3755.

Sindrup, J., Kastrup, J., Madsen, P., Christensen, H., Jørgensen, B., and Wildschoidtz, G. 1992. Nocturnal variations in human lower leg subcutaneous blood flow related to sleep stages. *J. Appl. Physiol.* 73:1246–1252.

Søeborg, T., Rasmussen, C., Mosekilde, E., and Colding-Jørgensen, M. 2009. Absorption kinetics of insulin after subcutaneous administration. *Eur. J. Pharm. Sci.* 36:78–90.

Søeborg, T, Rasmussen, C. H., Mosekilde, E., and Colding-Jørgensen, M. 2012. Bioavailability and variability of biphasic insulin mixtures. *Eur. J. Pharm. Sci.* 46:198-208.

Takeyama, M., Ishida, T., Kokubu, N., Komada, F., Iwakawa, S., Okumura, K., and Hori, R. 1991. Enhanced bioavailability of subcutaneously injected insulin by pretreatment with ointment containing protease inhibitors. *Pharm. Res.* 8:60–64.

Tarín, C., Teufel, E., Picó, J., Bondia, J., and Pfleiderer, H.-J. 2005. Comprehensive pharmacokinetic model of insulin glargine and other insulin formulations. *IEEE Trans. Biomed. Eng.* 52:1994–2005.

Thorisdottir, R. L., Parkner, T., Chen, J. W., Ejskjær, N., and Christiansen, J. S. 2009. A comparison of pharmacokinetics and pharmacodynamics of biphasic insulin aspart 30, 50, 70 and pure insulin aspart: A randomized, quadruple crossover study. *Basic Clin. Pharmacol. Toxicol.* 104:216–221.

Trajanoski, Z., Wach, P., Kotanko, P., Ott, A., and Skraba, F. 1993. Pharmacokinetic model for the absorption of subcutaneously injected soluble insulin and monomeric insulin analogues. *Biomed. Tecknik* 38:224–231.

Vague, P., Selam, J. L., Skeie, S., de Leeuw, I., Elte, J., Haahr, H. L., Kristensen, A., and Draeger, E. 2003. Insulin detemir is associated with more predictable glycemic control and reduced risk of hypoglycemia than NPH insulin in patients with type 1 diabetes on a basal-bolus regimen with premeal insulin aspart. *Diabetes Care* 26:590–596.

Vølund, A. 1993. Conversion of insulin units to SI units. *Am. J. Clin. Nutr.* 58:714–715.

Wach, P., Trajanoski, Z., Kotanko, P., and Skrabal, F. 1995. Numerical approximation of mathematical model for absorption of subcutaneously injected insulin. *Med. Bio. Eng. Comput.* 33:18–23.

Williams, G., Pickup, J. C., Bowcock, S., Cooke, E., and Keen, H. 1983. Subcutaneous aprotinin causes local hyperaemia: A possible mechanism by which aprotinin improves control in some diabetic patients. *Diabetologia* 24:91–94.

14 Registration of Peptides and Proteins

Niamh Kinsella

CONTENTS

14.1 INTRODUCTION

There are those in the biotech industry who would argue that biological medicinal products are too highly regulated and that the standards imposed for registration are too restrictive. It is certainly true that the number of regulations, guidelines, and other requirements relating to the registration of biological and biotechnological medicinal products has widened considerably in scope during the past two decades. However, this is due to the regulators seeking to keep pace with the rapid emergence of new technologies and product classes.

There have been isolated problems with many biotech products that have gained approval via the registration process, but these issues have usually been resolved pre- or postapproval without widespread, severe adverse events in the respective patient population. Those products showing an unacceptable level of risk when compared to the potential benefit (risk–benefit assessment) during early stage development have been halted before regulatory approval. In addition, unexpected postmarketing problems have not usually affected large numbers of patients. In recent years, there have been a few notable cases where there have been severe adverse reactions and deaths during the clinical development of biotech products. This has led to the development of new regulatory requirements for the development and approval of biotechnological products.

14.1.1 REGULATION OF MEDICINAL PRODUCTS IN EUROPE AND THE UNITED STATES

14.1.1.1 Regulation of Medicinal Products in Europe

14.1.1.1.1 Early Regulation of Medicinal Products in Europe

The development of regulations in the healthcare sector has been due to ethics and safety concerns related to early medicinal products. Certain major incidents have been the catalysts for the process of developing regulations for biological and medicinal products. Early medicinal products development programs had widespread problems with quality control, and many early biological medicinal products were associated with microbial and viral contamination. However, the thalidomide scandal in Europe and Canada in the late 1950s and early 1960s was a milestone in the development of regulations for pharmaceutical products in Europe. The thalidomide issues were the driving force for the first European directive on medicinal products: for the first time, detailed requirements for the marketing authorization of medicinal products for human use were described and were applicable across Europe.

Medicine control was an early activity of the European Economic Community (EEC). The first and basic EEC Directive to control medicines was introduced in 1965 (Directive 65/65/EEC). This directive reflected many of the requirements already established by the United States (U.S.) Food and Drug Administration (FDA).

In 1975, the Committee for Proprietary Medicinal Products (CPMP) was set up under Directive 75/319/EEC as a part of the initiative to create a single market in pharmaceuticals. The purpose of the CPMP was to assist member states in the review and approval of medicinal products. Directive 75/320/EEC set up the Pharmaceutical Committee with members from all the member states. The purpose of this committee was to provide expert advice to the Commission on the development of policies for the regulation of medicinal products.

Since the mid-1980s, a substantial body of harmonizing pharmaceutical legislation has been adopted throughout the European Union (EU). In 1987, the CPMP procedure was joined by a "concertation" procedure (Directive 87/22/EEC), which required the member states to consult each other before granting, refusing to grant, or withdrawing marketing authorizations for medicinal products produced by biotechnology processes. The aims of this procedure were to enable the member states to pool scarce resources when considering new and complex products, to encourage uniform authorization decisions, and to provide additional patent protection.

In addition to these new procedures, the late 1980s and early 1990s also saw the expansion of rules and requirements to specialized categories of vaccines and blood products that had not been covered earlier.

In 1995, a new European system for the authorization of medicinal products came into operation (Centralised Procedure), which was facilitated by a new subdivision of the European Commission (EC), the European Medicines Agency (EMA). This was the result of many years of the European Community working together toward a single EU-wide market for pharmaceuticals. The aim of a single market was to remove barriers to trade that previously existed so that a medicine marketed in one EU member state can also be made available in other EU member states. At the same time, the Community ensured that there are proper safeguards as to the safety, quality, and efficacy of medicinal products made available to consumers.

The EMA began its activities on February 1, 1995. Its main task is to coordinate the scientific evaluation of the safety, quality, and efficacy of medicinal products for human and veterinary use throughout the EU. The criteria and procedures for the approval of human medicines, as well as other important aspects of pharmaceutical legislation, have become harmonized within the EU. European pharmaceutical legislation now covers all industrially manufactured medicines, including biological products, biotechnological products, vaccines, and advanced therapy medicinal products. International developments in the pharmaceutical industry continually lead to the generation of new regulations and legislations. An overview of the organizational structure of the EMA is presented in Figure 14.1.

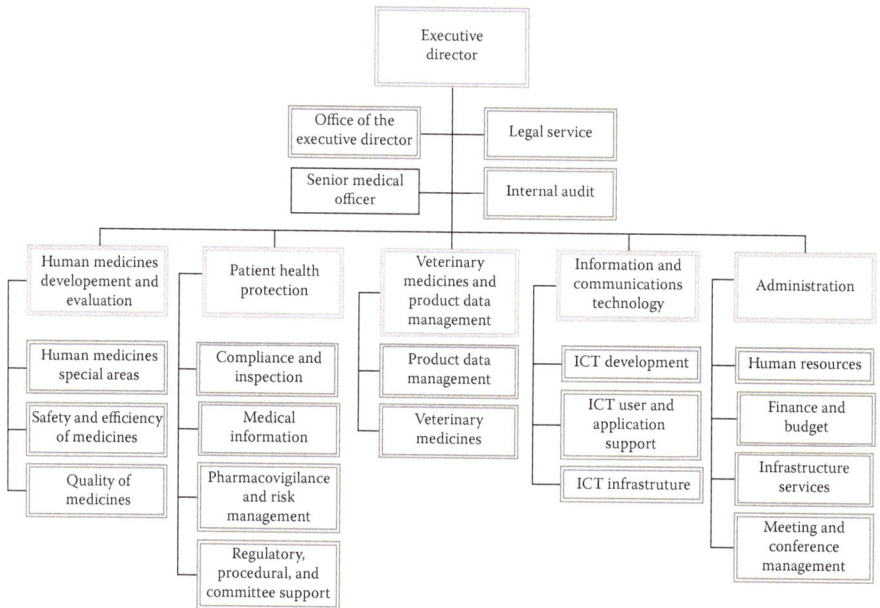

FIGURE 14.1   Organizational structure of the European Medicines Agency.

14.1.1.1.2   Regulation of Biological Medicinal Products in the EU

All biological products are defined as "medicinal products" in the EU. Regulation of these products falls under the auspices of Directive 2001/83/EC. Commission Directive 2003/63/EC, which amended Annex I to Directive 2001/83/EC, introduced a definition for "Biological Medicinal Products," as follows:

> A biological medicinal product is a product, the active substance of which is a biological substance. A biological substance is a substance that is produced by or extracted from a biological source and that needs for its characterisation and the determination of its quality a combination of physico-chemical-biological testing, together with the production process and its control. The following shall be considered as biological medicinal products: immunological medicinal products and medicinal products derived from human blood and human plasma as defined, respectively in paragraphs (4) and (10) of Article 1; medicinal products falling within the scope of Part A of the Annex to Regulation (EEC) No 2309/93; advanced therapy medicinal products as defined in Part IV of this Annex.

Complete harmonization of the basis of approval of biotechnological medicinal products was achieved through Regulation (EEC) No 2309/93/EC, which was issued in 1993 and implemented on January 1, 1995. Regulation (EEC) No 2309/93 was amended on March 31, 2004, through Regulation (EC) No 726/2004 and this included the extension of the product classes for which the Centralised Procedure is mandatory. As a result, the Centralised Procedure is mandatory for products derived from biotechnological processes. An overview of the Centralised Procedure is presented in Section 14.3.1.

14.1.1.2   Regulation of Medicinal Products in the United States

14.1.1.2.1   Early Regulation of Medicinal Products in the United States

As with the EU, the development of procedures for the regulation of medicinal products in the United States was driven by safety and ethical concerns. To prevent the marketing of ineffective smallpox vaccines, the U.S. Congress passed the first Federal Law to regulate any medicinal product, the Vaccine Act of 1813. However, this Act was later shown to be ineffective and was repealed.

The Biologics Control Act was passed in the United States on July 1, 1902, to help with standardization and quality control of new biological products after two incidents in 1901 involving the deaths of children caused by contaminated vaccines. First, 13 children died following the distribution of a diphtheria antitoxin from a horse ("Jim") that had contracted tetanus. The second involved contaminated smallpox vaccine, which killed nine children.

The U.S. Federal Food, Drug, and Cosmetic Act (FD&C) was rapidly enacted by Congress in 1938 following the public outcry over the 1937 Elixir Sulfanilamide tragedy, in which more than 100 people died after using this drug. This Act gave authority to the U.S. FDA to oversee the safety of food, drugs, and cosmetics. Following this, under the 1944 Public Health Service Act (PHS Act), biological products were subjected to licensing requirements. An overview of the organizational structure of the FDA is presented in Figure 14.2.

14.1.1.2.2   Regulation of Biological Medicinal Products in the United States

Drugs and biological products are regulated in the United States by a single Federal Agency within the Department of Health and Human Services: the U.S. FDA.

Biological products are defined separately in the PHS Act (42 U.S. Code §262 Regulation of Biological Products), which specifies product classes:

> In this section, the term "biological product" means a virus, therapeutic serum, toxin, antitoxin, vaccine, blood, blood component or derivative, allergenic product, or analogous product… applicable to the prevention, treatment, or cure of a disease or condition of human beings.

The separate legal frameworks of the FD&C Act and the PHS Act also provide different approval mechanisms for drugs and biological products. Drugs are approved under Section 505 of the FD&C Act, where the requirements for a New Drug Application (NDA) are set forth in Section 505(b). Most biological products are approved under Section 351 of the PHS Act via a Biologics License Application (BLA).

The review of applications relating to drugs is assigned to the FDA Center for Drug Evaluation and Research (CDER), whereas those for biological products were traditionally assigned to the Center for Biological Evaluation and Research (CBER). However, the FDA transferred responsibility for review of the following biological products to CDER in June 2003:

- Monoclonal antibodies for in vivo use
- Cytokines, growth factors, enzymes, immunomodulators, and thrombolytics

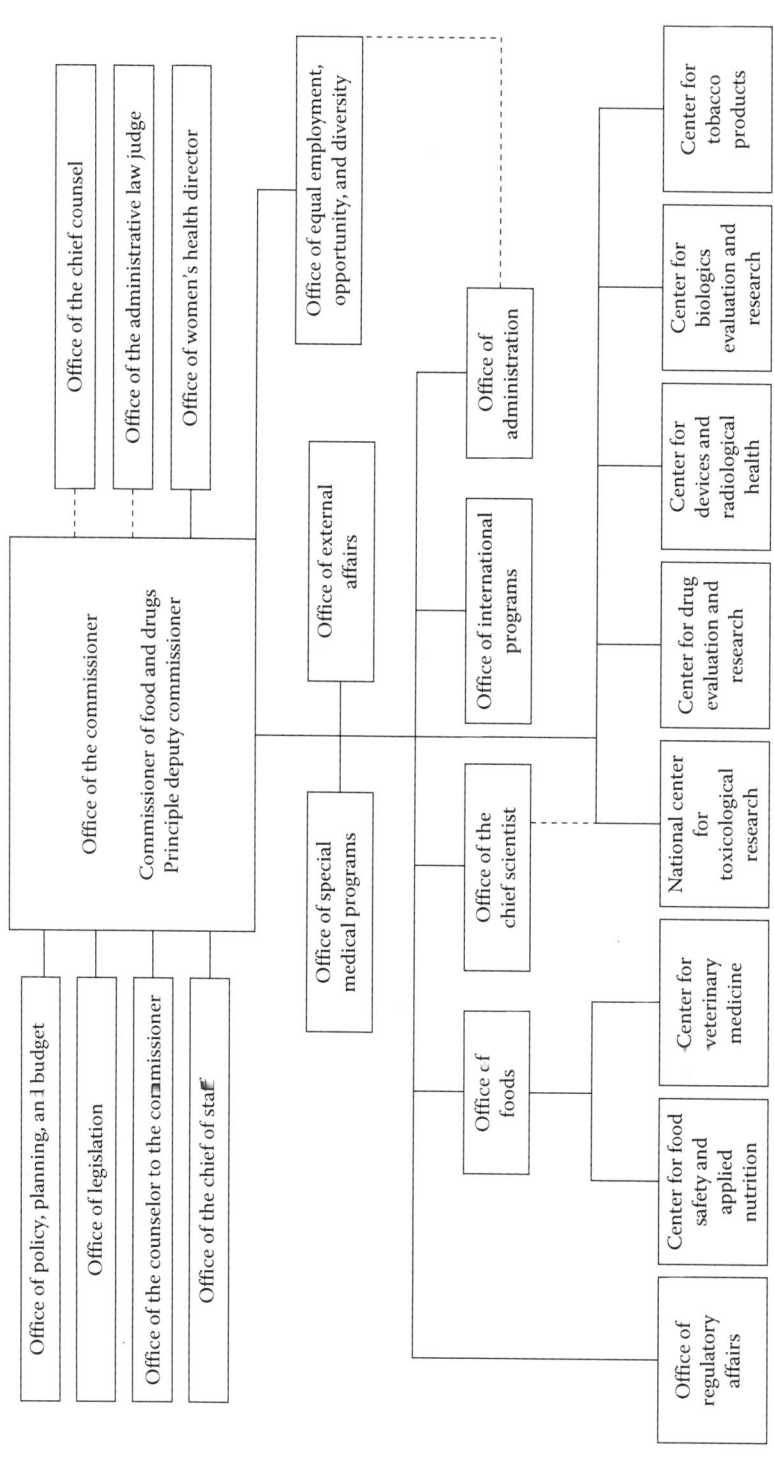

FIGURE 14.2 Organizational structure of the Food and Drug Administration.

- Proteins intended for therapeutic use, which are extracted from animals or microorganisms, including recombinant versions of these products (except clotting factors)
- Other nonvaccine immunotherapeutics

14.1.2 HISTORY OF THE APPROVAL OF BIOLOGICAL AND BIOTECHNOLOGICAL PRODUCTS

The first biological products to be approved were products prepared from human blood, and related products are still on the market today. These therapeutic proteins include a range of recombinant blood coagulation factors used to treat hemophilia, recombinant thrombolytics, and recombinant anticoagulants. Many of the initial therapeutic proteins approved for general medical use were replacement proteins (e.g., recombinant insulin and blood factors).

The manufacture of therapeutic proteins using recombinant DNA technology represented the first industrial application of this technology. Insulin, the first recombinant therapeutic protein, was approved for general medical use in 1978 in the United States and 1982 in Europe, and this recombinant product replaced the purified porcine and bovine insulin products previously approved for the treatment of diabetes.

Although some more recently approved products are also replacement proteins derived from natural sources, an increasing proportion of newly developed products is engineered. Protein engineering is increasingly used to tailor the functional attributes of commercially important proteins. It has been applied to achieve various objectives, including the alteration of the protein's immunogenicity, the alteration of biological half-life, the generation of faster or slower acting products, and the generation of novel hybrid/synthetic therapeutic proteins.

As of 2010, more than 100 recombinant therapeutic proteins have been approved in both the EU and the United States. Of these approved products, hormones and cytokines represent the largest product categories. In addition, approximately 20 therapeutic monoclonal antibodies have been approved, and more than 150 monoclonal antibody and monoclonal antibody-like products are in clinical development (Reichert et al., 2005). Biotechnological product categories also include vaccines, replacement proteins, and enzymes for the treatment of monogenetic diseases, and also includes proteins and enzymes for the treatment of other conditions. The regulatory guidelines developed alongside the development of recombinant DNA technology and there is now an extensive framework regulating these biotechnological products in both the EU and the United States.

More recently, advanced therapy medicinal products have become a focus of development and these products include cell therapy medicinal products, tissue-engineered products, and gene therapy products. To date, one tissue-engineered product, Tigenix®, has been approved in the EU and one cell therapy medicinal product, Provenge®, has been approved in the United States. Recently (July 2012), the first approval for a gene therapy product, Glybera, has been given by EMA. However, it is clear from the clinical trial websites and from news sources that many advanced therapy medicinal products are under development.

However, regulatory work has not kept apace with developments in the field. Development of regulatory guidelines for innovative or advanced medicinal products is in progress, and companies involved in the development of such products need to keep up to date with the changes in the regulatory environment, but should also be involved as stakeholders in the development of the legislation.

14.2 INFORMATION REQUIREMENTS FOR THE REGISTRATION DOSSIER

The registration dossier is arranged as a Common Technical Document (CTD) with the aim of preparing a dossier that would be acceptable for most regulatory regions globally. The organisation of the CTD is presented in International Conference on Harmonization (ICH) Guideline M4 (The CTD). The content requirements for the Marketing Authorisation Application (MAA) dossier for the EU are presented in the Notice to Applicants (NtA). However, different regional information may be required for registration in the various global regions and this is defined in regional legislation. The triangular representation of the CTD structure is presented in Figure 14.3.

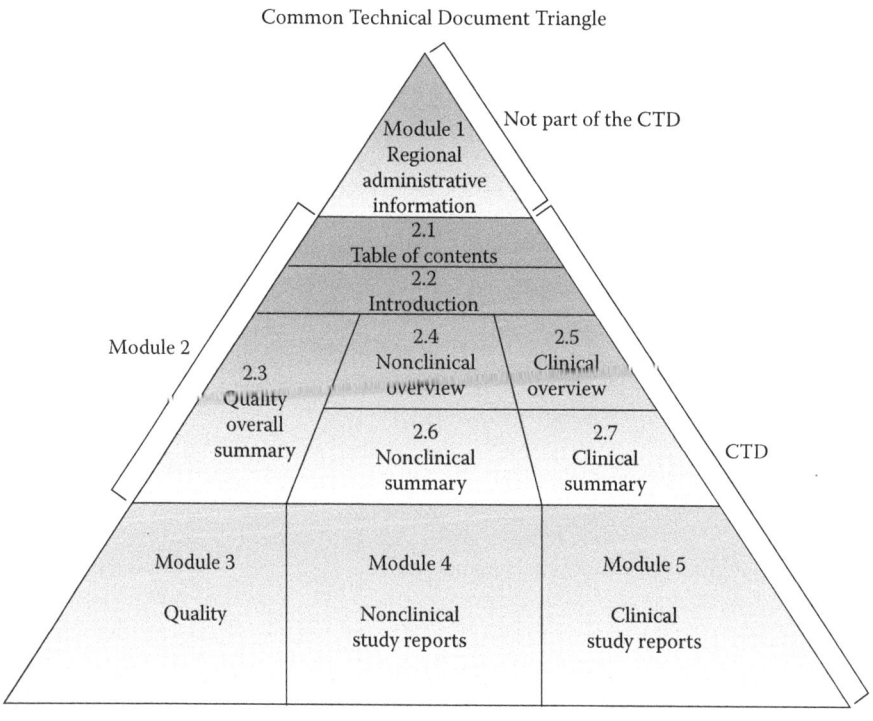

FIGURE 14.3 The Common Technical Document Triangle.

14.2.1 QUALITY REQUIREMENTS

For the review of the quality documentation for a biological or biotechnological product, the product is intrinsically linked to the process. The information supporting the process includes, but is not limited to, the cell bank from which the product is derived, the fermentation and purification process, the scale of manufacture, the control parameters of that process, defined raw materials used in the process, and the manufacturing facilities. This principle is stated and explained in a number of guidelines produced by the FDA, EMA, and ICH.

Regulators in the United States and at the EMA have consistently applied this principle to define product quality in terms of process consistency. Any changes to the process require an appropriate comparability exercise to determine the impact of the changes, if any, on the process as well as product attributes. Demonstrating comparability at the level of chemical, pharmaceutical, and biological documentation has always been considered a prerequisite for the approval of a process change. However, in some instances where comparability cannot be shown at the level of physicochemical and biological testing, additional nonclinical and/or clinical studies may be required.

In addition to the general guidelines on quality requirements for the registration dossier, the FDA and EMA have also published product-specific guidelines that should be consulted during product development.

Each product should be evaluated on a case-by-case basis and the CTDs of Module 2.3 (Quality overall summary [QOS]) and Module 3 (Quality) should be completed as appropriate for the product being developed.

The FDA will conduct a "bottom up" review, looking first at all protocols and reports, and will form conclusions based on the contents of these documents. Then, it will review Module 3 and the QOS. In the EU, the Committee for Medicinal Products for Human Use (CHMP) will conduct a "top down" review, initially looking at the QOS and then looking for more detail in Module 3; the Committee will only review protocols and reports if the information in Module 3 is insufficient or if any issues have been flagged.

14.2.2 NONCLINICAL REQUIREMENTS

Regulatory standards for the nonclinical testing of biotechnology-derived pharmaceuticals have generally been comparable among the EU, the United States, and Japan and these are outlined in ICH Safety Guideline ICH S6(R1). The principles outlined in this ICH guidance are applicable to products produced using recombinant DNA technology. However, this guideline can be considered to be an overarching guideline. Other ICH safety guidelines are available for the design of nonclinical studies. In addition, the EMA and FDA have published product-specific and/or indication-specific guidelines that should be consulted during the design of nonclinical studies.

Each product should be evaluated on a case-by-case basis and the CTDs of Module 2.4 (Nonclinical overview), Module 2.6 (Nonclinical summaries), and Module 4 (Nonclinical study reports) should be completed as appropriate for the product being developed. Where studies have not been conducted, an appropriate justification should be included in the relevant CTD sections.

14.2.3 CLINICAL REQUIREMENTS

The clinical development of biotechnological medicinal products is not significantly different to that of chemical compounds; for example, randomized controlled clinical trials are required for both. However, there are specific aspects of clinical development that are of particular importance for biological medicinal products. General guidance is provided under the ICH Efficacy Guidelines. As with nonclinical studies, the EMA and FDA have published product-specific and/or indication-specific guidelines that should be consulted during the design of nonclinical studies, and each product should be evaluated on a case-by-case basis. For the clinical development of more complex medicinal products, companies are strongly advised to seek scientific advice from the regulatory agencies to ensure that the planned clinical studies can be used to support marketing authorization.

Each product should be evaluated on a case-by-case basis and the CTD headings of Module 2.5 (Clinical overview), Module 2.7 (Clinical summaries), and Module 5 (Clinical study reports) should be completed as appropriate for the product being developed. Where studies have not been conducted, an appropriate justification should be included in the relevant CTD sections.

14.3 THE REGISTRATION PROCESS

14.3.1 AUTHORIZATION IN THE EU

The Centralised Procedure is compulsory for medicinal products derived from biotechnology and optional for other innovative new medicines according to the Annex of Regulation (EC) No 726/2004.

Under the procedure, applications are submitted directly to the EMA to be evaluated by the CHMP (formerly CPMP). Following review, the EMA then forwards its opinion to the EC, which makes the final decision on the granting of the European Community marketing authorization. An EC authorization is valid throughout the whole of the EU and European Economic Area (EEA) and is usually given for 5 years. Applications for extension must be made to the EMA 3 months before this 5-year period expires.

The centralized review process is initiated with the submission of a letter of intent by the applicant to the EMA. This letter of intent is sent up to 7 months in advance of the planned submission date and agreement of eligibility by the EMA. A Rapporteur and Co-Rapporteur are assigned to lead the review. Once the MAA has been submitted, the dossier undergoes a validation procedure to ensure that all required information is presented. Following successful validation, the EMA then initiates the review procedure.

This procedure entails an initial review period that takes place over 80 days by two Rapporteurs (one Rapporteur and one Co-Rapporteur). Following this 80-day review period, the Day 80 Draft Assessment Reports are sent to the applicant and these reports contain initial comments from the Rapporteur and the Co-Rapporteur. This is followed by an additional period of time to allow for comments from all CHMP members and their review teams. In addition, the Biologics Working Party (BWP) will review the quality modules. At Day 120, the CHMP provides the applicant with a Joint Assessment Report and a List of Questions. At this point in time, the assessment clock is stopped. The applicant then normally has to respond within 3 months, but this can be extended if there is sufficient justification. A second round of review is initiated once the applicant submits the additional information at Day 121, when the assessment clock is restarted. This second round of review takes 30 days. At Day 150, the second Joint Assessment Report prepared by the Rapporteur and the Co-Rapporteur is sent to the applicant. Following Day 150, there is an additional 30-day period for the CHMP to review the new information. At Day 180, the CHMP provides the applicant with a further List of Questions if there are still deficiencies, and the applicant has 30 days in which to respond. Again, the assessment clock is stopped. The applicant submits responses at Day 181 (point at which the assessment clock is restarted). There is also an opportunity at this stage for the applicant to have an "Oral Explanation" meeting with the CHMP at Day 181. The CHMP has a further 30 days to review the new information. Then, at Day 210, the CHMP will give its scientific opinion. An overview of this timeline is presented in Figure 14.4.

If the decision is positive, the applicant is then required to provide translations of the product information in all European languages (including the EEA). These, along with the opinion of the CHMP, are then transmitted to the EC who will issue the marketing authorization if in agreement with the opinion of the CHMP. Following a successful scientific review of a medicinal product for human use by the CHMP and authorization by the EC, the new product will automatically be authorized in all member states of the EU and the member states of the EEA.

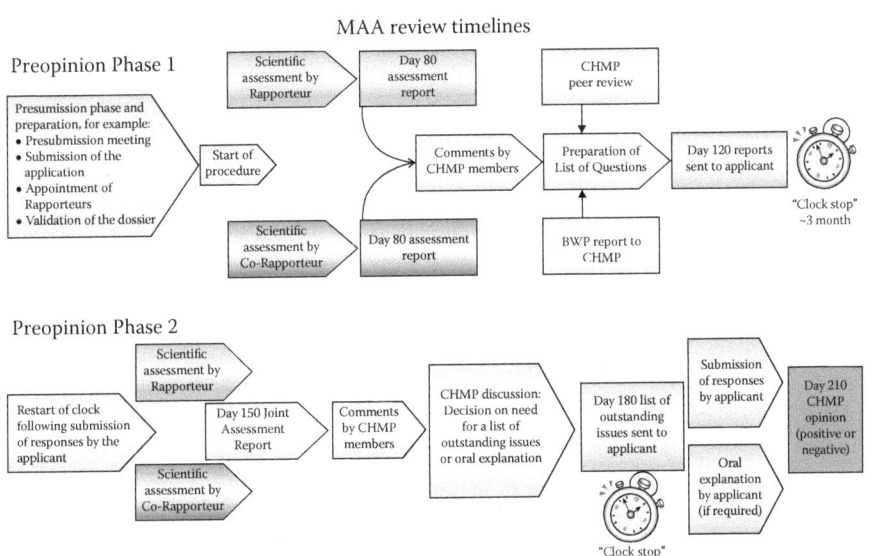

Clock stop: During the evaluation of a medicinal product, this refers to a period of time during which the "Clock" that calculates the length of a procedure "stops" while the applicant prepares answers to questions asked by the regulatory authority.

FIGURE 14.4 Marketing Authorisation Application review timelines.

14.3.2 REGISTRATION IN THE U.S.

During the BLA review process, there is considerable dialogue between FDA and the Sponsor. In addition, the procedure is less rigid with regard to both timelines and procedures than the EU Centralised Procedure. The FDA assessment timeline targets are that 90% of priority BLAs will be approved in 6 months and 90% of standard BLAs will be approved in 10 months.

Following notification of submission from the applicant, the review team is set up. An initial review is conducted to check for administrative completeness. A filing meeting is held and a filing letter is sent to the applicant. Any significant deficiencies noted up to that time are communicated to the applicant by Day 74 or a refuse to file (RTF) application is sent.

During the review process, review memos are sent to the applicant and these consist of information requests and discipline review letters. Information requests are issued while the review is in progress and these request further information that are needed to complete review.

Discipline reviews are issued when a particular discipline has finished its review, and the discipline review letter may contain comments or questions that might appear in an action letter and any responses are not necessarily reviewed prior to the issuance of action letter.

Typically, an Advisory Committee is held if the product meets the following criteria:

- It is first in class.
- There is a new indication.
- There are labeling issues.
- There are safety, efficacy, and risk/benefit questions.
- There are dosing concerns or concerns regarding the target population.

Advisory Committees are also held if FDA decisions are being appealed. The Advisory Committees are made up of preeminent scientists and specialists and also have a consumer representative and an industry representative, who is a nonvoting member. In addition, consultants (specialists and patient representatives) may also attend and these committee members are also nonvoting.

Following the initial round of review, a Review Committee meeting is held to discuss inspection findings. This meeting is held to discuss Advisory Committee recommendations, and to determine whether there are any outstanding issues and whether there should be agreements and commitments. Then the final decision is taken on whether the product is not ready for approval or whether it can be approved. Action letters are then sent to the applicant.

If the product is not ready for approval, a Complete Response letter is sent, which outlines all the deficiencies in the application that must be corrected or addressed prior to approval. The review process starts again as described above when the additional information is submitted.

If the product is approvable, the approval letter is sent to the applicant, which grants permission to distribute the product.

On approval, a compliance check is conducted, the Summary Basis of Approval (SBA) is prepared, and the approval letter and the license are then issued.

14.3.3   AUTHORIZATION OF BIOSIMILAR MEDICINAL PRODUCTS

14.3.3.1   Authorization of Biosimilar Medicinal Products in the EU

14.3.3.1.1   The EU Omnitrope Story

In 2001, Sandoz, the generics subsidiary of Novartis, first attempted to obtain EU approval for Omnitrope. They submitted an application to have the product considered for generic authorization based on a detailed scientific bibliography, under Article 10 of Directive 2001/83/EC, which legislates the approval of generic medicinal products. The application was accompanied by studies aimed at showing comparability with the reference product, Genotropin.

In June 2003, the CHMP issued a positive opinion. However, the EC decided not to follow the opinion and did not give approval for the product. In March 2004, after legal action by Sandoz, the EC published a statement indicating that the CHMP had improperly accepted Sandoz's application as a bibliographical application based on the well-established use of the medicine, because at the same time the company had required comparability studies to be performed. Sandoz appealed the EC decision with the European Courts of Justice, contesting that the performance of comparability studies implied that the legal conditions for the application of the bibliographical application procedure were not met.

At that time, a new regulatory framework was put in place. Article 10(4) of Directive 2001/83/EC was amended in Directive 2004/27/EC to refer to biosimilar medicinal products as follows:

> Where a biological medicinal product which is similar to a reference biological product does not meet the conditions in the definition of generic medicinal products, owing to, in particular, differences relating to raw materials or differences in manufacturing processes of the biological medicinal product and the reference biological medicinal product, the results of appropriate preclinical tests or clinical trials relating to these conditions must be provided. The type and quantity of supplementary data to be provided must comply with the relevant criteria stated in Annex I and the related detailed guidelines. The results of other tests and trials from the reference medicinal product's dossier shall not be provided.

Sandoz then resubmitted the MAA in July 2004 under this new regulatory framework, obtained a positive opinion from CHMP in January 2006, and received EC approval in April 2006.

14.3.3.1.2   Process for the Approval of Biosimilar Medicinal Products in the EU

Following the initial Omnitrope MAA, Article 10(4) of Directive 2001/83/EC was amended in Directive 2004/27/EC to introduce the concept of biosimilar medicinal products. In particular, the Directive notes the fact that the manufacturing process, including raw materials and other variables for the manufacture of the biosimilar product, is unlikely to be identical to those used for the manufacture of the innovator product. As such, MAAs for biosimilar products require appropriate pre-clinical studies and clinical trials.

This is further reinforced in the aforementioned Annex I of Directive 2001/83/EC, as amended in Directive 2003/63/EC, which includes the following reference in Part II(4) to "similar biological medicinal products":

> The provisions of Article 10(1)(a)(iii) may not be sufficient in the case of biological medicinal products. If the information required in the case of essentially similar products (generics) does not permit the demonstration of the similar nature of two biological medicinal products, additional data, in particular, the toxicological and clinical profile shall be provided.

Thus, MAAs for biosimilar products in the EU are expected to be "full applications." The degree to which the MAA dossiers can be abridged in any way by including information from literature or providing justification from the results of comparability exercises is decided on a case-by-case basis.

Since the publication of the directives, the EMA has published a number of guidelines on the development of biosimilar products. There is one overarching guideline that defines the philosophy and principles of biosimilar development (CHMP/437/04). There are also general guidelines, which provide an overview of the principles for quality, nonclinical, and clinical development of biosimilar medicinal products (CHMP/49348/05 and CHMP/42832/05). The adopted product-specific guidelines include guidelines for the development of erythropoietin, low-molecular-weight heparins, interferon alpha, granulocyte-colony stimulating factor, somatropin (human growth hormone), and human insulin. More recently, draft guidelines for the development of biosimilar medicinal products containing recombinant follicle stimulation hormone, recombinant interferon beta, and monoclonal antibodies have been released for consultation.

14.3.3.2 Authorization of Biosimilar Medicinal Products in the U.S.

In the U.S., the FDA can approve generic drugs via mechanisms contained in Sections 505(b)(2) and 505(j) of the FD&C Act. As such, these apply only to products regulated by the Act, that is, to products classified as drugs. There is no corresponding mechanism for the approval of generic versions of biologics ("biosimilar" or "follow-on biologic") under the PHS Act. However, polypeptide hormones such as somatropin, insulin, and calcitonin are regulated as drugs under the FD&C Act and this potentially allows for biosimilar versions of these polypeptide hormones to be approved.

For an application to be considered under Section 505(j), an important complication for the review of a biosimilar medicinal product would be the requirement to demonstrate "sameness."

> An abbreviated application for a new drug shall contain... information to show that the active ingredient of the new drug is the same as that of the listed drug...

The other possible pathway for a follow-on biologic application under the FD&C Act is the 505(b)(2) route.

An application... for which the investigations... relied upon by the applicant for approval of the application were not conducted by or for the applicant and for which the applicant has not obtained a right of reference or use from the person by or for whom the investigations were conducted shall also include -

A certification, in the opinion of the applicant and to the best of his knowledge, with respect to each patent which claims the drug for which such investigations were conducted or which claims a use for such drug for which the applicant is seeking approval under this subsection and for which information is required to be filed under paragraph (1) or subsection (c) -

    i.   that such patent information has not been filed,

    ii.   that such patent has expired,

    iii. of the date on which such patent will expire, or

    iv. that such patent is invalid or will not be infringed by the manufacture, use, or sale of the new drug for which the application is submitted

In this case, the requirements shared with the 505(j) mechanism come down to just patent information and the "sameness" criteria are no longer applicable. The FDA released a draft Guidance (Applications covered by Section 505(b)(2)) on this topic in October 1999.

Sandoz resubmitted a BLA for Omnitrope in July 2003 under Section 505(b)(2) of the FD&C Act. In 2004, the FDA declared that a decision could not be made due to "unresolved scientific and legal issues." In September 2005, Sandoz filed a lawsuit against the FDA for failure to take action on its pending application and in April 2006, the Court ordered the FDA to comply with its statutory obligation to act on the application. As a result, the Omnitrope application was finally approved in June 2006.

Subsequently, there has been only one other submission for a biosimilar product in the U.S. Teva Pharmaceutical Industries filed for approval of its filgrastim product, which is a biosimilar to Amgen's Neupogen® (granulocyte colony-stimulating factor [G-CSF]), in February 2010. Teva has simply filed for approval of its product using a traditional application route via a BLA with supporting clinical data. This application is still under review and the FDA issued a Complete Response letter in October 2010 to request additional information needed to complete the review of the application for product approval.

Despite Teva's strategy, Amgen has filed a patent infringement claim in the U.S. Federal Court against Teva trying to block the move. Teva had sought a declaration from the same court that its product does not infringe Amgen's patents. However, the District Court in Pennsylvania has entered final judgment and a permanent injunction against Teva prohibiting them from infringing Amgen's patents relating to human G-CSF and methods for its use. The judgment was accompanied by Teva's admission that its filgrastim product infringes the two Amgen patents at issue in the litigation and that those patents are valid and enforceable. The Court's injunction extends until November 10, 2013, after which date Teva may sell its filgrastim product in the United States.

Over the last 4 to 5 years, the U.S. government has been attempting to introduce biosimilar legislation. At the beginning of 2006, the legislation for biosimilar medicinal products, also called follow-on biologics, was blocked by U.S. biotechnology lobbyists. The proposal was reintroduced by Congressman Waxman at the beginning of 2007 to create a regulatory pathway for the approval of biosimilar medicinal products. However in May 2007, President Bush officially opposed any legislation paving the way for biosimilar medicinal products. Therefore, in September 2007, the bill changing FDA provisions approved by Congress did not include a regulatory pathway for biosimilar medicinal products.

On March 17, 2009, the Pathway for Biosimilars Act was introduced in the House of Representatives. The FDA was given the authority to approve biosimilars, including interchangeable products that are substitutable with their reference product, as part of the Patient Protection and Affordable Care Act signed by President Obama on March 23, 2010. This Act included three significant components. First, the Act established a licensure pathway for competing versions of previously marketed biologics. In particular, the legislation established a regulatory pathway for two sorts of products, termed "biosimilar" and "interchangeable" biologics. The FDA was afforded a prominent role in determining the particular standards for biosimilarity and interchangeability for individual products. Second, the Act created FDA-administered periods of data protection and marketing exclusivity for certain brand-name drugs and follow-on products. Brand-name biologic products receive 4 years of marketing exclusivity and 12 years of data protection. The Act also provides for a term of marketing exclusivity for the applicant who is the first to establish that its product is interchangeable with the brand-name product. Finally, the Act created a patent dispute resolution procedure for use by brand-name and follow-on biologic manufacturers. In February 2012, FDA released a draft Guidance for Industry "Biosimilars: Questions and Answers Regarding Implementation of the Biologics Price Competition and Innovation Act of 2009." This guidance provides answers to common questions from sponsors interested in developing proposed biosimilar products, BLA holders, and other interested parties regarding FDA's interpretation of the Biologics Price Competition and Innovation Act. At the same time, two further draft Guidance for Industry documents were published providing input on scientific and quality considerations for the development of biosimilar medicinal products (Guidance for Industry: Scientific Considerations in Demonstrating Biosimilarity to a Reference Product and Guidance for Industry: Quality Considerations in Demonstrating Biosimilarity to a Reference Protein Product). The release of these draft documents is a step forward in biosimilar medicinal product development in the United States.

14.4   TYPICAL PITFALLS DURING THE REGISTRATION PROCESS

Although regulatory standards and procedures in Europe and the United States have improved, the number of major issues with MAAs and BLAs for biotechnological products remains high. For example, the pivotal clinical trials of some late-stage failures have been found not to meet the regulatory guidelines of the EU, and regulators are increasingly concerned that attempts to accelerate the process of biotechnological product development leads to the neglect of important issues (Schneider and Schäffner-Dallmann, 2008).

In order to help companies developing medicinal products in the EU, it has been a requirement that details of both positive and negative decisions are made available to the public since late 2005. Previously, only information on positive opinions was made available as the European Public Assessment Report. Since 2005, information on the negative opinions (refusals) or, if the company withdraws the MAA prior to receipt of a negative opinion, information on the official assessment conducted to the point of withdrawal are published on the EMA website.

Similarly in the United States, regulatory standards and procedures are continually improving but there are still a substantial number of failures for registration dossiers submitted to FDA. This information is more difficult to locate as there is no centralized filing location for information on withdrawals and refusals as there is on the EMA website.

It is however important not to assume that products that are approved by the FDA will be automatically approved by EMA and *vice versa*. Some decisions from these two regulatory regions are provided in Sections 14.4.1 through 14.4.3 to provide an overview of differences in the review of registration dossiers in the two regions. Though not all examples are for biotechnological or biological products, similar issues can be seen for all product types.

14.4.1 EXAMPLES OF MEDICINAL PRODUCTS APPROVED IN THE UNITED STATES BUT NOT IN THE EU

14.4.1.1 Cimzia (Certolizumab Pegol) for the Treatment of Crohn's Disease

Cimzia was approved by the FDA for the treatment of Crohn's disease in April 2008. However, the EMA refused to authorize this product in November 2007. CHMP was concerned that there was insufficient evidence to show a benefit of Cimzia. The CHMP was also concerned over Cimzia's safety as there was some concern over a possible increased risk of bleeding in patients receiving Cimzia. In addition, the CHMP was concerned that the company had not demonstrated that it would have been able to monitor the quality of the medicine to an acceptable level. In March 2008, following a reexamination procedure, the CHMP removed its concern regarding the ability to monitor the medicine's quality. It also removed its concern over the possible increased risk of bleeding, but maintained a general concern over Cimzia's safety. Therefore, at that point in time, the CHMP was of the opinion that the benefits of Cimzia in the treatment of severe, active Crohn's disease did not outweigh its risks, and therefore it recommended that Cimzia be refused marketing authorization.

Since this initial refusal by EMA, Cimzia is now approved in both the EU and the United States for the treatment of rheumatoid arthritis, but at the time of writing this chapter, Cimzia had not yet been approved for the treatment of Crohn's disease by EMA.

14.4.1.2 Mylotarg (Gemtuzumab Ozogamicin)

The FDA approved Mylotarg in May 2000. However, in September 2007, CHMP adopted a negative opinion, recommending the refusal of the marketing authorization for Mylotarg. The CHMP was concerned that the studies of Mylotarg had not shown a benefit of the medicine, because of the way they were designed. The CHMP noted that there were side effects associated with Mylotarg, which included severe and long-lasting bone marrow suppression causing low levels of white blood cells and platelets, liver problems, and side effects related to the infusion, such as chills, fever, and low blood pressure. At that point in time, the CHMP was of the opinion that there was insufficient evidence to establish the effectiveness of Mylotarg in the treatment of acute myeloid leukemia, and therefore the medicine's benefits did not outweigh its risks. Hence, the CHMP recommended that Mylotarg be refused marketing authorization; the refusal was confirmed after reexamination.

14.4.1.3 Natalizumab Elan Pharma

Although natalizumab is approved for the treatment of multiple sclerosis in the EU (marketed as Tysabri by Biogen Idec), the CHMP recommended refusal of the marketing authorization for Natalizumab Elan Pharma for the treatment of Crohn's disease in July 2007, and this opinion was confirmed following reexamination in November 2007. During the initial review, the CHMP was concerned that there was insufficient evidence to show the effectiveness of Natalizumab Elan Pharma. The CHMP also considered that there was insufficient evidence for maintenance of the medicine's effects, and it also had concerns over the safety of Natalizumab Elan Pharma in patients with Crohn's disease, because of the risk of serious infections, including the brain infection progressive multifocal leucoencephalopathy. Following the reexamination, the CHMP removed its concern regarding the effectiveness of the medicine in patients starting treatment. However, all other concerns remained and the CHMP was of the opinion that the benefits of Natalizumab Elan Pharma in the treatment of Crohn's disease did not outweigh its risks. In contrast, in January 2008, the FDA approved the extension of indication for Tysabri to include the treatment of Crohn's disease.

14.4.1.4 Gemesis (Growth-Factor Enhanced Matrix)

GEM21S is a dental bone filling device with a biological component (platelet-derived growth factor [PDGF]) that is used to treat patients who have bone defects due to periodontal disease. This product was approved by FDA in 2005 as a device for the treatment of intrabony periodontal defects, furcation periodontal defects, and gingival recession associated with periodontal defects. Reflecting the difference in review processes, the EMA reviewed Gem21S as a medicinal product with a device component under the trade name Gemesis. Following the review, the CHMP was of the opinion that the main study failed to show that Gemesis was effective in treating periodontal defects. The CHMP noted that at this time, the company did not have sufficient information on how strongly becaplermin (the active substance) binds to PDGF receptors and did not sufficiently demonstrate that Gemesis used in clinical studies was comparable to the product intended to be placed on the market. The CHMP was also concerned about the level of product-related impurities present. Therefore, at that point in time, the CHMP was of the opinion that the benefits of Gemesis did not outweigh its risks and recommended that Gemesis be refused marketing authorization. This opinion was confirmed after reexamination.

14.4.2 EXAMPLES OF MEDICINAL PRODUCTS APPROVED IN THE EU BUT NOT IN THE U.S.

On occasion, some products may be approved in the EU by EMA but not by the FDA.

14.4.2.1   Ruconest (Conestat Alfa)

Although initially refused (under the trade name Rhucin), the EMA granted marketing authorization for Ruconest in October 2010. However, in February 2011, the FDA issued a "refusal to file" indicating that the clinical study results submitted did not provide sufficient data to support the proposed dose, and that the results lacked sufficient information on the scale used in measuring the clinical effects. The applicant, Pharming, initiated studies in February 2011 to respond to the FDA issues and should resubmit to FDA when the results of this clinical study are available.

14.4.3   MEDICINAL PRODUCTS REFUSED IN EU AND U.S.

There are also examples of products that were refused in both regions but on different grounds.

14.4.3.1   Surfaxin

Surfaxin is an Orphan Drug consisting of a liposomal formulation containing a target of 30 mg/mL phospholipids (dipalmitoylphosphatidylcholine and palmitoyl-oleoyl phosphatidylglycerol), palmitic acid, and a synthetic 21-amino acid peptide (sinapultide) in an aqueous medium. The MAA was withdrawn in the EU at Day 180 in July 2006. The following concerns were raised:

- Quality issues concerning manufacturing and stability.
- The design and the results of the clinical studies were not sufficient to establish that the product will be at least as effective and safe as other surfactant agents currently used to prevent respiratory distress syndrome (RDS).
- Data to support the indication for the treatment of RDS were not sufficient.

The FDA refused authorization in July 2009 due to problems with the biological activity tests. No clinical concerns were raised.

14.4.3.2   Advexin

Advexin is an Orphan Drug and is an adenoviral vector containing the p53 gene (Ad5CMV-p53) intended for administration into tumors of Li-Fraumeni syndrome patients. The MAA was withdrawn in the EU at Day 179 in December 2008. The following concerns were raised:

- Not enough evidence was available to show that the injection of Advexin into Li-Fraumeni tumors led to benefits for patients.
- The CHMP also had concerns over what happened to the medicine in the body, how it should be given, and how safe it was.
- The company had not supplied enough evidence to demonstrate that Advexin could be made in a reliable manner, or that it would not be harmful to the environment or to people in close contact with the patient.

The FDA also refused to file in September 2008, on the grounds that an incomplete BLA had been submitted. As no review has taken place, we cannot know if the FDA findings would be the same as those listed by CHMP.

14.5 REGULATORY LEGISLATION IN THE EU

In the EU, regulation of biological and peptide products is controlled by Directives, Regulations, and Guidelines. A directive is a legislative act of the EU that requires member states to achieve a particular result without dictating the means of achieving that result. Directives normally leave member states with a certain amount of leeway as to the exact rules to be adopted. A regulation is a legislative act of the EU that becomes immediately enforceable as law in all member states simultaneously. Guidelines are enacted by the EMA (formerly EMEA). Guidelines are intended to provide a basis for practical harmonization of the manner in which the EU member states and the EMA interpret and apply the detailed requirements for the demonstration of quality, safety, and efficacy contained in the directives. Scientific guidelines do not have legal force but companies are urged to follow the guidelines unless strong justification for deviation is available.

14.6 CONCLUSION

The development of regulations for the control of medicinal products in both the EU and the United States has primarily been due to ethics and safety concerns related to early medicinal products. As a result, the number of regulations, guidelines, and other requirements relating to the registration of biological and biotechnological medicinal products have widened considerably in scope during the past two decades in the two regions, in response to a rapidly developing field.

However, the pace of development in the field of biological medicinal products has been increasing rapidly over the last decade and the regulations appear not to be keeping up with these developments. Occasionally, the generation of new regulations appear reactive rather than proactive, particularly in the EU. In one particular instance, the Guideline on Strategies to Identify and Mitigate Risks for First-in-Human Clinical Trials with Investigational Medicinal Products (EMEA/CHMP/SWP/28367/07) was developed in the wake of the TGN1412 incident, where the product (also known as CD28-SuperMAB) was withdrawn from development after inducing severe inflammatory reactions in the first human subjects to receive the drug. Typically, however, the agencies start developing regulatory documentation in conjunction with the "stakeholders" (i.e., the interested parties in the particular field of development) when new and innovative products are entering into development. This can be seen in a number of directives, regulations, guidelines, and reflection papers that have been developed for advanced therapy medicinal products over the last few years.

Although it can be argued that biological medicinal products are too highly regulated and that the standards imposed for registration are too restrictive, this level of regulation is necessary to ensure patient safety throughout the life cycle of the biological product.

These guidance documents should be seen as a source of reference that, if followed, will aid companies in gaining approval for their medicinal products rather than becoming a major hurdle to product development. In addition, the agencies are not closed to considering alternative development pathways for novel biotechnological products as long as solid justifications can be provided where guidelines have not been followed. However, ongoing discussion with the agencies should be conducted in these instances.

DEFINITIONS

FDA

Biological product (biologics): Biological products include a wide range of products such as vaccines, blood and blood components, allergenics, somatic cells, gene therapy, tissues, and recombinant therapeutic proteins. Biologics can be composed of sugars, proteins, or nucleic acids or complex combinations of these substances, or may be living entities such as cells and tissues. Biologics are isolated from a variety of natural sources—human, animal, or microorganism—and may be produced by biotechnology methods and other cutting-edge technologies. Gene-based and cellular biologics, for example, often are at the forefront of biomedical research and may be used to treat a variety of medical conditions for which no other treatments are available.

Biosimilar (previously known as follow-on biologic): Biological products that are demonstrated to be "highly similar" to or "interchangeable" with an FDA-approved biological product.

EMA

Biological medicinal product: A biological medicinal product is a medicinal product whose active substance is made by or derived from a living organism.

Biosimilar medicinal product: A biosimilar medicinal product is a medicinal product which is similar to a biological medicinal product that has already been authorized (the "biological reference medicinal product"). The active substance of a biosimilar medicinal product is similar to one of the biological reference medicinal product.

ABBREVIATIONS

BLA	Biologics License Application
BWP	Biologics Working Party
CBER	Center for Biological Evaluation and Research
CDER	Center for Drug Evaluation and Research
CHMP (formerly CPMP)	Committee for Medicinal Products for Human Use
CPMP	Committee for Proprietary Medicinal Products
CTD	Common Technical Document
DPPC	Dipalmitoylphosphatidylcholine

(Continued)

DNA	Deoxyribonucleic acid
EC	European Commission
EEA	European Economic Area
EEC	European Economic Community
EMA (formerly EMEA)	European Medicines Agency
EPAR	European Public Assessment Report
EU	European Union
FDA	Food and Drug Administration
FD&C Act	Food, Drug and Cosmetic Act
G-CSF	Granulocyte colony-stimulating factor
ICH	International Conference on Harmonization
LFS	Li-Fraumeni syndrome
MAA	Marketing Authorisation Application
NDA	New Drug Application
NtA	Notice to Applicants
POPG	Palmitoyl-oleoyl phosphatidylglycerol
PHS Act	Public Health Service Act
PDGF	Platelet-derived growth factor
PML	Progressive multifocal leucoencephalopathy
QOS	Quality Overall Summary
RDS	Respiratory Distress Syndrome
RTF	Refusal to file
SBA	Summary Basis of Approval
U.S.	United States

REFERENCES

Reichert, J. M., Rosensweig, C. J., Faden, L. B., and Dewitz, M. C. 2005. Monoclonal antibody successes in the clinic. *Nat. Biotech.* 23, 1073–1078.

Schneider, C. K., and Schäffner-Dallmann, G. 2008. Typical pitfalls in applications for marketing authorization of biotechnological products in Europe. *Nat. Rev. Drug Discov.* 7, 893–899.

FURTHER READING

Relevant Websites

EMA: http://www.ema.europa.eu/
FDA: http://www.fda.gov/
CDER: http://www.fda.gov/Drugs/default.htm
CBER: http://www.fda.gov/BiologicsBloodVaccines/default.htm
ICH: http://www.ich.org/products/guidelines.html
EC Directives and Regulations: http://ec.europa.eu/health/documents/eudralex/vol-1/index_en.htm
NtA: http://ec.europa.eu/health/documents/eudralex/vol-2/index_en.htm

EMA Public Assesstment Reports

Refusal of Cimzia (certolizumab pegol) for the treatment of Crohn's disease (http://www.ema.europa.eu/docs/en_GB/document_library/EPAR_-_Public_assessment_report/human/000740/WC500070613.pdf)

Authorization of Cimzia (certolizumab pegol) for the treatment of rheumatoid arthritis (http://www.ema.europa.eu/docs/en_GB/document_library/EPAR_-_Public_assessment_report/human/001037/WC500069735.pdf)

Refusal of Mylotarg (gemtuzumab ozogamicin) (http://www.ema.europa.eu/docs/en_GB/document_library/EPAR_-_Public_assessment_report/human/000705/WC500070677.pdf)

Refusal of Natalizumab Elan Pharma (http://www.ema.europa.eu/docs/en_GB/document_library/EPAR_-_Public_assessment_report/human/000624/WC500070715.pdf)

Refusal of Gemesis (Growth-factor Enhanced Matrix) (http://www.ema.europa.eu/docs/en_GB/document_library/EPAR_-_Public_assessment_report/human/000997/WC500091372.pdf)

Approval of Ruconest (conestat alfa) (http://www.ema.europa.eu/docs/en_GB/document_library/EPAR_-_Public_assessment_report/human/001223/WC500098546.pdf)

Withdrawal of Surfaxin (http://www.ema.europa.eu/docs/en_GB/document_library/Application_withdrawal_assessment_report/2010/01/WC500069610.pdf)

Withdrawal of Advexin (http://www.ema.europa.eu/docs/en_GB/document_library/Application_withdrawal_assessment_report/2010/01/WC500063080.pdf)

Regulation of Medicinal Products

EC Directives

Council Directive 65/65/EEC of 26 January 1965 on the approximation of provisions laid down by law, regulation, or administrative action relating to medicinal products (http://www.echamp.eu/fileadmin/user_upload/Regulation/Directive_65-65-EEC__-__Consolidated_Version.pdf)

Council Directive 75/319/EEC of 20 May 1975 on the approximation of provisions laid down by law, regulation, or administrative action relating to proprietary medicinal products (http://www.echamp.eu/fileadmin/user_upload/Regulation/Directive_75-319-EEC__-__Consolidated_Version.pdf)

Council Directive 87/22/EEC of 22 December 1986 on the approximation of national measures relating to the placing on the market of high-technology medicinal products, particularly those derived from biotechnology (http://ec.europa.eu/health/files/pharmacos/docs/doc2001/may/directive_87-22_en.pdf)

Council Directive 2001/83/EC of the European Parliament and of the Council of 6 November 2001 on the Community code relating to medicinal products for human use (http://www.ema.europa.eu/docs/en_GB/document_library/Regulatory_and_procedural_guideline/2009/10/WC500004481.pdf)

Council Directive 2003/63/EC of 25 June 2003 amending Directive 2001/83/EC of the European Parliament and of the Council on the Community code relating to medicinal products for human use (http://ec.europa.eu/health/files/eudralex/vol-1/dir_2003_63/dir_2003_63_en.pdf)

Council Directive 2004/27/EC of the European Parliament and of the Council of 31 March 2004 amending Directive 2001/83/EC on the Community code relating to medicinal products for human use (http://ec.europa.eu/health/files/eudralex/vol-1/dir_2004_27/dir_2004_27_en.pdf)

EC Regulations

Council Regulation (EEC) No 2309/93 of 22 July 1993 laying down Community procedures for the authorization and supervision of medicinal products for human and veterinary use and establishing a European Agency for the Evaluation of Medicinal Products (http://ec.europa.eu/health/files/eudralex/vol-1/reg_1993_2309/reg_1993_2309_en.pdf)

Council Regulation (EC) No 726/2004 of the European Parliament and of the Council of 31 March 2004 laying down Community procedures for the authorization and supervision of medicinal products for human and veterinary use and establishing a European Medicines Agency (http://ec.europa.eu/health/files/eudralex/vol-1/reg_2004_726_cons/reg_2004_726_cons_en.pdf)

EC Decisions

Council Decision 75/320/EEC of 20 May 1975 setting up a pharmaceutical committee (http://eur-lex.europa.eu/LexUriServ/LexUriServ.do?uri=OJ:L:1975:147:0023:0023:EN:PDF)

EMA Guidelines

CHMP/437/04: Similar Biological Medicinal Products (http://www.ema.europa.eu/docs/en_GB/document_library/Scientific_guideline/2009/09/WC500003517.pdf)

CHMP/49348/05: Similar Biological Medicinal Products Containing Biotechnology-Derived Proteins as Active Substance: Quality Issues (http://www.ema.europa.eu/docs/en_GB/document_library/Scientific_guideline/2009/09/WC500003953.pdf)

CHMP/42832/05: Similar Biological Medicinal Products Containing Biotechnology-Derived Proteins as Active Substance: Non-Clinical and Clinical Issues (http://www.ema.europa.eu/docs/en_GB/document_library/Scientific_guideline/2009/09/WC500003920.pdf)

EMEA/CHMP/SWP/28367/07: Guideline on Strategies to Identify and Mitigate Risks for First-in-Human Clinical Trials with Investigational Medicinal Products (http://www.ema.europa.eu/docs/en_GB/document_library/Scientific_guideline/2009/09/WC500002988.pdf)

ICH Guidelines

ICH S6(R1): Preclinical Safety Evaluation of Biotechnology-Derived Pharmaceuticals (http://www.ich.org/fileadmin/Public_Web_Site/ICH_Products/Guidelines/Safety/S6_R1/Step4/S6_R1_Guideline.pdf)

U.S. Acts

Vaccine Act of 1813 (http://biotech.law.lsu.edu/cases/vaccines/vac_act_1813.pdf)

The U.S. Federal Food, Drug, and Cosmetic Act of 1938 (http://www.fda.gov/RegulatoryInformation/Legislation/FederalFoodDrugandCosmeticActFDCAct/default.htm)

The Public Health Service Act of 1944 (http://www.ncbi.nlm.nih.gov/pmc/articles/PMC1403520/pdf/pubhealthrep00059-0006.pdf

Biologics Price Competition and Innovation Act, 2009 (http://www.fda.gov/downloads/Drugs/GuidanceComplianceRegulatoryInformation/UCM216146.pdf)

U.S. Guidance for Industry

Guidance for Industry—Biosimilars: Questions and Answers Regarding Implementation of the Biologics Price Competition and Innovation Act of 2009 (http://www.fda.gov/downloads/Drugs/GuidanceComplianceRegulatoryInformation/Guidances/UCM273001.pdf)

Guidance for Industry: Quality Considerations in Demonstrating Biosimilarity to a Reference Protein Product (http://www.fda.gov/downloads/Drugs/GuidanceComplianceRegulatoryInformation/Guidances/UCM291134.pdf)

Guidance for Industry: Scientific Considerations in Demonstrating Biosimilarity to a Reference Product (http://www.fda.gov/downloads/Drugs/GuidanceComplianceRegulatoryInformation/Guidances/UCM291128.pdf)

Index